Ambient Intelligence

W. Weber J.M. Rabaey E. Aarts (Eds.)

Ambient Intelligence

With 143 Figures

 Springer

Werner Weber
Infineon Technologies
Corporate Research
Otto-Hahn-Ring 6
81739 Munich
Germany
e-mail: Werner.Weber@infineon.com

Emile Aarts
Philips Research
Prof. Holstlaan 4 (WAY2.11)
5656 AA Eindhoven
The Netherlands
e-mail: emile.aarts@philips.com

Jan M. Rabaey
Dept. of Electrical Engineering
and Computer Sciences
University of California
511 Cory Hall #1770
Berkeley, CA 94720-1770
USA
e-mail: jan@eecs.berkeley.edu

Library of Congress Number: 2004114852

ISBN 3-540-23867-0 Springer Berlin Heidelberg New York

Springer is a part of Springer Science+Business Media

springeronline.com

Typesetting: by the authors and TechBooks using a Springer LaTeX macro package
Cover design: *design & production* GmbH, Heidelberg

Printed on acid-free paper   SPIN 10948439   57/3141/jl   5 4 3 2 1 0

Preface

The emergence of a new technical concept that profoundly affects human life is a relatively rare event. Yet, over the last centuries and decades, we have witnessed acceleration in the introduction of new society-affecting technologies. While it took a long time for book printing, electricity, the car and television to put their stamp on the world and to affect the daily living patterns, information-technology related technologies have been introduced at an ever-increasing pace. It took the internet and the mobile phone just two or one decades to profoundly change the way we communicate and interact.

It is our belief that yet another paradigm-shifting technology is now on the horizon. "Ambient intelligence" is the most commonly used descriptor of the phenomenon, although in the United States it is often dubbed as sensor (and actuator) networks. As a third wave of computing, it presents a true departure from the way electronic devices and humans interact. In the first wave of computing – mainframes, super-computers, mini-computers, desk- and laptops, the meaning of computation was centered in a single device, and users mostly interacted in batch-mode or through a limited interface. With the advent of the internet and the world-wide web in the 1990s, the picture changed dramatically. Computation and information access became a globally distributed concept with data and computation scattered over a wide range of servers and data storage devices located all over the world. However, the interface remained quite similar to the previous era. To obtain data or information, a user has to proactively initiate an exchange.

The "Ambient Intelligence" paradigm differs in two major ways from these previous generations. First of all, the user interface has become reactive, that is actions are not explicitly requested but are the result of the mere presence of people or their avatars (of course, with their explicit or implicit goals and constraints). Secondarily, the meaning of computation can no longer be associated to a single device or a set of connected devices, but is located in the "collection of devices". This means that the failure of a single component does not mean that the goal cannot be accomplished.

Yet, new technologies do not appear all of a sudden. After initially being in the domain of a mere few (just think about the ARPANET of the 1970s and early 1980s), they gradually emerge into the global market. For this to happen, potential users have to prepared, convincing benefits have to be conveyed, and concerns regarding negative side-effects have to be resolved. "Ambient Intelligence" is such a new concept. To be successful, it is up to the advocates of this vision to preach: that is, to make it known, provide understanding, create demands, shine light onto the concept from various angles, and discuss various peripheral aspects. Only when this is accomplished will the technology find broader penetration and begin to accelerate.

These concerns provided the ultimate motivation for the editors to assemble this book. As true believers, it is our goal to "preach": convey the opportunities and benefits of the ambient intelligence concept, evaluate the current status and identify challenges and concerns.

To that effect, we have organized the book in three major parts:

- Part I discusses a number of potential applications of ambient intelligence, and describes a set of scenarios. The part starts by addressing social, economic and ethical implications. It discusses electronics integrated into textiles, in smart rooms and intelligent buildings that could make our environment more friendly and enjoyable, more user-friendly, more effective, and in addition more energy efficient.
- Part II gives an overview of the networking and infrastructure issues involved in the realization and the implementation of an ambient intelligence environment. A networked infomechanical system, an operating system, a service-based application interface and a locationing and timing service are discussed for peer-to-peer ad-hoc wireless sensor networks. Furthermore, the security issue is discussed. This part concludes by presenting an alternative architecture, namely a network with a star topology, for improved power consumption and efficient data communication.
- Part III describes the basic components and technologies needed for the low-cost, low-power, small-size implementation of these ad-hoc ubiquitous networks of communication and computation nodes. Issues such as programming environment, energy supply, privacy and security, packaging and algorithms for various applications are addressed as well.

This book should appeal to wide range of audiences including the technologists, the system developers, the application programmers, and the potential users. As such, it can be used as a reference document for practicing engineers, but also as a text book for graduate courses that explore the avant-garde of the information technology age.

Munich   *Werner Weber*
Berkeley   *Jan Rabaey*
Eindhoven   *Emile Aarts*
November 2004

Contents

List of Contributors

Emile Aarts
Philips Research
Prof. Holstlaan 4 (WAY2.11)
5656 AA Eindhoven
The Netherlands
emile.aarts@philips.com

Josie Ammer
Dept. of Electrical Engineering
and Computer Sciences
University of California
511 Cory Hall #1770
Berkeley, CA 94720-1770, USA
mjammer@eecs.berkeley.edu

Edward Arens
Center for the Built Environment
University of California
390 Wurster Hall #1839
Berkeley, CA 94720-1839,USA
earens@berkeley.edu

Jürgen Bohn
ETH Zürich
Swiss Federal Institute of Technology
Department of Computer Science
Institute for Pervasive Computing
ETH-Zentrum
CH-8092 Zürich, Switzerland
bohn@inf.ethz.ch

Eric Brewer
Dept. of Electrical Engineering
and Computer Sciences
University of California

511 Cory Hall #1770
Berkeley, CA 94720-1770, USA
brewer@cs.berkeley.edu

Fred Burghardt
Dept. of Electrical Engineering
and Computer Sciences
University of California
511 Cory Hall #1770
Berkeley, CA 94720-1770, USA
flb@eecs.berkeley.edu

Vlad Coroama
ETH Zürich
Swiss Federal Institute of Technology
Department of Computer Science
Institute for Pervasive Computing
ETH-Zentrum
CH-8092 Zürich, Switzerland
coroama@inf.ethz.ch

David Culler
Dept. of Electrical Engineering
and Computer Sciences
University of California
511 Cory Hall #1770
Berkeley, CA 94720-1770, USA
culler@cs.berkeley.edu

Deborah Estrin
UCLA
Computer Science Department
3531H Boelter Hall, Box 951596
Los Angeles, CA 90095-1596, USA
destrin@cs.ucla.edu

Clifford C. Federspiel
Center for the Built Environment
University of California
390 Wurster Hall #1839
Berkeley, CA 94720-1839,USA
cliff_f@berkeley.edu

David Gay
Intel Research Berkeley
2150 Shattuck Avenue, Suite 1300
Berkeley, CA 94704, USA
dgay@intel-research.net

Jana van Greunen
Dept. of Electrical Engineering
and Computer Sciences
University of California
511 Cory Hall #1770
Berkeley, CA 94720-1770, USA
janavg@eecs.berkeley.edu

Y. Gsottberger
Infineon Technologies
Corporate Research
Otto-Hahn-Ring 6
81739 Munich, Germany
yvonne.gsottberger
@infineon.com

S. Guttowski
IZM FhG
Gustav-Mayer-Allee 25
Geb. 17.2
13355 Berlin, Germany
stephan.guttowski
@izm.fraunhofer.de

Jason Hill
JLH Labs
35231 Camino Capistrano
Capistrano Beach, CA 92624, USA
jhill@jlhlabs.com

Alireza Hodjat
UCLA
Electrical Engineering Department
56-125B Engineering IV Building
Box 951594
Los Angeles, CA 90095-1594,USA
ahodjat@ee.ucla.edu

Charlie Huizenga
Center for the Built Environment
University of California
390 Wurster Hall #1839
Berkeley, CA 94720-1839,USA
huizenga@berkeley.edu

David Hwang
UCLA
Electrical Engineering Department
56-125B Engineering IV Building
Box 951594
Los Angeles, CA 90095-1594,USA
dhwang@ee.ucla.edu

Wijnand Ijsselsteijn
University of Technology Eindhoven
Den Dolech, P.O. Box 513
MB 5600 Eindhoven
The Netherlands
W.A.IJsselsteijn@tue.nl

Stefan Jung
Infineon Technologies
Wearable Technology Solutions
Balanstr. 73
81541 Munich, Germany
stefan.jung@infineon.com

William J. Kaiser
UCLA
Electrical Engineering Department
56-125B Engineering IV Building
Box 951594
Los Angeles, CA 90095-1594,USA
kaiser@ee.ucla.edu

C. Kallmayer
IZM FhG
Gustav-Mayer-Allee 25
Geb. 17.2
13355 Berlin, Germany
Kallmayer@izm.fhg.de

Jan Korst
Philips Research
Prof. Holstlaan 4 (WAY2.11)
5656 AA Eindhoven
The Netherlands
jan.korst@philips.com

Bo-Cheng Lai
UCLA
Electrical Engineering Department
56-125B Engineering IV Building
Box 951594
Los Angeles, CA 90095-1594,USA
dhwang@ee.ucla.edu

Marc Langheinrich
ETH Zürich
Swiss Federal Institute of Technology
Department of Computer Science
Institute for Pervasive Computing
ETH-Zentrum
CH-8092 Zürich, Switzerland
langheinrich@inf.ethz.ch

Christl Lauterbach
Infineon Technologies
Corporate Research
Lab for Emerging Technologies
Otto-Hahn-Ring 6
81730 Munich, Germany
christl.lauterbach@infineon.com

Philip Levis
Dept. of Electrical Engineering
and Computer Sciences
University of California
511 Cory Hall #1770
Berkeley, CA 94720-1770, USA
pal@cs.berkeley.edu

En-yi Lin
Dept. of Electrical Engineering
and Computer Sciences
University of California
511 Cory Hall #1770
Berkeley, CA 94720-1770, USA
enyi@eecs.berkeley.edu

Sam Madden
Dept. of Electrical Engineering
and Computer Sciences
University of California
511 Cory Hall #1770
Berkeley, CA 94720-1770, USA
madden@cs.berkeley.edu
and
Intel Research Berkeley
2150 Shattuck Avenue, Suite 1300
Berkeley, CA 94704; USA
madden@intel-research.net
and
MIT
CSAIL
32 Vassar St.
Cambridge, MA 02139, USA
madden@csail.mit.edu

Friedemann Mattern
ETH Zürich
Swiss Federal Institute of Technology
Department of Computer Science
Institute for Pervasive Computing
ETH-Zentrum
CH-8092 Zürich, Switzerland
mattern@inf.ethz.ch

Panos Markopoulos
University of Technology Eindhoven
Den Dolech, P.O. Box 513
MB 5600 Eindhoven
The Netherlands
p.markopoulos@tue.nl

M. Niedermayer
IZM FhG
Gustav-Mayer-Allee 25
Geb. 17.2
13355 Berlin, Germany
Michael.Niedermayer
@izm.fraunhofer.de

Brian Otis
Dept. of Electrical Engineering
and Computer Sciences
University of California
511 Cory Hall #1770
Berkeley, CA 94720-1770, USA
botis@uclink.berkeley.edu

Joseph Polastre
Dept. of Electrical Engineering
and Computer Sciences
University of California
511 Cory Hall #1770
Berkeley, CA 94720-1770, USA
polastre@cs.berkeley.edu

Gregory J. Pottie
UCLA
Electrical Engineering Department
56-125B Engineering IV Building
Box 951594
Los Angeles, CA 90095-1594,USA
pottie@ee.ucla.edu

Jan M. Rabaey
Dept. of Electrical Engineering
and Computer Sciences
University of California
511 Cory Hall #1770
Berkeley, CA 94720-1770, USA
jan@eecs.berkeley.edu

H. Reichl
IZM FhG
Gustav-Mayer-Allee 25
Geb. 17.2
13355 Berlin, Germany
Reichl@izm.fraunhofer.de

Michael Rohs
ETH Zürich
Swiss Federal Institute of Technology
Department of Computer Science
Institute for Pervasive Computing
ETH-Zentrum
CH-8092 Zürich, Switzerland
rohs@inf.ethz.ch

Shad Roundy
Faculty of Engineering
and Information Technology
Building 32, Baldessin Precinct
Building
The Australian National University
ACT 0200 Australia
shad.roundy@anu.edu.au

Boris de Ruyter
Philips Research
Prof. Holstlaan 4 (WAY2.11)
5656 AA Eindhoven
The Netherlands
boris.de.ruyter@philips.com

Alberto Sangiovanni-Vincentelli
Dept. of Electrical Engineering
and Computer Sciences
University of California
511 Cory Hall #1770
Berkeley, CA 94720-1770, USA
alberto@eecs.berkeley.edu

Marco Sgroi
DoCoMo Euro-Labs
Landsberger Strasse 312
80687 Munich, Germany
MarcoSgroi@aol.com

Rahul Shah
Dept. of Electrical Engineering
and Computer Sciences
University of California
511 Cory Hall #1770
Berkeley, CA 94720-1770, USA
rahul_shah@ieee.org

Mike Sheets
Dept. of Electrical Engineering
and Computer Sciences
University of California
511 Cory Hall #1770
Berkeley, CA 94720-1770, USA
msheets@eecs.berkeley.edu

X. Shi
Infineon Technologies
Corporate Research
Otto-Hahn-Ring 6
81739 Munich, Germany
xiaolei.shi@infineon.com

Fred Snijders
Philips Research
Prof. Holstlaan 4 (WAY2.11)
5656 AA Eindhoven
The Netherlands
fred.snijders@philips.com

Mani Srivastava
Dept. of Electrical Engineering
and Computer Sciences
University of California
511 Cory Hall #1770
Berkeley, CA 94720-1770, USA
mbs@ee.ucla.edu

Marc Strasser
Infineon Technologies
MP PDT CS SIM
Balanstr. 73
81541 Munich, Germany
marc.strasser@infineon.com

G. Stromberg
Infineon Technologies
Corporate Research
Otto-Hahn-Ring 6
81739 Munich, Germany
guido.stromberg@infineon.com

T.F. Sturm
University of the
Federal Armed Forces
85577 Neubiberg
Germany
thomas.sturm@unibw-muenchen.de

Gaurav S. Sukhatme
Computer Science (MC 0781)
University of Southern California
941 W. 37th Place
Los Angeles, CA 90089-0781, USA
gaurav@usc.edu

Robert Szewczyk
Dept. of Electrical Engineering
and Computer Sciences
University of California
511 Cory Hall #1770
Berkeley, CA 94720-1770, USA
szewczyk@cs.berkeley.edu

Ingrid Verbauwhede
UCLA
Electrical Engineering Department
56-125B Engineering IV Building
Box 951594
Los Angeles, CA 90095-1594,USA
ingrid@ee.ucla.edu

Wim F.J. Verhaegh
Philips Research
Prof. Holstlaan 4 (WAY2.11)
5656 AA Eindhoven
The Netherlands
wim.verhaegh@philips.com

John Villasenor
UCLA
Electrical Engineering Department
56-125B Engineering IV Building
Box 951594
Los Angeles, CA 90095-1594,USA
villa@icsl.ucla.edu

Danni Wang
Center for the Built Environment
University of California
390 Wurster Hall #1839
Berkeley, CA 94720-1839,USA
wangdn@berkeley.edu

Werner Weber
Infineon Technologies
Corporate Research
Otto-Hahn-Ring 6
81739 Munich, Germany
Werner.Weber@infineon.com

Matt Welsh
Intel Research Berkeley
2150 Shattuck Avenue, Suite 1300
Berkeley, CA 94704, USA
mdw@intel-research.net
and
Division of Engineering
and Applied Sciences
Harvard University
Cambridge, MA 02138, USA
mdw@eecs.harvard.edu

Kamin Whitehouse
UCLA
Electrical Engineering Department

56-125B Engineering IV Building
Box 951594
Los Angeles, CA 90095-1594,USA
kamin@cs.berkeley.edu

Adam Wolisz
Technical University of Berlin
Dept. of Electrical Engineering
Telecommunication Networks
Einsteinufer 25
10587 Berlin, Germany
wolisz@tkn.tu-berlin.de

Alec Woo
UCLA
Electrical Engineering Department
56-125B Engineering IV Building
Box 951594
Los Angeles, CA 90095-1594,USA
awoo@cs.berkeley.edu

Paul K. Wright
Dept. of Mechanical Engineering
University of California
5133 Etcheverry
Berkeley, CA 94720-1740, USA
pwright@robocop.me.berkeley.edu

Introduction

W. Weber, J. Rabaey, and E. Aarts

Why does mankind always want to predict the future? The science of futurology has invented various methods such as Delphi studies and scenario techniques that try to predict various aspects of upcoming societies. In the past, these forecasts, however, have had limited success. Important breakthroughs such as the success of the Internet came as a surprise to this field of science. However, it has been observed that as soon as ideas become common to opinion leaders, they develop into self-fulfilling prophecies and materialize as powerful and universal trends. "Ambient Intelligence" is a concept that has this potential.

Ambient intelligence is the vision that technology will become invisible, embedded in our natural surroundings, present whenever we need it, enabled by simple and effortless interactions, attuned to all our senses, adaptive to users and context-sensitive, and autonomous. High-quality information access and personalized content must be available to everybody, anywhere, and at any time. Ambient intelligence is characterized by an environment

- where technology is embedded, hidden in the background
- that is sensitive, adaptive, and responsive to the presence of people and objects
- that augments activities through smart non-explicit assistance
- that preserves security, privacy and trustworthiness while utilizing information when needed and appropriate.

Ambient intelligence is used to support human contacts and accompany an individual's path through the complicated modern world. It offers him information and guidance whenever needed. In other words, it improves the control of his surrounding. Control of human health and security functions is another goal of ambient intelligence. Various logistics applications such as tracking of bags in airports or of books in libraries are also sometimes considered ambient intelligence, although a highly developed interface to humans is not key to those. Moreover, ambient intelligence will enable people to express themselves in ways that are unprecedented as natural interaction and augmented environments will become available. This may result in enhanced expressiveness, productivity, and well-being that may improve the quality of life substantially.

From the technical standpoint, ambient intelligence means distributed electronic intelligence. The necessary hardware vanishes into the background. Devices used for ambient intelligence are small, low-power, low weight and (very importantly) low-cost; they collaborate or interact with each another; and they are redundant and error-tolerant. That means that the failure of one device will not cause failure of the whole system. Often wired connections do not exist; thus radio methods play an important role for data transfer.

Some believe that ambient intelligence is only a vision of a very distant future. Indeed, some of these visions lie at least 20 years ahead of us. The realization of intelligent electronic wall paper or smart dust forming distributed intelligence, for instance, will require additional technological achievement. For some of those not even basic feasibility has been demonstrated today. Examples are large-size paper-like display technologies, or printable, ultra-low-cost, ultra-low-power electronics. Experience tells that the introduction to the market of new basic hardware technologies takes a long time even after a basic proof of functionality has been demonstrated.

On the other side, indications of a trend towards ambient intelligence are already evident today. While in the 1970s one computer served many users, and in the 1990s the personal computer served humans on a one-to-one basis, individuals are already served by many computational devices today. Peripheral devices that were dependent on central computing a few years ago now have become independent and contain powerful controllers. It is our vision that this trend towards distributed electronics and intelligence will only accelerate in the near future. In addition, many of the technologies that would enable this revolution are becoming available today. Technologies such as energy-scavenging, low-power and low-cost digital and wireless components, and integrated sensors are making their way into the market. In addition, self-configuring, robust networking techniques are being conceived making it possible to connect 100's of devices seamlessly and effortlessly in a scalable fashion.

The research work on ambient intelligence strives to demonstrate applications and technologies that will enable semiconductor solutions for the technology lifestyle of the individual in the 21st century. In this work, the individual is the subject, whereas electronics is the object. In some sense, this concept is the opposite of virtual reality, in which the user is embedded in an electronic environment that provides sensory inputs. In virtual reality, the user is controlled by the system and becomes the object.

The users of the targeted applications want their daily needs to be met. They want an excellent interface between technology and themselves. Neither specific operational knowledge, nor awareness of the functioning of the appliances or applications should be necessary. It is our vision to provide human individuals with better control of their surroundings, assisted by invisible, highly functional electronic appliances that support personal expression, well-being, and productivity, eventually leading to a better quality of life.

Part I

Applications

Social, Economic, and Ethical Implications of Ambient Intelligence and Ubiquitous Computing

J. Bohn, V. Coroamă, M. Langheinrich, F. Mattern, and M. Rohs

Summary. Visions of ambient intelligence and ubiquitous computing involve integrating tiny microelectronic processors and sensors into everyday objects in order to make them "smart." Smart things can explore their environment, communicate with other smart things, and interact with humans, therefore helping users to cope with their tasks in new, intuitive ways. Although many concepts have already been tested out as prototypes in field trials, the repercussions of such extensive integration of computer technology into our everyday lives are difficult to predict. This contribution is a first attempt to classify the social, economic, and ethical implications of this development.

1 Introduction

The increasing miniaturization of computer technology will, in the foreseeable future, result in processors and tiny sensors being integrated into more and more everyday objects, leading to the disappearance of traditional PC input and output media such as keyboards, mice, and screens. Instead, we will communicate directly with our clothes, watches, pens, and furniture – and these objects will communicate with each other and with other people's objects.

More than 10 years ago, Mark Weiser foresaw this development and described it in his influential article "The Computer for the 21st Century" [1]. Weiser coined the term "ubiquitous computing," referring to omnipresent computers that serve people in their everyday lives at home and at work, functioning invisibly and unobtrusively in the background and freeing people to a large extent from tedious routine tasks. In its 1999 vision statement, the European Union's Information Society Technologies Program Advisory Group (ISTAG) used the term "ambient intelligence" in a similar fashion to describe a vision where "people will be surrounded by intelligent and intuitive interfaces embedded in everyday objects around us and an environment recognizing and responding to the presence of individuals in an invisible way" [2].

This contribution is based on an earlier journal article "Living in a World of Smart Everyday Objects – Social, Economic, and Ethical Implications", Human and Ecological Risk Assessment, Vol. 10, No. 5, October 2004.

The vision of a future filled with smart and interacting everyday objects offers a whole range of fascinating possibilities. For example, parents will no longer lose track of their children, even in the busiest of crowds, when location sensors and communications modules are sewn into their clothes. Similar devices attached to timetables and signposts could guide blind people in unknown environments by "talking" to them via a wireless headset [3]. Tiny communicating computers could also play a valuable role in protecting the environment, for example as sensors the size of dust particles that detect the dispersion of oil spills or forest fires. Another interesting possibility is that of linking any sort of information to everyday objects, allowing for example future washing machines to query our dirty clothes for washing instructions.

While developments in information technology never had the explicit goal of changing society, but rather did so as a side effect, the above-mentioned visions expressly propose to transform society by fully computerizing it. It is therefore very likely that this will have long-term consequences for our everyday lives and ethical values that are much more far-reaching than the Internet with all its discussions about spam e-mails, cyber crime, and child pornography. With its orientation towards the public as well as the private, the personal as well as the commercial, it aspires to create technology that will accompany us throughout our entire lives, day in and day out. And if Mark Weiser's vision of "invisible computing" actually materializes, we won't even notice any of it.

It seems to be clear that with these technical developments – pushed through largely unnoticed by the general public and extending quite rapidly into our everyday lives – unanticipated (if not unacceptable) standards could soon be set for the rest of our lives. In the following, we examine the driving factors behind the visions of ubiquitous computing and ambient intelligence – from a technical as well as an economic perspective – and we try to illustrate the social and ethical implications of a "smart world" that connects everything to everything else, where anywhere can potentially be contacted from anywhere else, and where everybody could conceivably interact with anybody (and anything) else.

2 Technology Trends

The driving force behind continuing technological progress in the field of information technology is the long-term trend in microelectronics: Moore's Law [4], drawn up in the late 1960s by Gordon Moore and roughly stating that the power of microprocessors doubles about every 18 months, has held true with astonishing accuracy and consistency. A similarly high increase in cost-efficiency can be observed for some other technological parameters such as storage capacity and communications bandwidth. To put it another way, prices for microelectronic functionality with an equivalent amount of computing power are falling radically over time. This trend, which is expected to continue for at least another 15 years, means that computer processors and storage components will become much more powerful, smaller, and cheaper in the future, so that there will be an almost unlimited supply of them.

Even more important are the results of microsystems technology and nanotechnology. These could lead, for example, to flexible displays or electronic paper. Another interesting development is radio sensors that can report their readings within a few meters distance without an explicit energy supply – such sensors obtain the necessary energy from the environment or directly from the measuring process itself.

Electronic labels (so-called "smart labels" or RFID tags) also operate without their own energy supply. Depending on their construction, these are less than a square millimeter in area and thinner than a piece of paper. In some ways, this is a further development of the well-known anti-theft technology involving security gates in department stores. However, this is not just about the binary information "paid/stolen"; within milliseconds, several hundred characters could be read and written "wirelessly" up to a distance of a few meters [5].

What is interesting about such remote-inquiry electronic markers is that they enable objects to be clearly identified and recognized, and therefore linked in real time to an associated data record held on the Internet or in a remote database. This ultimately means that specific data can be associated with any kind of object. If everyday objects can be uniquely identified from a distance and furnished with information, this opens up application possibilities that go far beyond the original purpose of automated warehousing or supermarkets without cashiers.

Significant advances have also been made in the field of wireless communications. Especially interesting are recent short-range communications technologies that require very little energy, making it possible to produce designs that are much smaller and cheaper than today's mobile phones. Intensive research is also being carried out into improved options for indicating the position of mobile objects. As well as increased accuracy (currently around ten meters for the GPS system), the aim is also to make the devices much smaller.

If you summarize these technology trends and developments – tiny, cheap processors with integrated sensors and wireless communications capability, attaching information to everyday objects, the remote identification of objects, the precise localization of objects, flexible displays based on polymers, and electronic paper – it becomes clear that the technological basis for a strange new world has been created: everyday objects that are in some respects "smart," and with which we can even communicate under certain circumstances.

There are various ways of implementing such communication with things. As one example, imagine everyday objects such as furniture, packaged food, medication, clothing, or toys being equipped with an electronic label containing a specific Internet address as digital information. If you can then read this Internet address with a portable device just by pointing it at the object, this device can, independently and with no further assistance from the object in question, access and display the associated information from the Internet via the mobile phone network. The user has the impression that the object itself has "transmitted" the information, although in fact it has been supplied to the display device via the Internet. The information could be, for example, operating instructions, or cooking instructions for a ready-to-serve meal, or the information leaflet for medication. The details of what is displayed may depend on the "context" – for example, whether the user is a good customer and paid a lot of money for the product, whether he is over 18 years of age, what language he speaks, or his current location, – but also maybe whether he has paid his taxes on time . . .

The foreseeable technological developments will therefore add an additional new quality to everyday objects – these might be able not only to communicate with people and other "smart" objects, but also to discover where they are, which other objects are in their vicinity, and what has happened to them in the past, for example. Objects and devices could thus behave in a context-sensitive manner and appear to be "smart," without actually being "intelligent."

While technological advances such as miniaturization, increasing computing power, and wireless connectivity open up the possibility of new applications, critics [6–8] argue that it is not yet clear how these possibilities are actually going to be put into practice: we are "brilliant on means, but pretty hopeless when it comes to ends" [9]. However, this *innovation dilemma* – we may know *how* we can create incredible things, but we don't know *what* needs they are supposed to meet – is only superficial. The following section describes potential economic benefits that ubiquitous-computing technology offers when it comes to industrial processes – benefits that will be a prominent driver for the proliferation of ambient intelligence, perhaps even more than the above-mentioned technological progress itself.

3 The Ambient Economy

"The most profound revolutions are not the ones trumpeted by pundits, but those that sneak in when we are not looking" [10]. What Mark Weiser formulated over ten years ago accurately describes the current atmosphere surrounding the fields of ambient intelligence and ubiquitous computing. While personal gadgetry in the form of smartphones and Internet fridges continues to bedazzle the press, industry has quietly begun setting its sights on the enormous business potential that technologies such as wireless sensors, RFID tags, and positioning systems have to offer. Analysts call it the *real-time economy* or *now-economy* [11], where more and more entities in the economic process, such as goods, factories, and vehicles, are being enhanced with comprehensive methods of monitoring and information extraction. Ultimately, the whole lifecycle of products, beginning with the "birth" of their components and ending with their complete consumption (or recycling), can be witnessed (and, to some extent, even controlled) in real time.

Two important technologies form the core of these new economic processes and applications: the ability to *track* real-world entities, and the *introspection capabilities* of smart objects. Tracking objects in real-time allows for more efficient business processes, while objects that can monitor their own status via embedded sensors allow for a range of innovative business models.

3.1 Staying on Track: Smart Products

Inventory management obviously benefits from accurate, real-time information on the location and condition of goods, equipment, and manpower. If a company does not know the location and condition of its stock, and how long it has been in the warehouse, significant costs are incurred. Missed profits, oversized inventories, and the devaluation of goods depreciating in the warehouse are possible consequences of a lack of information. The stocktaking required for business or legal reasons also typically requires a considerable amount of effort. Stocktaking is not only expensive; it is inherently error-prone as well. A factory floor or warehouse equipped with technologies such as indoor localization and automatic identification can largely automate the stocktaking process, thereby reducing costs.

If several companies along a supply chain simultaneously use such precise inventory data in addition to real-time order information, they can achieve additional savings by significantly attenuating the so-called "bullwhip effect" [12]. This effect, often noticed in practice, describes the following phenomenon: although consumer demand for a product remains almost constant over time, small changes in this demand amplify along the supply chain and ultimately result in either excess production (and associated storage costs) or sudden interruptions to supply (and associated missed sales). However, the more information transparency there is along the supply chain, the more these undesirable effects are attenuated [13]. By making comprehensive information available along the supply chain, a significant reduction in the bullwhip effect can be achieved.

A further step towards the *now economy* is the constant monitoring of critical product parameters (e.g., of temperature-sensitive goods such as chemicals or groceries) by tiny wireless sensors. Equipped with communications capabilities, such "introspective" goods are not only able to monitor themselves, but can also communicate relevant parameters to the outside world [14]. Smart goods could observe their condition while in transit and trigger an alarm in the event of excessive temperatures, which could – if appropriate – lead to an automatic reordering of damaged goods. Alternatively, the goods could also attempt to take corrective action, for example by controlling the temperature of their container: "As sensors improve and always-on connectivity becomes a reality, products will be able to do something about their condition" [15]. In this way, "self-conscious" products (i.e., products that perceive their condition, analyze it, and attempt to change their situation if they are dissatisfied with it) would lower the total transaction costs by reducing the time necessary to procure replacements for damaged goods.

3.2 Anything, Anywhere: Innovative Business Models

The benefits of a world full of smart objects do not stop at the factory floor. Once comprehensive infrastructures for tracking goods and facilitating communication among "self-conscious" objects are in place, ambient intelligence throughout our environment – in private homes, cars, trains, and public places – would facilitate a range of new applications and business models. The prospect of their revenue streams might become another substantial driver for the deployment of ambient-intelligence technologies in the near future.

3.2.1 Real-Time Shopping

"See a great sweater on someone walking by? Find out the brand and price, and place an order. Or maybe you'll be wearing the sweater and earning a commission every time someone near you sees and buys." This vision [15] describes a future, maybe not that far off, in which the boundary between the real world and the world of information has become blurred. Invisible tags embedded in most products would allow consumer devices to read out the unique identification number of an item and use this to access the object's virtual representation in the information world, which in turn could provide the user with a wealth of background information (e.g., ingredients, product reviews, etc.) as well as direct links to online (or even real-world) shops selling this item. People could shop on the move – on the streets, in buses, or whilst chilling out in their favorite bar at night – and every item sold could in turn become a new sales channel.

In a more proactive fashion, smart products could begin to subtly advertise themselves, or even use cross-marketing to advertise their "friends." A smart refrigerator could, for example, recommend recipes based on both the groceries it contains, and on the items currently discounted at the local supermarket. It could also accrue reward points every time goods of a promoted brand are stored in it. With the help of ambient-intelligence technology and the detailed profiles it enables, prices of everyday goods could even be adjusted to suit individual customers. In the vendor's ideal world "each consumer would be quoted an individual price, which [. . .] exactly corresponded to his readiness to pay" [16] – economists call this "perfect price discrimination." Such individual prices have to be introduced carefully, however: the online-bookseller Amazon discovered that such perfection might not be universally appreciated, after their trial of individual DVD prices had to be suspended almost immediately due to massive customer criticism [17,18].

The ultimate in shopping may be achieved when all the decision-making is removed from people, and things do the shopping themselves (see the section on social challenges below for a discussion of ubiquitous computing and consumer control). The business consultancy Accenture has already coined a phrase for this – "silent commerce" [19]. Their vision of "autonomous purchasing objects" not only includes photocopiers responsible for ordering their own paper, but also Barbie dolls[1] that delight children (and their parents ...) by ordering new clothes with their own pocket money: "Barbie detects the presence of clothing and compares it with her existing wardrobe – after all, how many tennis outfits does a doll need? The toy can buy straight from the manufacturer via the wireless connection ... She can be constantly and anonymously shopping, even though the owner might not know it" [20].

Apart from such rather questionable ways of increasing sales revenues, however, the "smartness" of objects might also facilitate trading at a more fundamental level, by reducing *information asymmetries* [21]: goods could not only "talk" about their price, ingredients, and availability, but also provide a detailed history of their production, use, and repair. A used car could give a detailed list of the parts that have been replaced or repaired over the course of its lifetime, thus reducing the amount of trust the buyer must have in the seller. Organic food could provide consumers with a comprehensive history of its cultivation, fertilization, and processing, potentially increasing their willingness to pay a premium for it. Obviously, in order for consumers to trust this information, it would either require verification from trusted third parties, or some sort of technical means to prevent tampering (much like odometers today). However, by increasing the overall market transparency, both for new and used goods, an environment supporting trustworthy ambient-intelligence systems could potentially increase overall market activities by reducing the uncertainty inherent in many transactions today.

3.2.2 Pay-Per-Weight?

In a future full of smart objects, we may not only be tempted to buy (and have the ability to shop) just about anywhere at any time, we may also *have* to buy just about everywhere, all the time. Digital rights management systems that have recently been developed for distributing digital music and video are a first step in this direction. These systems make it possible for owners of digital content to exert control over the access to digital information even after it has been sold – Buena Vista's Disney home video unit, for example, recently started selling DVDs that render themselves unplayable 48 hours after unpacking [22]. Digital rights management systems could even be programmed

---

[1] Apparently, the toy manufacturer Mattel, owner of the brand, was not at all happy about this and sent in the lawyers right away – Accenture's Web site now refers only to "a doll", with no mention of Barbie.

to require a *continuous payment* while listening to digital music or watching digital video. Such *pay-per-use* models have traditionally been used for phone calls or public utilities (e.g., electricity, gas), but could now – in an ambient-intelligence world – be implemented using everyday objects equipped with sensors and communications capabilities. Furniture, for instance, could monitor its usage (e.g., a sofa could count the number of persons that sit on it, the persons' weight and seating time) and create a monthly itemized billing statement. While private homes might still prefer owning their beds rather then being billed for sleeping in them, corporate buyers with a high turnover of furniture (such as hotels or offices) could potentially provide a sufficiently large customer base for such a business model to develop. However, as with the market transparencies described above, mechanisms would need to be in place in order to assure both parties that the sensor readings had not been tampered with.

3.2.3 No Risk, No Premium

Information asymmetry is not only a hindrance for trading, but also affects the risk management (moral hazard) of insurance companies. Typically, insurers use broad categories such as age or location in order to judge the risk involved in providing individual coverage, e.g., for health or car insurance. An ambient-intelligence environment, however, would allow for a much more fine-grained calculation of the individual risks involved. A smart car, for example, could provide detailed information about the driving style and parking habits of its owner, thus providing the insurer with a much better assessment of the likelihood of an accident or theft. For "safe" drivers, this would result in reduced insurance premiums as they would be exempted from paying for the insurer's uncertainty regarding their driving style.

Insurance rates could even become entirely dynamic, reducing or increasing the premium in a "pay-per-use" fashion in real time, e.g., during driving. Taken to the extreme, the current insurance rate, similar to today's indicators showing current gas consumption, could be calculated on the spot and displayed in an "insurance meter", based on the time of day, the route traveled, weather and road conditions, and of course driving style. Similar adjustments could be made for home owners' insurance (new furniture automatically registers its net worth when placed inside the home) or health insurance (smoking a single cigarette would increase the rate, a walk in the park decrease it). Even in the case of an actual insurance claim, smart objects could significantly lower the insurer's cost by observing the circumstances of an accident or by noticing which items were actually missing after a break-in.

Obviously, such detailed activity records would constitute a serious privacy hazard for many consumers. While technical remedies might be possible (e.g., data would be encrypted before being sent to the insurer, who would need an authorized key from the customer – or law enforcement – to read it in case of an accident), both the increased complexity and the limited usefulness to the insurer of such encrypted information would most likely favor the privacy-threatening alternatives. Even if customers had the choice of opting out, drivers who did not wish to pass their details on to their insurer would most likely have to pay a considerably higher premium, as the insurer's risk would be spread among fewer and fewer non-participating customers. This would not only limit the desired freedom of choice, particularly for low-income drivers, but also implicitly accelerate the move away from fixed rates to cheaper dynamic (but privacy-invading) solutions.

3.3 An Economy on Autopilot

Despite the number of potential economic advantages described above, there are also substantial risks involved when relying on ubiquitous-computing technologies for large parts of an economy. The increasing automation of economically relevant aspects and the exclusion of humans as decision makers could certainly become a cause for concern. Under "normal" circumstances, automated control processes increase system stability – machines are certainly much better than humans if they have to devote their whole attention to a particularly boring task. But situations that have not been anticipated in the software can easily have disastrous consequences if they are not directly controlled by humans (as tragic accidents involving airplane autopilots have shown in the past). Other problems might arise from the intricate interplay of several automated processes, which might quickly escalate into an unanticipated feedback loop that gets out of control. For example, the stock market crash of 1987 was partly caused by newly implemented trading software [11], which was designed in such a way that it would trigger the sale of shares whenever a certain pattern appeared in the daily fluctuations of their prices. Since a large number of traders were using the same software, the first appearance of this pattern caused a flood of sales, which in turn reinforced the software's selling pattern and thus triggered the crash.

Another risk stems from the increased efficiency of an economy supported by ambient-intelligence technologies. While detailed tracking and inventory systems, along with smart, "self-conscious" products, allow a company to trim production and stock keeping as much as possible, this lack of "slack" in the production process also increases the risk that unforeseen interruptions can have grave consequences. In the case of supply chain management, for example, the attenuation of the bullwhip effect permits a reduction in storage capacity. However, if all companies along the delivery chain drastically reduce their stocks, one small unexpected interruption in supply by the weakest member would immediately halt all production along the whole chain.

The automation and acceleration of the economy increases not only the potential for possible savings, but also influences the associated risks in these types of complex and sensitive systems. Nevertheless, the increased flexibility and control that a world of ambient intelligence offers will certainly be an important incentive for businesses to drive its deployment in the near future.

Therefore, it is important that the ambient-intelligence landscapes we build should be both *reliable* and *socially acceptable*. We will come back to these issues in a later section. For now, we want to take a closer look at what probably constitutes the most visible implication of an economy driven by ambient-intelligence technologies, namely its threat to personal privacy.

4 Privacy in Ambient Intelligence

Intelligent fridges, pay-per-use scenarios, and dynamic insurance rates paint a future in which all of our moves, actions, and decisions are recorded by tireless electronic devices, from the kitchen and living room of our homes to our weekend trips in our cars. Not surprisingly, many critics see this as "an attempt at a violent technological penetration of everyday life" [7], as the "feverish dream of spooks and spies – to plant a 'bug' in every object" [23] or even as "a project that aims at totality and, of course, verges on the totalitarian" [6].

By virtue of its very definitions, the vision of ambient intelligence has the potential to create an invisible and comprehensive surveillance network, covering an unprecedented share of our public and private life: "The old sayings that 'the walls have ears' and 'if these walls could talk' have become the disturbing reality. The world is filled with all-knowing, all-reporting things" [8]. And with the economic possibilities described above, such a comprehensive coverage seems more likely to be put into place: shopping without participating in comprehensive profiling, buying instead of renting items in a pay-per-use scheme, as well as fixed insurance schemes that do not constantly transmit information to insurers – all of this might become an expensive luxury for well-off citizens, while the population at large must trade in their privacy for increased productivity and market transparency. This might very well be self-inflicted: given the immediate economic returns of consumer loyalty programs or low insurance rates, the rather vague threats of future privacy violations are easily enough ignored.[2] The following sections try to add a differentiated view to this problem, especially with respect to ubiquitous-computing technology, by first examining *why* personal privacy is desirable, describing *when* we feel that it has been violated, and then assessing *how* the deployment of future ambient-intelligence systems will affect all that.

---

[2] The continuing media attention that privacy issues are receiving has nevertheless prompted industry to tread carefully: "A very cautious approach is needed [. . .] with this kind of monitoring, otherwise newspaper headlines such as 'Spy in the Kitchen' would soon appear, killing the intelligent appliance before it takes off" [55].

4.1 The Many Facets of Personal Privacy

Even though critics continue to argue that "all this secrecy is making life harder, more expensive, dangerous and less serendipitous" [24], privacy is still predominantly seen as a fundamental requirement of any modern democracy [25]. It is only when people are free to decide what to do with their lives, according to their interests and beliefs, and without fear of repression from their fellow citizens, that the necessary plurality of ideas and attitudes can develop that will prevent society being subjugated under a charismatic leader. Harvard law professor Lawrence Lessig [26] takes this requirement a step further and distinguishes between a number of motives for the protection of privacy in today's laws and standards:

- *Privacy as Empowerment.* Seeing privacy mainly as informational privacy, its aim is to give people the power to control the publication and distribution of information about themselves [27]. A recent legal discussion surrounding this motivation revolved around the question of whether personal information should be seen as private property (which would entail the right to sell all or part of it as the owner sees fit) or as intellectual property (which would entitle the owner to certain inalienable rights, preventing him for example from selling the rights to his own name to anybody else).
- *Privacy as Utility.* From the viewpoint of the person involved, privacy can be seen as a utility providing more or less effective protection against nuisances such as unsolicited phone calls or emails. This view probably best follows Brandeis' definition of privacy as "The right to be left alone," where the focus is on minimizing the amount of disturbance for the individual [28].
- *Privacy as Dignity.* Dignity not only entails being free from unsubstantiated suspicion (for example being the target of a wire tap, where the intrusion is usually not directly perceived as a disturbance), but also focuses on the *equilibrium* of information available between two people: as in a situation where you are having a conversation with a fully dressed person when you yourself are naked, any relationship where there is a significant information imbalance will make it much more difficult for those with less information about the other to keep their composure.
- *Privacy as a Regulating Agent.* Privacy laws and moral norms to that extent can also be seen as a tool for keeping checks and balances on the powers of a decision-making elite. By limiting information gathering of a certain type, crimes or moral norms pertaining to that type of information cannot be effectively controlled.

Depending on what kind of motives one assumes for preserving privacy, ambient-intelligence technology can become the driving factor for changing the scope and impact of privacy protection as it exists today, and creating substantially different social landscapes in the future. This is because ambient-intelligence technology influences two important design parameters relating to privacy: the ability to monitor and the ability to search [26].

4.2 Ambient Intelligence and Surveillance

The conscious observation of the actions and habits of our fellow men is as old as mankind itself. However, observations using automated systems differ from our nosy neighbors in one important aspect: while in the "good old days", anything out of the ordinary would attract the attention of our fellow citizens, it is now the *ordinary*, the everyday routine, that can be (and often is) the sole focus of tireless computerized monitoring. With ambient–intelligence technologies, today's monitoring capabilities can obviously be extended far beyond credit-card records, call logs, and news postings. Not only will the *spatial* scope of such monitoring activities be significantly extended in ambient-intelligence landscapes, but their *temporal* coverage will also greatly increase: starting from pre-natal diagnostics data stored on babies' hospital smart cards, to activity patterns in kindergarten and schools, to workplace monitoring and senior citizen's health monitoring.

Such comprehensive monitoring (or surveillance) techniques create new opportunities for what MIT professor emeritus Gary T. Marx calls *border crossings*: "Central to our acceptance or sense of outrage with respect to surveillance ... are the implications for crossing personal borders" [29]. He goes on to define four such border crossings that form the basis for perceived privacy violation:

- *Natural Borders.* Physical borders of observability, such as walls and doors, clothing, darkness, and also sealed letters and phone conversations. Even facial expressions can represent a natural border against the true feelings of a person.
- *Social Borders.* Expectations with regard to confidentiality in certain social groups, such as family members, doctors, and lawyers. This also includes the expectation that your colleagues do not read personal fax messages addressed to you, or material that you leave lying around the photocopier.
- *Spatial or Temporal Borders.* The expectation by people that parts of their lives can exist in isolation from other parts, both temporally and spatially. For example, a previous wild adolescent phase should not have a lasting influence on the current life of a father of four, nor should an evening with friends in a bar influence his coexistence with work colleagues.
- *Borders due to Ephemeral or Transitory Effects.* This describes what is best known as a "fleeting moment," a spontaneous utterance or action that we hope will soon be forgotten, or old pictures and letters that we put out in our trash. Seeing audio or video recordings of such events subsequently, or observing someone sifting through our trash, would violate our expectations of being able to have information simply pass away unnoticed or forgotten.

Putting ambient-intelligence systems into place will most certainly allow far greater possibilities for such border crossings in our daily routines. Consider the popular vision of a wearable "memory amplifier" [30, 31], allowing its wearer to constantly record the events of her daily life in a lifetime multimedia diary. While at first sight such technology promises assistance to those of us who frequently tend to forget small details, it also has substantial consequences for our privacy borders stemming from *ephemeral and transitory effects*: any statement I make during a private conversation could potentially be played back if my conversation partner gave others access to her multimedia diary. Even if this information were never disclosed to others, the very thought of dealing with people who have a perfect memory (and thus would *never* forget anything) would probably have a considerable effect on our interpersonal relationships.

The problem of *spatial and temporal borders,* on the other hand, is well known from the field of consumer profiles. Although such profiles are often the subject of public debate, the social and legal attitudes towards them have, until now, been relatively relaxed. Consumer acceptance is also much higher than the frequent negative news coverage might indicate, mostly because their negative consequences are often perceived as being rather minor (such as unsolicited spam) compared to their advantages (e.g., monetary incentives in the form of discounts or rewards). However, there are well-known risks associated with profiles, and their adoption as the basis for ambient intelligence would only exacerbate such problems. Besides the obvious risk of accidental leaks of information [32], profiles also threaten universal equality, a concept central to many constitutions, basic laws, and human rights, where "all men are created equal" [33]. Even though an extensively customized ambient-intelligence future where I only get the information that is relevant to my profile holds great promise, the fact that at the same time a large amount of information might be deliberately *withheld* from me because I am not considered a valued recipient of such information, would constitute a severe violation of privacy for many people.

Applying ambient-intelligence technology in areas with primarily *social borders* – for example where a close social group interacts only among itself, such as families [34, 35] or co-workers – might appear to alleviate some of the above concerns. Most participants would already share close relationships and tend to know a great deal about each other, without needing a system to compile a profile of their communication partner. Such systems, however, also raise the ante as to what *type* of information they handle. While a communication whiteboard for families may facilitate social bonding between physically and temporally separated members, it would also increase the risk of unwanted social border crossings by accidentally allowing Mum to read a message you left for your sister, or a visiting friend to be recorded in the house activity log even though you told grandma you would spend the weekend alone.

Natural borders, then, might be the easiest to respect when designing ambient-intelligence systems. Here, the concept of surveillance is well known and usually fairly straightforward to spot, after all: if others are able to watch your actions behind closed doors, they are most certainly intruding on your privacy. Proponents of wearable computing systems often cite the fact that information could both be gathered and stored *locally* (i.e., on the user's belt, or within her shirt) as a turnkey solution for privacy-conscious technologists [36]. Border crossings, however, are not only about *who* does something, but also *what* is happening. Even though a context-aware wearable system might keep its data to itself, its array of sensors nevertheless probe deep into our personal life, and the things they might find there could easily startle (and trouble) us, once such systems start anticipating our future actions and reactions. The feeling of having someone (or something) constantly looking over our shoulder and second-guessing us would certainly constitute a natural border crossing for most of us. And the temptation of law enforcement subpoenaing such information not only to determine your physical data (were you at the crime scene?) but also to guess your *intentions* (by assessing the data feed from our body sensors) would certainly motivate legislation that would make the deletion of such information a crime (just as recent legislation against cybercrime [37] does for computer log files).

4.3 The Power of Searching and Combining Information Bits

All these examples serve to show that ambient-intelligence landscapes, even when created for the greater good and with the best of intentions, will run a high risk of involuntarily threatening the personal borders that separate our private and public lives, simply because their monitoring capabilities will facilitate more of the border crossings described above. However, whether such crossings ultimately occur, given the opportunities created, will depend very much on the type of *searching* capabilities such ambient-intelligence systems might offer.

Search technology is traditionally a topic in the fields of information retrieval or databases, rather than that of ambient intelligence. However, the chances are high that such technology will be a basic building block of future ambient-intelligence landscapes, as many of the envisioned applications in the field require precisely these capabilities. An automated diary collecting 24/7 audio and video data would not be much use unless it was combined with powerful search and retrieval technology that allowed us to comb large amounts of data for very specific information. And the ability to combine different information sources, especially large, innocuous ones such as walking patterns or eating habits, is the backbone of any "smart" system, which must make the best use of a large variety of different sensor input to take decisions that make it appear to *understand* what is happening around us.

Having thus both monitoring and search capabilities at the very core of their architecture, ambient-intelligence systems will very likely provide their developers, owners, and regulators with a significant tool for driving the future development of privacy concepts within society. Depending on the actual systems deployed, some of the motivating aspects of privacy as discussed above might become more or less prominent, thus influencing corresponding legal and social norms. However, as important as privacy is, it is merely the tip of the iceberg that constitutes the social implications of ambient intelligence. In the following section we want to explore a bit more of the edges that lurk just below the waterline, where they seem to be of no immediate threat but where they could potentially become much more dangerous as the deployment of ambient intelligence picks up speed.

5 Social Challenges and Implications

Life without computers is unimaginable for most of us today – embedded processors monitor the condition of high-risk patients around the clock, they control central heating in buildings, air conditioning in tunnels, and they safely guide airplanes between continents. The potential economic benefits of ubiquitous computing are certainly key factors for the further proliferation of information technology, such as novel indoor and outdoor positioning systems, ubiquitous communication platforms, and unobtrusive monitoring installations. This technology will form and shape the foundations of future ambient-intelligence landscapes.

As more and more objects and environments are being equipped with ambient-intelligence technology, the degree of our dependence on the correct, reliable functioning of the deployed devices and microcomputers including their software infrastructures is increasing accordingly. Today, in most cases, we are still able to decide for ourselves whether we want to use devices equipped with modern computer technology (e.g., by choosing manual control for our central heating, or by deciding not to carry a mobile phone if we dislike the constant accessibility its usage implies). But in a largely computerized future, it might not be possible to escape from this sort of technologically induced dependence, which leads to a number of fundamental social challenges for future ambient-intelligence systems. Privacy is just one of these challenges, though probably the most prominent one. However, the more thoroughly "computerized" our environment becomes, the more basic attributes of the world we live in will subtly change, such as its reliability, accessibility, and transparency. In the following, we attempt to identify these concerns and try to address additional ethical and social implications of future ambient-intelligence landscapes.

5.1 Reliability

The vision of ambient intelligence describes systems that work completely in the background, discreetly and unobtrusively helping us to carry out our tasks. Since our needs and circumstances can change over time, such systems must be able to adapt themselves dynamically to the current situation. In doing so, one crucial basic requirement is reliability in the broadest sense of the word. In addition to ensuring dependability from a technological point of view, a complex and highly dynamic system must also remain manageable and controllable, and must retain the ability to predict (and, to a certain extent, verify) that the system is behaving correctly:

- *Manageability.* It is far from clear that implementing large-scale ambient-intelligence scenarios involving potentially millions of smart, adaptive devices, is simply a question of scaling up existing toy examples. Will these services and applications still be able to meet their original requirements, even with a massive increase in the number of tiny interacting objects? And, above all, how will we be able to understand and control such a highly dynamic world involving such large numbers of individual objects?
- *Predictability.* Today's technical infrastructures, such as the phone system, television, and electricity, are relatively easy to use, even for people with no special qualifications. This also entails the ability to detect malfunctions: for example, if you lift a telephone receiver and do not hear a dial tone, it is immediately evident that the phone (either the handset or the landline) is not working properly. However, this type of predictability of system behavior can no longer be taken for granted in an ambient-intelligence landscape, as systems are expected to function without users noticing their presence. This will make fault detection and diagnosis fundamentally difficult, especially for the layman [38]. Additionally, users might continue to rely on a failed service (e.g., an automated backup service or the self-diagnostics of a smart product) without noticing, thus increasing the damage done until the problem is finally discovered.
- *Dependability.* Incorporating computing and communication technology into everyday artifacts requires ever-decreasing form factors and minimal energy consumption. This makes it difficult to use hardware redundancy in such systems, even though their envisioned unobtrusive and ubiquitous use implies much harsher surroundings than, say, an everyday indoor environment. This calls for alternative concepts and mechanisms in order to overcome service interruptions and device failures, such as an explicit diversification of system functions. Such a diversification can be achieved by providing fully independent ways of carrying out the same task, preferably based on separate sets of system resources wherever feasible. A communications connection, for instance, can be diversified if the system provides different communications mechanisms in parallel, such as GSM, Bluetooth, and wireless LAN.

The power outages that affected not only large parts of the USA and Canada but also Italy and some other countries in 2003 have demonstrated our dependence on existing technical infrastructures, in this case the power grid. With the constant goal of saving costs, any industry-built ambient-intelligence infrastructure will run a high risk of forgoing safety for the sake of efficiency, resulting in brittle systems that will work only sporadically.

5.2 Delegation of Control

In order to minimize the need for human intervention in complex, highly dynamic environments, new concepts for delegating control are necessary – automatic processes should take care of routine tasks in a dependable manner, but also provide accounting mechanisms for monitoring complex control flows. Control and accounting mechanisms are important tools for determining who is in control of an autonomous system, and who is responsible if something goes wrong. At the same time, however, the autonomy of artifacts is also limited by their reliance on the technical infrastructure:

- *Content Control.* If smart objects provide information about themselves, this raises the question of who guarantees the objectivity and accuracy of the statements made. For example, smart products might be used to tie customers more closely to traders by recommending they purchase other goods produced by that same trader. In a certain sense, smart objects are becoming media representing a particular "ideology" (e.g., that of the product's manufacturer, or the politically motivated opinion of a consumer protection organization). Could a consumer protection institute use its own electronic directory to map a smart product label onto information other than that which the producer intended (for example, to warn of allergies to ingredients)? And maybe more importantly: who will decide what a smart toy tells the children, potentially shaping the children's opinions without their parents' knowledge? Tempting children to buy additional toys would only be the most obvious strategy – a much more serious threat would be the moral values induced by smart toys during play.
- *System Control.* It is similarly conceivable that automobiles or other products, as components of an ambient-intelligence network, would no longer feel completely "loyal" to their owners, but would instead enforce the guidelines of insurance companies, manufacturers, or the judiciary. For example, a smart car might refuse to open the door for its driver because he or she has stopped in a no-parking zone. But when should an intelligent device obey human orders, and when should it follow its own "convictions"? While such a no-parking system might be desirable for congested cities, some kind of manual override mechanism would obviously be needed for emergency situations, e.g., when rushing a seriously injured person to hospital (and trying to park in front of it). Even if a system were designed to only make suggestions, it would still find itself treading a fine line between inspiration and frustration, between obliging helpfulness and pig-headed patronization [39].

- *Accountability.* If autonomous objects such as the previously mentioned smart doll start taking decisions on their own (e.g., buying new clothes), legal guidelines need to be drawn up in order to resolve who is ultimately responsible for these business transactions. Smart assistants might order unwanted plane tickets, smart fridges excessive amounts of food – in both cases the automated system might have performed according to specification, though neither the original programmers nor the layman user would be able to understand its reasoning. Providing the user with a detailed explanation of completed transactions is only part of the solution, especially when monetary damages are involved. It may look appealing to simply shift the responsibility and liability onto the end user by changing the license agreements of smart objects accordingly in the small print, but it is questionable whether such a procedure would prove tenable if taken to court.

Similar discussions involving the questions of accountability and content control are already taking place in the context of the World Wide Web. For example, questions regarding the right to possess and use certain prestigious domain names [40, 41] can be compared to the issue of content control (i.e., who is allowed to resolve a certain URL stored in the RFID tag of a product), while national laws trying to control digital copyright as well as freedom of speech [42] might already set standards regarding the future "freedom" of smart devices to obey their owners.

5.3 Social Compatibility

Another fundamental challenge for ambient-intelligence systems is their social compatibility. If we, as humans, want to be capable of participating in highly dynamic systems, their parameters will have to be adjusted accordingly. System behavior relating to particular aspects should retain a certain transparency and inertia, allowing humans to detect and adjust to changes. On the other hand, it should also be taken into account that an all-encompassing ambient-intelligence landscape must also meet the needs and requirements of as broad a section of society as possible, especially if participation is practically mandatory.

- *Transparency.* With the pay-per-use model discussed above, perusing an ambient-intelligence environment might incur a large number of micropayments, e.g., for bus or theater seats we have sat on, pages of books or newspapers we have read, or clothing we have worn. Irrespective of technical feasibility, this prompts the question of how we could keep track of the resulting number of short-term contracts and the countless associated micropayments, let alone retrospectively check the legitimacy of these transactions. Not only would it be extremely tedious and unrealistic to manually check thousands of transactions, it is also questionable to what extent inappropriate items could be identified and rejected, and to what extent

legitimate payments could be unambiguously and indisputably allocated
to the responsible party. Dynamic insurance rates that vary according to
the style of driving (as discussed in an earlier section) constitute another
example of a potential loss of transparency, especially if the underlying
assessment methods changed dynamically with no warning, or if they were
unknown or too complicated to be understood by the user.

- *Knowledge Sustainability.* Most information in our everyday life today remains valid for an extended period of time, e.g., food prices in our favorite
  supermarket, or prices for public transport. It is this inertia of information
  that permits us to use acquired knowledge and prior experiences to cope
  with future situations and tasks. In a highly dynamic world, the sustainability of knowledge risks being lost – an experience that was valid and
  useful one minute could become obsolete and unusable the next. Such a
  loss or accelerated devaluation of long-term experiences could, in the long
  term, contribute to an increased uncertainty and lack of direction for people
  in society.
- *Fairness.* Detailed cross-marketing based on ambient intelligence promises
  tailor-made offers that virtually eliminate unwanted advertisements. However, a specific offer may be withheld from a particular consumer for one of
  two reasons: either the offer was not worth the consumer, or the consumer
  was not worth the offer. David Lyon, Professor of Sociology at Queen's
  University in Canada, calls this process "social sorting" – "Categorizing
  persons and groups in ways that appear to be accurate and scientific, but
  which in many ways accentuate differences and reinforce existing inequal-
  ities" [43]. People not matching a certain "desirable" profile might have
  to pay much higher prices, as they do not qualify for any of the existing
  discounts, which might in turn reinforce the non-matching patterns.
- *Universal Access.* The natural interfaces envisioned in ambient-intelligence
  scenarios certainly have the opportunity to overcome many of today's accessibility problems, such as the small screens and keypads of modern mobile phones that often prevent elderly people from using them [44]. Many
  projects in the field target elderly and physically disabled people in particular, for example with electronic "memory aids," reading aids and navigation systems [45], which might pave the way for a universal design [46] that
  considers the needs of minorities and marginal groups early on in the design stage. Intelligent interfaces and the concept of ubiquitous information
  access are often seen as key developments for bridging the *digital divide*,
  where different sections of the population have different abilities to participate in the information society. However, having more information opportunities does not necessarily mean more justice or freedom, simply because
  the potential dependencies and opportunities for manipulation would be so
  numerous they could overwhelm individuals, making it even more difficult
  to assess the trustworthiness of the information's source. Information that
  was uncritical or sponsored by advertisers (and therefore one-sided) could

become available free of charge, while independent, high-quality information would cost money, thus widening the digital divide even further. Since ubiquitous computing is not just about information itself, but is inherently linked to real-world objects, these new means of access and content control could easily lead to the digital divide becoming a real and perceivable rift in our everyday lives.

In history, the development of regulatory, social, and ethical standards tends to lag considerably behind the rapid proliferation of pioneering technological inventions, as was the case with the invention of the assembly line and mass production at the beginning of the 20th century, and with the appearance of the global Internet in the 1980s, for example. In an emerging future of ambient-intelligence systems, one exciting question is whether we will be aware of the impending pitfalls and tackle them in an early (design) phase, where we still have the means to shape the envisaged systems according to fundamental social and ethical requirements, or if there is a need for yet another social revolution that subsequently brings about necessary adaptations by force.

5.4 Acceptance

The fundamental paradigm of ambient intelligence, namely that computers disappear from the user's consciousness and recede into the background, is sometimes seen as an attempt to have technology infiltrate everyday life unnoticed by the general public in order to circumvent any possible social resistance [7]. Yet beyond any perceived sinister motives (which might be easy enough to counter), a widespread public acceptance of ambient intelligence also rests on issues of an almost philosophical nature, such as the fundamental nature of smart objects or our changing relationship with our environment.

- *Feasibility and Credibility.* Many philosophers and social scientists identify a prevailing self-confident and technophile attitude among scientists in the field of ambient intelligence and ubiquitous computing [47], where the non-critical anticipation of future technological developments almost attains the characteristics of a metaphysical prophecy. Others doubt the credibility of the envisioned scenarios, e.g., when ambient intelligence is said to simplify our lives, help us save time, and relieve us of laborious tasks. While this assertion has been constantly repeated throughout the twentieth century by the consumer goods industry, adding "smart machines" everywhere will not help to overcome the existing pattern of hurry, rush, stress, and separation from other people, but will only increase their efficiency [48]. Such criticism may build up and induce a serious credibility gap, reducing the acceptability of ubiquitous-computing technologies.
- *Artifact Autonomy.* Networked everyday objects embedded in an ambient-intelligence landscape lose part of their autonomy and, with this, exhibit

an increased dependence on the infrastructure. For users, this reduces the "object constancy" of the objects that surround them, as the example of electronic books made from smart paper shows: reading such a book may presuppose a regular connection to a server (license server, accounts server, etc.). Because of this, it appears to be more error-prone and less autonomous than a "normal" book, which can always be read, whereas the electronic one can only be read if the infrastructure is functioning.

- *Impact on Health and Environment.* It is hard to predict the impact that a large-scale use of ubiquitous computing and communication technology would have on our environment in terms of raw material consumption, energy consumption, and disposal. For example, if all supermarket goods were equipped with smart labels in the future, billions of these tiny and individually quite harmless chips would end up in the household garbage. On the other hand, the remote identification capability provided by smart labels would enable information on products to be made available throughout their entire lives, permitting the different materials in waste products to be efficiently identified and separated. It is also not yet fully understood whether, and to what extent, electromagnetic radiation (e.g., produced by wirelessly communicating smart objects) could affect our physical health. A vision involving myriads of everyday objects and wearable "information appliances" that communicate wirelessly with each other thus gives due cause for concern, as its potential adverse environmental effects could permanently influence the lives of future generations [49].
- *The Relationship between Man and the World.* From a philosophical point of view, the vision of ambient intelligence fundamentally changes the environment in which we live: "By this weaving of extensions of ourselves into the surroundings, significant parts of the environment lose important aspects of their otherness and the environment as a whole tends to become more and more a subservient 'artifact'. This artifact, which the world immediately surrounding us becomes, is almost entirely 'us' rather than 'other'. In this sense, the surrounding world has almost disappeared." [7] Similarly, Adamowsky stipulates that our inability to handle the physical world in a flexible enough way will force us to replace it by digital surrogates – equivalents of particular aspects of the real world in the digital world, implemented in the form of models, simulations, and virtual counterparts – which will ultimately lead to a transformation, dislocation, substitution, and the loss of fundamental properties relating to the world [6].

Dryer et al. [50] conducted two empirical studies to examine the theoretical relationships between system design for mobile computing, human behaviors, social attributions, and interaction outcome. In their conclusion, they express "doubt that our inevitable future is to become a machinelike collective society. How devices are used is not determined by their creators alone. Individuals influence how devices are used, and humans can be tenaciously

social creatures." They conclude "Given the importance of social relationships in our lives, we may adopt only those devices that support, rather
than inhibit, such relationships." With the substantial amount of skepticism
related to technology, such findings seem to counterbalance the immediate
threat that a thoroughly computerized future appears to hold. However, apart
from personal prejudices, the wide range of social consequences that ambient intelligence may have will certainly need to be addressed in future systems and debates. These challenges are of fundamental importance and may
ultimately even have a decisive influence on the large-scale acceptance of
ambient-intelligence technologies and environments.

6 Brave New World?

"Everything will be connected to everything else," but "no one has any idea
what all those connections will mean" [8]. This criticism can be taken as
a perceived lack of focus when it comes to ambient-intelligence applications,
but also as a deficiency in terms of understanding the consequences of deploying ubiquitous-computing systems in the real world: how will we use "smart
things" in our everyday lives? When should we switch them on or off? What
should smart things be permitted to hear, see, and feel? And whom should
they be allowed tell about it? Whether these consequences concern the protection of personal data, the implications for the macro-economy, or social
acceptance – developers of ubiquitous-computing systems can profit greatly
from a careful evaluation of the consequences of such technology within the
framework of established concepts from the fields of sociology, economics,
and jurisprudence.

Although predicting the future is difficult, if not impossible, the above
discussion allows us to guess at a few of the possible implications of wide-
scale use of ambient-intelligence technology: new business models will increase
profits, possibly at the expense of safety margins; the balance of political and
economic power could shift; economic developments will accelerate and initiate long-term changes in our social values and motives; personal borders
could be violated by new surveillance and search technology; and, not least,
there is the danger that we will lose confidence in our environment, thus fundamentally and unfavorably changing our attitude towards the world that
surrounds us. The intention of this article was to throw light on the interdisciplinary fields of ubiquitous computing and ambient intelligence, in order
better to understand how far these visions can and should influence our everyday lives. By identifying and addressing the great challenges of technical and
social change, as well as their environmental sustainability, it may be possible to steer this development in a direction that has more in common with
Weiser's optimistic vision of the 21st century than with the depressing mix
of consumer terror and police state conjured up by Steven Spielberg in his
movie "Minority Report" [51].

Acknowledgements

This work has been partially funded by the Gottlieb Daimler- and Karl Benz-Foundation, Ladenburg, Germany, as part of the interdisciplinary research project "Living in a Smart Environment – Implications of Ubiquitous Computing". The authors also wish to thank Jochen Jagob, TU Darmstadt, Germany, for fruitful discussions on the economic issues.

References

1. Weiser, M. (1991) The Computer for the 21st Century. Scientific American, 265(3):66–75, September 1991.
2. Ahola, J. (2001) Ambient Intelligence, ERCIM News, No 47, October 2001.
3. Coroama, V., Röthenbacher, F. (2003) The Chatty Environment – Providing Everyday Independence to the Visually Impaired. In: The 2nd International Workshop on Ubiquitous Computing for Pervasive Healthcare Applications (UbiHealth2003), October 2003. Available at www.vs.inf.ethz.ch/publ/papers/ubicomp2003-hc.pdf.
4. Moore, G.E. (1965) Cramming more components onto integrated circuits. Electronics, 38:114–117, April 1965.
5. Finkenzeller, K. (2003) RFID-Handbook. John, Wiley & Sons, 2nd edition.
6. Adamowsky, N. (2000) Kulturelle Relevanz. Ladenburger Diskurs "Ubiquitous Computing", February 2000. Available at www.vs.inf.ethz.ch/events/slides/adamowldbg.pdf.
7. Araya, A.A. (1995) Questioning Ubiquitous Computing. In: Proceedings of the 1995 ACM 23rd Annual Conference on Computer Science. ACM Press, 1995. Available at doi.acm.org/10.1145/259526.259560.
8. Lucky, R. (1999) Everything will be connected to everything else. Connections. IEEE Spectrum, March 1999. Available at www.argreenhouse.com/papers/rlucky/spectrum/connect.shtml.
9. Thackara, J. (2001) The design challenge of pervasive computing. Interactions, 8(3):46–52, May 2001. Available at doi.acm.org/10.1145/369825.369832.
10. Weiser, M. (1993) Ubiquitous Computing. IEEE Computer, 26(10):71–72, October 1993.
11. Siegele, L. (2002) How about now? A survey of the real-time economy. The Economist, 362(8257):3–18, January 2002.
12. Lee, H.L., Padmanabhan, V., Whang, S. (1997) The Bullwhip Effect in Supply Chains. MIT Sloan Management Review, 38(3):93–102, Spring 1997.
13. Joshi, Y.V. (2000) Information Visibility and its Effect on Supply Chain Dynamics. Master's Thesis, MIT, June 2000.
14. Fleisch, E. (2002) Von der Vernetzung von Unternehmen zur Vernutzung von Dingen. In: Schögel, M., Tomczak, T., Belz, C. (eds) (2002) Roadm@p to E-Business – Wie Unternehmen das Internet erfolgreich nutzen. Thexis, St. Gallen, pp 270–284.
15. Ferguson, G.T. (2002) Have Your Objects Call My Objects. Harvard Business Review, 80(6):138–144, June 2002.

16. Skiera, B., Spann, M. (2002) Preisdifferenzierung im Internet. In: Schögel, M., Tomczak, T., Belz, C. (eds) (2002) Roadm@p to E-Business – Wie Unternehmen das Internet erfolgreich nutzen. Thexis, St. Gallen, pp 270–284.
17. Siegele, L. (2002) Tante Emma lebt. Die Zeit, (42):27, October 2002.
18. USA Today (2000) Amazon May Spell End for "Dynamic" Pricing. Associated Press. September 29, 2000.
19. Accenture Technology Labs (2001) Silent Commerce. Available at www. accenture.com/xd/xd.asp?it=enweb&xd=services/technology/vision/silent_commerce.xml.
20. Maeder, T. (2002) What Barbie Wants, Barbie Gets. Wired Magazine, 10(1), January 2002.
21. Akerlof, G. (1970) The Market for Lemons: Qualitative Uncertainty and the Market Mechanism, The Quarterly Journal of Economics, 84(3), pp 488–500.
22. Reuters (2003) This DVD Will Self-Destruct. Wired News, May 16, 2003. Available at www.wired.com/news/technology/0,1282,58883,00.html.
23. Talbott, S. (2000) The Trouble With Ubiquitous Technology Pushers, or: Why We'd Be Better Off without the MIT Media Lab. NetFuture: Technology and Human Responsibility. Available at www.netfuture.org/2000/Jan0600_100.html#3.
24. Cochrane, P. (2000) Head to Head. Sovereign Magazine, pp 56–57, March 2000.
25. Rotenberg, M. (2001) Testimony and Statement for the Record. Hearing on Privacy in the Commercial World before the Subcommittee on Commerce, Trade, and Consumer Protection, U.S. House of Representatives, March 2001. Available at www.epic.org/privacy/testimony_0301.html.
26. Lessig, L. (1999) Code and Other Laws of Cyberspace. Basic Books, New York NY, 1999.
27. Westin, A.F. (1967) Privacy and Freedom. Atheneum, New York NY, 1967.
28. Warren, S., Brandeis, L. (1890) The Right to Privacy. Harvard Law Review, 4(1):193–220, December 1890.
29. Marx, G.T. (2001) Murky Conceptual Waters: The Public and the Private. Ethics and Information Technology, 3(3):157–169, July 2001.
30. Mayo, R.N. (2001) The Factoids Project. July 2001. Available at www.research.compaq.com/wrl/techreports/abstracts/TN-60.html.
31. Rhodes, B. (1997) The Wearable Remembrance Agent: A System for Augmented Memory. Personal Technologies Journal. Special Issue on Wearable Computing, 1:218–224, January 1997.
32. O'Harrow, Jr. R. (2001) Prozac Maker Reveals Patient E-Mail Addresses. The Washington Post, July 2001.
33. The Declaration of Independence and the Constitution of the United States. Bantam Books, New York. August 1998.
34. Nagel, K.S., Kidd, C.D., O'Connell, T., Day, A., Abowd, G.D. (2001) Family Intercom: Developing a Context-Aware Audio Communication System. In: Abowd, G.D. et al [52], pp 176–183.
35. Westerlund, B., Lindquist, S., Sundblad, Y. (2001) Cooperative Design of Communication Support for and with Families in Stockholm, September 2001. Available at interliving.kth.se/papers.html.
36. Rhodes, B., Minar, N., Weaver, J. (1999) Wearable Computing Meets Ubiquitous Computing – Reaping the Best of Both Worlds. In: Proceedings of the Third International Symposium on Wearable Computers (ISWC '99), San Francisco CA, October 1999, pp 141–149.

37. Council of Europe (2001) Convention on Cybercrime. ETS No 185. Available at conventions.coe.int/Treaty/en/Treaties/Html/185.htm, November 2001.
38. Estrin, D., Culler, D., Pister, K., Sukhatme, G. (2002) Connecting the Physical World with Pervasive Networks. IEEE Pervasive Computing – Mobile and Ubiquitous Systems, 1(1):59–69, January 2002.
39. Satyanarayanan, M. (2001) Pervasive Computing: Vision and Challenges. IEEE Personal Communications, 8(4):10–17, August 2001.
40. Bitlaw.com (2000) Some well publicized examples of domain names disputes. See www.bitlaw.com/internet/domain.html#disputes.
41. Domain Name Handbook (2003) Domain Dispute Index. See www.domainhandbook.com/dd.html.
42. Electronic Frontier Foundation EFF (2003) Blue Ribbon Campaign for free speech online. See www.eff.org/br/.
43. Lyon, D. (2001) Facing the Future: Seeking Ethics for Everyday Surveillance. Ethics and Information Technology, 3(3):171–180, July 2001.
44. Gonçalves, D.J. (2001) Ubiquitous Computing and AI Towards an Inclusive Society. In: Heller et al [53], pp 37–40.
45. Makris, P. (2001) Accessibility of Ubiquitous Computing: Providing for the Elderly. In: Heller et al [53].
46. Stephanidis, C. (2001) Towards Universal Access in the Information Society. In: Heller et al [53].
47. Adamowsky, N. (2003) Smarte Götter und magische Maschinen – zur Virulenz vormoderner Argumentationsmuster in Ubiquitous-Computing-Visionen. In: Mattern, F. [54], pp 231–247.
48. Winner, L. (1999) The Voluntary Complexity Movement. NetFuture: Technology and Human Responsibility, September 1999. Available at www.netfuture. org/1999/Sep1499_94.html#3.
49. Hilty, L., Behrendt, S., Binswanger, M., Bruinink, A., Erdmann, L., Fröhlich, J., Köhler, A., Kuster, N., Som, C., Würtenberger, F. (2003) Das Vorsorgeprinzip in der Informationsgesellschaft. Auswirkungen des Pervasive Computing auf Gesundheit und Umwelt. Studie des Zentrums für Technologiefolgen-Abschätzung TA-SWISS, TA 46/2003, August 2003.
50. Dryer, D.C., Eisbach, C., Ark, W.S. (1999) At what cost pervasive? A social computing view of mobile computing systems. IBM Systems Journal, 38(4):652–676.
51. Dick, P.K. (1956) Minority Report. Fantastic Universe, January 1956.
52. Abowd, G.D., Brumitt, B., Shafer, S. (eds) (2001) Ubicomp 2001: Ubiquitous Computing. Prooceedings Series: Lecture Notes in Computer Science, Vol. 2201, Springer-Verlag.
53. Heller, R., Jorge, J., Guedj, R. (eds) (2001) Proceedings of the 2001 EC/NSF Workshop on Universal Accessibility of Ubiquitous Computing: Providing for the Elderly, Alcácer do Sal, Portugal, May 2001, ACM Press. See also virtual.inesc.pt/wuauc01/.
54. Mattern, F. (eds) (2003) Total vernetzt: Szenarien einer informatisierten Welt. Springer-Verlag, 2003.
55. IBM Global Services (2001) Transforming the appliance industry. Switching on revenue streams in services. White Paper, 2001.

Integrated Microelectronics for Smart Textiles

C. Lauterbach and S. Jung

1 Introduction

Of all materials we are directly interacting with in our daily life, textile structures traditionally belong to the most widely spread and popular ones. We are in touch with textiles in many ways, be it as garments which we use to protect ourselves from cold and heat, fabrics covering the surfaces of floors, or the upholstery of a car seat. Because of their unique properties, textiles also appear in many technical applications. They are fundamental, yet often disguised components in every part of our artificially built environment. The world-wide consumption of technical textiles and nonwovens reached 16.7 million tonnes in the year 2000 [1]. An annual growth of 3 to 5% is predicted, depending on the field of application.

Today, the interaction of human individuals with electronic devices demands specific user skills. In future, improved user interfaces can largely alleviate this problem and push the exploitation of microelectronics considerably [2]. In this context the vision of smart clothes promises greater user-friendliness, user-empowerment, and more efficient services support. Wearable electronics recognizes and responds to the presence of individuals in a more or less invisible way. It serves the human individual in his needs for a very convenient lifestyle [3]. However the integration of smart devices seamlessly integrated into textiles requires novel approaches both on the electronic and textile parts. We believe that today, the cost level of important microelectronic functions is low enough and enabling key technologies are mature enough to exploit this vision to the benefit of society.

In the following we present key technologies for smart textiles, as well as examples for their relevance in daily life applications. Table 1 gives an overview over the twelve main groups to which the technical textiles are divided according to their field of application and shows specific examples for additional functionalities through microelectronic integration: Large-area textiles within our environment for drainage, irrigation or slope stabilization may be improved by electronic defect detection. Sensor systems within textile reinforced concrete could detect cracks within supporting construction elements after an earthquake. Within buildings smart textiles could control the air-conditioning, or function as alarm plants or guiding systems in case of emergency. Industrial textiles combined with electronics may improve the

Table 1. Technical textiles, their applications, and possible microelectronic functionality

Segment	Application	Microelectronic functionality
Agrotech	drainage, irrigation, silos, sun shields	defect detection, temperature control
Buildtech	concrete reinforcement, isolation, sound absorption, interior fittings, tents	crack detection, air conditioning, humidity measurement, alarm plants, robot navigation
Clothtech	outdoor clothes, functional clothes, protective clothes, boots	navigation systems, fire/safety garment, authentication, medical monitoring, fun/entertainment
Geotech	shoreline/slope stabilization, drainage, reservoirs, tunnels, road construction	defect detection, fire/smoke/humidity sensors, traffic control
Hometech	floor/wall coverings, sun shields, molding, interior decoration	air conditioning, alarm plants, large area displays
Indutech	power transmission belts, conveyor, absorber, roll unit cover, filtration, gaskets	breakdown prevention, abrasion/wear out control, production parameter control
Medtech	hygiene products, bandages, prosthesis, implants, hospital sheets, surgery sheets	inflammation diagnosis, continuous medication, medical monitoring, emergency call
Mobiltech	truck tarps, car seats, airbags, safety belts, tires, interior/floor coverings, power transmission belts	alarm plants, airbag control, air conditioning, active sound suppression, tire pressure control, authentication
Ecotech	ground sealing, erosion protection, watercourse protection, air purification	defect detection, wear out control
Packtech	packages, container (bags), canvas covers, silos, package nets	logistics, temperature/humidity sensoring, expiration date, cool chain control
Protech	heat/fire/cold/radiation/ antistatic protection, protective clothes, camouflage nets	medical monitoring, navigation, localization, communication,
Sporttech	aviation, sails, cycling, protective clothing, mountaineering, boots	defect detection, energy conversion, medical monitoring, emergency call, localization

homogeneity of production parameters and prevent breakdowns due to abrasion or wear out of conveyor bands and transmission belts. Wearable electronics can improve the safety of workers in hazardous environments or support navigation for outdoor activities. Even monitoring of vital functions of at-risk patients becomes possible. Possible applications of electronic textiles are almost as manifold as the usage of textiles itself.

Several approaches to smart clothes and intelligent textiles for the new era of the information society have been presented recently [2–4]. For integration into everyday's clothing, electronic components should be designed in a functional, unobtrusive, robust, small, and inexpensive way [5, 6]. With the ongoing progress of miniaturization, many complex and large-size electronic systems will soon be replaced by tiny silicon microchips measuring just a few square millimeters.

2 Key Technologies for Microelectronic Integration in Textiles

For unobtrusive active functionality of textiles the seamless integration of enabling microelectronic devices is a big challenge. An interdisciplinary approach on both, the microelectronic and the textile industry is necessary to develop robust and marketable systems. In research, the materials have to be selected not only according to their electrical properties, but also their feasibility during textile production, interconnect properties and endurance in daily life.

2.1 Conductive Fibres

The textile industry already developed the weaving technology for implementing conductive fibers into fabrics for textile shields against electromagnetic radiation or the prevention of electrostatic discharge (ESD). Many trademarks of fabrics that are commercially available, contain different amounts of conductive filaments such as carbon [7], polymers [8], polymers with additional nickel, copper and silver coatings of varying thicknesses [9], steel [10] or copper wires [11]. Their mechanical, chemical and electrical properties vary with the core material, diameter of the fiber, coating and doping materials. Therefore a direct comparison between the different ommercially available fibers is almost impossible. They have to be chosen according to the intended application. The main goal is to keep the conductive fabrics lightweight, durable, flexible, cost competitive and compatible with standard braiding equipment. With well defined characteristics of the conductive materials, they may also be used for conducting electrical signals and power for integrated microelectronics.

The most suitable material for power and signal transmission are copper wires due to the high electrical conductivity of copper. For their implementation in textiles, these copper wires are usually silver-plated to improve the corrosion resistance. For their use in textiles exposed to humidity like wearables, they are insulated by an additional polyester or polyamide coating. Depending on the diameter of the interwoven wires, they are used as single filaments or spun together with materials such as polyester to yarns. The suitability of such textile constructions for a reliable data transmission up to rates of 100 MHz was shown in [12]. Copper wires are more fragile than the very flexible carbon or polymer filaments, which may have consequences for a reliable implementation into products.

2.2 Packaging and Interconnect Technologies

Electrical contacts to the conductive fibers, which are woven in or embroidered [13] to the surface of the textile fabric, can be formed by crimping, soldering or by using conductive adhesives. However, the difference in size and flexibility of textile fabrics and silicon-based microelectronic integrated circuits (ICs) must be considered: The contact dimensions of ICs are approximately 80 to 100 microns, leading to a contact pitch of 160 to 200 microns. Considering that several conductive fibers within the fabric are needed for one contact to ensure failure tolerance, the resulting pitch of contacts in the textile is in the range of millimeters. There are different possibilities to bridge this gap and provide a smooth adaptation of the soft and flexible textile to the solid microelectronic components.

In [13] a special "Plastic Threaded Chip Carrier" (PTCC) is described, which is designed to be stitched into an embroidered fabric circuit and for this purpose has long, flexible conducting leads.

An interconnect and packaging technology, based on a wire bonding technology is demonstrated in [14]. The polyester narrow fabric (Fig. 1a) contains several copper wires which are coated with silver and polyester. For the electrical connections, the coating of the wires and the surrounding textile material are removed by laser treatment. The wires within the resulting openings in the fabric are then soldered to tiny contact plates and thin bonding wires (Fig. 1a). Then the module and the whole contact area are encapsulated, using polyurethane adhesive for mechanical protection (Fig. 1b).

A very straightforward interconnect technology uses flexible printed circuit boards (PCBs), typically finding applications in cameras, printers, portable computers and antennas. Kapton®(Dupont) is one of the most commonly used polyimide-based films used as flexible PCBs, which can sustain the high temperatures of conventional soldering processes. Again, before soldering the flexible board to the fabric the insulation of the copper wires has to be removed by laser treatment (Fig. 2a) and contact area is encapsulated, using polyurethane or silicone adhesive [14].

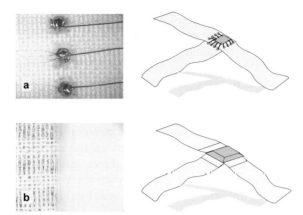

Fig. 1. Textile integration of electronic circuitry using a wire bonding concept. A polyester fabric has been used with warp threads locally replaced by electrically isolated copper wires (**a**). The contact area is encapsulated by polyurethane adhesive (**b**)

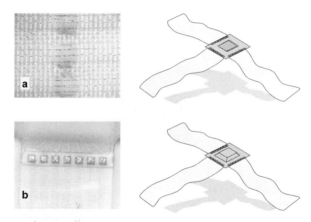

Fig. 2. Alternative interconnect concept using a thin flexible circuit board. The contact areas in the fabric are prepared by laser treatment (**a**). The contact area of the flexible board is encapsulated by silicone (**b**)

The flexible PCB can be further miniaturized by using multichip modules (MCMs) instead of conventionally packaged ICs [13]. However the flexibility of the PCBs is still reduced in the region where the chips are mounted.

An interesting approach for gaining flexibility of the complete microelectronic system is the use of ultra-thin silicon chips [15]. Silicon wafers become flexible like foils, when thinned below 50 microns. The handling of these silicon chips, which have to be mounted directly on the conductive fibers, their fixation and encapsulation in the textile environment is still a challenge.

The interconnect technologies discussed above are suitable for comparatively small microelectronic systems. Large-area applications like smart carpets need interconnect technologies, that lead to producible systems in a reel-to-reel process. Manufacturing time and cost should be optimized for adjusting the modules to the conductive fibers and forming a reliable interconnect. Expensive technologies like laser treatment for stripping insulating coatings will not suceed in the market. If an insulating coating of the conductive fibers is required, its application after mounting the modules should be considered. To avoid difficult adjustment of the microelectronic system in a large area substrate, the ICs or the silicon chips are fixed preferably on a submount, which bridges the gap between the contact pitch of the microelectronic module and the pitch of the conductive fibers in the textile fabric (Fig. 3). Conventional soldering processes for mounting the chips on the submount or the submount to the textile require expensive flexible PCBs and should therefore be avoided. A promising approach is the adoption of the technologies commonly used for mounting chips on plastic foils for RFID tags. The interconnect by conductive adhesive (CA), anisotropic conductive adhesive (ACA) or non-conductive adhesives (NCA) facilitate the use of low-cost submount substrates like aluminium and PET laminates [16].

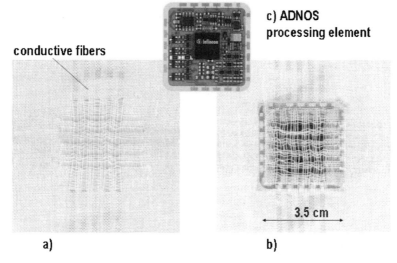

conductive fibers

c) ADNOS
processing element

3.5 cm

a)   b)

Fig. 3. Textile for large-area applications using copper wires for warp and fill threads in a regular grid. Within the cross over areas a jacquard technique makes the conductive fibers accessible (**a**). The modules are mounted by non-conductive adhesive (**b**)

3 Smart Textile Applications

When developing smart textile applications, the main task is to meet the demands of the prospective users. For wearable electronics, the usability and convenience of the human computer interface (HCI) is a crucial part of the whole system. While head mounted displays [17] are fully acceptable for professional users like service technicians or military, most users may find them too obtrusive for leisure activities.

Tight fitting sensor vests for monitoring vital signals may be accepted as a necessity for at-risk patients [18], but could be very inconvenient during sports. Identification chips within clothes may act as life-saver in case of an emergency, however many people worry about deliberate infringements on privacy and liberty [19]. Developers of smart textile applications are well advised to take the costumers needs and objections seriously, because success in the market will depend on their demand, only.

3.1 Wearable Electronics: MP3 Jacket

In this section, the deep integration of a speech-controlled digital music player system [14] into a textile environment is described, as an example for wearable electronics. Attention has been paid to an appropriate textile design for the tailoring smart clothes. Damage of the components by washing processes and daily use must be avoided. Most solutions known so far require removal of complex electronics before washing. The aim of this work was to demonstrate technological solutions for the interconnect between the textile structures and electronics, as well as for robust electronic packaging, which does not impair wearing comfort and offer easy handling. All materials are chosen according to maximum wear comfort and environmental compliance. E.g., the audio module is fully covered by garment. The wearer still has a comfortable textile touch in case the electronic module gets into direct contact with the skin. The supply voltages of the integrated electronics are as harmless as the ones of a standard music player or comparable device. An overview of the demonstrator system is depicted in Fig. 4. The audio module is directly covered by the cloth, resulting in a textile touch. The combined battery and data storage card pack also has a cover made of a textile fabric. The earpieces are connected to the narrow fabric by using textile cords that contain electrically conductive threads. In cooperation with the German Fashion School in Munich (Deutsche Meisterschule für Mode, München) several garments designs have been developed using the presented hardware. As an example, the design of a hooded jacket is shown in Fig. 5.

The core of the digital music player is the audio module [6]. Its basic overview is shown in Fig. 6. The module contains an audio signal processing microchip which is programmable to perform various functions. An 8051 microcontroller is used for the control functions and a 16-bit OAK DSP for signal processing. Possible input devices are switches, temperature sensors,

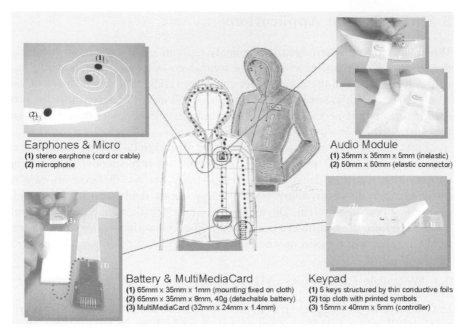

Fig. 4. Overview the digital music player system with the proposed textile design-in. The *dotted* lines drawn in the garment indicate the textile wiring realized by a narrow fabric with conductive stripes

Fig. 5. The audio processing architecture integrated into a sports jacket (design by Petra Spee, German Fashion School, Munich) The conductive ribbons are cleverly used as design elements

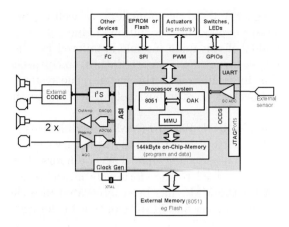

Fig. 6. Basic overview of the Infineon audio processor chip [6]

microphones or fingertip sensors for security purposes. Potential actuators are for example heating elements and loudspeakers. Sufficient general-purpose input/output lines (GPIOs) support the simple actuators. More complex ones may be connected to DC-analogue to digital converters. Furthermore, the processor system maintains a power management functionality for an optimized standby time.

Besides this audio processor chip (size $7 \times 7\,\mathrm{mm}^2$), further components such as a program memory, and a few auxiliary devices are implemented. All together, the module measures approximately $25 \times 25 \times 3\,\mathrm{mm}^3$ (see Fig. 7). Microphones, earphones, storage media, keypads, displays, sensors, actuators, and a battery can be directly connected to its interfaces. The functionality of the module is determined by the built-in software, such as: speakerphone, speech compression and decompression, music synthesis (e.g., MIDI), music

Fig. 7. Photograph of the audio module (size $25 \times 25\,\mathrm{mm}^2$) before encapsulation. The HiFi DAC and the voltage regulators are placed on the back side

decompression (e.g., MP3), software-modem, as well as, both speaker independent and speaker dependent recognition engines.

The functionality of this main module is tailored for the typical requirements in wearable electronics and smart clothes, since speech recognition allows for hands-free interaction with the system. The system can be extended to establish a connection to standard networks like the Internet (software modem function).

The battery and MultiMediaCard (MMC) module features a rechargeable lithium-ion polymer battery to supply the necessary electric energy. Its capacity is sufficient for an operating time of several hours. The battery module weighs 50 grams and is fixed to the clothes by means of a simple one-piece connector. A slot for the MMC has been integrated into the housing of the battery module. Both battery and MMC can thus be detached fast and easily. The MMC stores up to 128 Mb of digital music or audio data. When detached from the clothes, battery and MMC module can be plugged to a PC in order to recharge the battery and download new music to the MMC.

The keypad module consists of thin metallized foils brought onto the conductive narrow fabric. The metal foils are fixed using melt adhesives commonly used in garment production. The metal foils are connected to a small sensor chip module which detects whether a finger is close to a specific pad. By means of the keys, the user can activate the music player, control the volume or activate voice control. Speaker independent voice recognition ("stop!", "start!", "volume!") is activated as soon as a specific button of the keypad has been touched. This measure avoids unintentional activation of control functions. The audio module recognizes the spoken words, e.g., the number or title of the music track.

The ear- and microphone module simply consists of a piece of narrow fabric connecting the audio module and conventional earpieces and microphones.

In future, the system is expandable by further modules, such as fingerprint sensor chips for authentication, sensors for monitoring vital signals in health or safety applications, or wireless data transceivers. The generation of electrical power from ambient energy is a possible solution to the power supply problem for wearables. For ambient energy conversion, flexible and washable solar cells [24], powerboots [25] and piezoelectric transmitters [26] are commercially available. Thermogenerators use a very reliable power source, the body heat [20]. They may be used for low-energy applications like sensor systems and wristwatches [23].

A challenge for the production of wearable electronics is the necessary combination of at least two completely different industries with distinct differences within their value chains. The fixed and permanent integration of a complete MP3 player module, like the one shown in Fig. 4, during the fabrication of the clothes introduces the microelectronics directly into the textile value chain, leading to extremely high sales price for the wearable electronics product. While a direct integration of a convenient keypad is desirable, a detachable MP3 player, which can be used in other garments or even as a standalone device, can offer more freedom and flexibility to the user. In contrast, the deep textile integration of sensors which need to be in direct and reliable contact to certain parts of the body, seems very reasonable.

3.2 Large-Area Application: Smart Carpet

Microelectronics integrated in a regular grid within large-area textiles allow for huge flexible sensor or actuator areas (see Table 1). A schematic of such an embedded network of processing elements (PE) in a textile fabric, forming a regular grid is shown in Fig. 8. The PEs are equipped with sensors or actuators and mounted at the cross-over points of the conductive fibers.

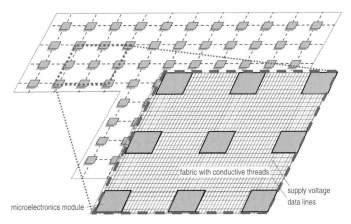

Fig. 8. Schematic of an embedded network of processing elements (PE) in a textile fabric, forming a regular grid. The PEs are equipped with sensors or actuators and mounted at the cross-over points of the conductive fibers

However, for the practical realization of such an embedded network, the following questions have to be considered: How is it possible to exploit the functionality of the integrated PEs, sensors and actuators at low installation cost? Is it possible to use a reel-to-reel process for the microelectronic integration? What happens, if the smart textile has to be cut to fit e.g., as a functional floor covering in an arbitrarily shaped room? Will a single destroyed or defective module or wire lead to a complete failure of the smart textile system?

To solve these problems we applied a self-organizing technique based on newly invented algorithms [21], locally performed by the integrated PEs.

In Fig. 9 the concept demonstrator of a smart carpet is shown. Threads indicate the supply wires, data and sensor lines. Small microelectronic modules

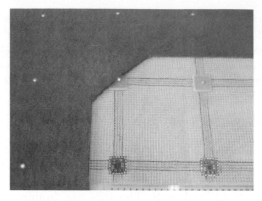

Fig. 9. Concept demonstrator of a smart carpet: microelectronics modules with integrated light emitting diodes connected to a coarsely meshed fabric, with interwoven conductive wires, before and after encapsulation (background); a carpet forms the top layer

are connected at the crossover points of the wires. Light emitting diodes can be integrated, when piercing the top layer of the carpet. Main task for the integration of the smart textile within floor covering is the planarization of the surface and the prevention of mechanical stress during use. A lot of work has been done on self-organization of microelectronic networks in large areas with an arbitrary distribution of PEs [22,27,28]. In our case, the regular grid of the PEs and their interconnection by wires simplifies the self-organization considerably.

During an initializing self-organization of the integrated network, information routes are established, based on heuristic rules. These allow for an efficient data transport from and to the portal on the edge of the network, circumventing defective regions and even tolerating arbitrary shapes.

Within the active area of the smart textile each PE exchanges control messages or data with its four nearest neighbors, respectively, and controls and drives a specific region. All PEs are identical 32-bit microcontrollers, featuring four UARTs (Universal Asynchronous Receiver and Transmitter) for serial communication with their neighbors. No prior knowledge about their position and orientation in the grid is used. The control or in case of a display application the image data are fed into the network by a functional block referred to as portal, which is connected to one or more initial PEs on the array edges. Before operation, a set-up phase takes place in which each PE self-dependently determines its orientation and position. After the self-organization, directed and explicit routing paths are established, beginning at the initial PEs and proceeding across the array. As a result, each PE is connected to the PC interface via a reasonably short path. If new defects occur, the self-organizing routine will be repeated during a new booting cycle and new routing paths will emerge in the network.

The entire network is connected via the so called portal to the PC using the RS232 communication port. Software running on the PC is called SCM (smart carpet monitor) and is based on Java Swing technology. It controls and monitors the network.

The specifications to this architecture differ from typical network situations, see e.g., [22], and demand dedicated solutions.

During self-organization, the network discovers its own geometry, size, and functional elements, and it constructs routes to each PE. The organization is divided into a sequence of phases as shown in Fig. 10. Every phase is initiated by the SCM and consists of relatively simple algorithms which are implemented into each PE and are based on messages being interchanged between neighboring PEs.

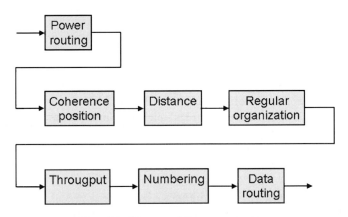

Fig. 10. Phases of the organization

The first phase is the power routing phase, where starting from the portal each PE checks its neighbors for leakage currents, disables defective neighbors and switches the power through to all functional PEs until the whole network has its power supply. During the second phase, the network is booted. The sent boot frames contain data to identify the orientation of each PE (coherence), its position and the distance to the portal.

Based on those elements of information, data routes are established within the network by an algorithm called backward organization.

These algorithms tolerate defects in the array structure, like missing PEs, damaged interconnects or irregularly shaped display borders (see Fig. 11). In the following the general ideas of these algorithms are illustrated for selected phases.

Depending on the manufacturing process of the smart textile, the PEs have no initial information about their positions inside the array and do not even have any knowledge about a common orientation in the plane. A quadratic PE tile, for example, may have a random north direction out of

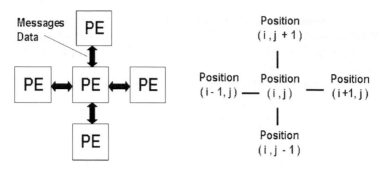

Fig. 11. Messages for position determination

four possibilities. Therefore, in the first phase of the organization, a common "coherence" of all connected PEs and a unique position in a regular coordinate space is determined. This is done by communicating special messages between the nodes. The process is started by the portal which invokes its direct neighbor PEs. The portal sends one of four possible direction number over the appropriate communication lines and thus enables its neighbors to calculate their north direction. After a PE has been set coherent, it sends appropriate coherence messages to its neighbors and so on.

Next, each PE computes the positions of its neighbors using its own temporary position which is set to $(0, 0)$ at the beginning. As shown in Fig. 11, each neighbor is informed about these computed coordinates. On reception of a position message, every PE checks its own position (i, j) against the communicated coordinates (r, c), exchanges i by r if $r > i$, and exchanges j by c if $c > j$. If the coordinates are changed, the PE computes the new positions of its direct neighbors by incrementing or decrementing (i, j) appropriately, and sends position messages to them, respectively. Again, the portal starts this process by feeding the initial PEs with their absolute position values. Figure 12 shows, that upon accomplishment of the position determination phase a consistent coordinate space is generated.

In the next phase, the minimum distance of every PE to the portal is computed in an analogous way by the network. Each PE sends messages to its neighbors about its own estimated distance to the portal. On reception of such messages, each PE selects the neighbor with the smallest distance d and sets its own distance to $d + 1$. Again, this information is communicated to its neighbors.

Based on the previously computed data, routes can be generated. Figure 12 shows channels that are assigned to specific communication lines where the data streams will flow later. The algorithm for route generation, called regular backward organization, attempts to generate regular structured routes with minimal distance to the portal.

The main criterion for evaluating the routes is the transmission of sensor or actor data. Typically, this is a system dependent function of route lengths

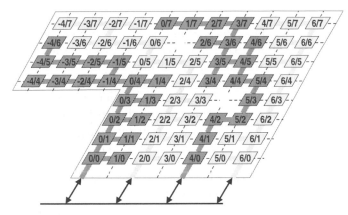

Fig. 12. Coordinates and routing paths created on the basis of regular backward organization

and throughput numbers. The throughput of a PE is defined as the number of routes leading through the PE. Again the network computes the throughput numbers by simple algorithms.

Finally, the portal decides on the optimum result for the actual configuration. The portal starts an automated numbering based on the established routes, enabling the PE to determine a unique address number. The generated address number format delivers sufficient routing information for every PE. Therefore, with the numbers of its neighbors, no additional routing tables are needed inside the network.

All algorithms are implemented in Java as software simulator ADNOS (Algorithmic Device Network Organization System). Figure 13 depicts the application of the smart carpet as a monitor system for tracking persons.

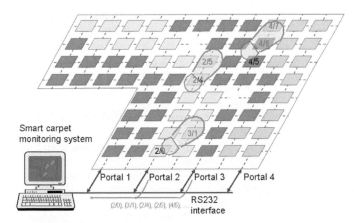

Fig. 13. Application of the smart carpet as a monitor system for tracking persons

The implemented capacitve sensors generate signals, when a person steps on the sensitive areas. The information is lead out of the network by the PEs through the portals. An RS232 interface is used as communication port to the PC. The smart carpet monitoring software on the PC evaluates the data from the network. Within this application it is decided wether steps on the carpet are an alarm case, like a burglar intruding into the room at night from the window side, or it is normal use during daytime. With intelligent data mining it even might be possible to detect an emergency case, when a person falls on the carpet and does not move afterwards. This could be an interesting application for making life more safe for elderly people living alone.

For implementation of the smart textile sensor or display area, the information gained within the network has to be transmitted to the application on a PC. We decided to implement an RS232 interface. However, this can be changed easily to another serial interface or to wireless transmission. The PC application software is adapted to the specific functions the smart textile should support. Within this application the customized features are defined, e.g., how the sensed data are processed and evaluated, or how specific data lead to an indication of light-emitting diodes.

References

1. Divid Rigby Associates, 2002, "Technical textiles and industrial nonwovens: world market forecast to 2010, www.dratex.co.uk.
2. S. Ditlea; "The PC goes ready to wear", IEEE spectrum, Oct. 2000, pp. 35–39.
3. Philips, "New Nomads - an Exploration of Wearable Electronics"; 010 publishers, 2001.
4. W.D. Hartmann, K. Steilmann, A. Ullsperger; "High-tech-fashion"; Heimdall Verlag, 2000.
5. P. Lukowicz and G. Tröster, "Packaging Issues in Wearable Computing"; CPD2000, pp. 19–22
6. B. Burchard, S. Jung, A. Ullsperger, and W.D. Hartmann; "Devices, Software, their Applications and Requirements for Wearable Electronics"; ICCE, 2001, pp. 224–225.
7. Shakespeare, "Conductive Fibers, Technical Data", URL: www.shakespearemonofilaments.com/Pages/Products/Resistat/
8. R.V. Gregory, W.C. Kimbrell, H.H. Kuhn, "Conductive Textiles", Synthetic Metals, 28, No. 1, 2, (1989), pp. C823–C835.
9. Dupont, "About Aracon", URL: http://www.dupont.com/afs/aracon/html/about.htm.
10. R-Stat, " R-Stat yarns and fibers for antistatic protection from Swicofil", URL: http://www.r-stat.com/GB/index.html.
11. Elektro-Feindraht, "Textile wire", URL: http://www.textile-wire.ch/.
12. D. Marculescu, R. Marculescu, N. H. Zamora, P. Stanley-Marbell, P. K. Khosla, S. Park, S. Jayaraman, S. Jung, C. Lauterbach, W. Weber, T. Kirstein, D. Cottet, J. Grzyb, G. Tröster, M. Jones, T. Martin, Z. Nahkad, "Electronic Textiles: A Platform for Pervasive Computing", Proceedings of the IEEE, Vol. 91, NO. 12, Dec. 2003, pp.

13. E. R. Post, M. Orth, P. R. Russo, N. Gershenfeld, "E-broidery Design and fabrication of textile-based computing", IBM SYSTEMS JOURNAl, Vol. 39, NOS 3&4, 2000, pp. 840-860.
14. S. Jung, C. Lauterbach, and W. Weber, "A Digital Music Player Tailored for Smart Textiles: First Results", Proceedings of Avantex Sympoium 2002, pp.
15. C. Landesberger, S. Scherbaum, G. Schwinn, H. Spöhrle, "New Process Scheme for Wafer Thinning and Stress-free Separation of Ultra-Thin ICs", H. Reichl: Micro System Technologies, VDE-Verlag, Berlin, 2001, 2001ISBN 3-8007-2601-7, pp. 431–436.
16. Delo "ACABOND", URL: http://www.delo.de
17. T. E. Starner, "The Enigmatic Display", IEEE pervasive computing, Vol. 2 NO. 1, 2003, pp. 15–18.
18. The Georgia Tech wearable motherboard: The intelligent garment for the 21st century (1998), [Online]. URL: http://ww.smartshirt.gatech.edu.
19. A. Stone, "The Dark Side of Pervasive Computing", IEEE pervasive computing, Vol. 2 NO. 1, 2003, pp. 4–8.
20. C. Lauterbach, M. Strasser, S. Jung, and W. Weber, "Smart Clothes Self-Powered by Body Heat", Proceedings of the Avantex Symposium, 2002.
21. T.F. Sturm, S. Jung, G. Stromberg, and A. Stöhr: "A Novel Fault-Tolerant Architecture for Self-organizing Display and Sensor Arrays", SID Symp. Digest of Technical Paper, Vol. XXXIII, NO. II, 2002, pp 1316–1319.
22. J. Duato, "Interconnection Networks: an Engineering Approach", IEEE Computer Society Press, Los Alamitos (1997).
23. M. Strasser, R. Aigner, C. Lauterbach, T.F. Sturm, M. Franosch, G. Wachutka, "Micromachined CMOS Thermoelectric Gmerators as On-Chip Power Supply", Dig. Tech. Papers Transducers '03 Boston, USA, 2003, pp. 45–48.
24. M. de Lainsecq, "Hauchdünne Solarzellen für Trekkerinnen und Astronauten", URL: http://www.ch-forschung. ch/index.php?artid=53.
25. A. Farrelly, "These Boots are made for electricity", (2000), URL: http:// www.anu.edu.au/mail-archives/link/link0006/0669.html.
26. EnOcean, "Radio Transmission without Batteries", URL: http://www.enocean.de/indexe.html.
27. J. Lifton, D. Seetharam, M. Broxton, J. Paradiso,"Pushpin Computing System Overview: A Platform for Distributed, Embedded, Ubiquitous Sensor Networks", Proc. Of Pervasive 2002, LNCS 2414, pp. 139–151.
28. W. Butera, "Programming a Paintable Computer", MIT Media Laboratory, doctoral dissertation, 2002.

Ambient Intelligence Research in HomeLab: Engineering the User Experience

B. de Ruyter, E. Aarts, P. Markopoulos, and W. Ijsselsteijn

Abstract. As technologies in the area of storage, connectivity and displays evolve rapidly and business developments point to the direction of the experience economy, the vision of Ambient Intelligence is positioning human needs central to technology development. This chapter describes concerted research efforts surrounding the HomeLab, a special research instrument that supports scientific investigations of the interaction between humans and technology. Such investigations reach beyond traditional usability and technology acceptance, aiming to characterize the nature of user experiences, measuring them and designing them. Starting from a historical view upon the vision of Ambient Intelligence, this chapter presents the HomeLab, the rationale for its set-up and three related studies that were conducted within it. These studies have focused on technology use at home for leisure purposes; more specifically they have explored the experience of immersion from displays that extend beyond screen boundaries to encompass the lighting in a room, and the experience of social presence and connectedness with remote friends and family as they result from peripheral awareness displays.

Keywords: Ambient Intelligence, HomeLab, User Experiences

1 Introduction

The vision of Ambient Intelligence [2] requires considerable advances in several technologies, but crucially takes a human-centric perspective. A cornerstone to Ambient Intelligence research is the design and creation of new user experiences and the scientific investigation of their nature. Contrasting traditional research in human-computer interaction and usability engineering against user research in the era of Ambient Intelligence, we note how it is necessary to move beyond the Tayloristic model that was relevant when optimizing the ergonomics of video display to support work (cf. ISO 9241 standard). Established organizations such as the Usability Professional's Organization have recognized the need for new approaches to usability testing [6].

Although research methodologies such as User Centered Design [8] remain important, much more radical departures are required. Rather than seeing interactive products as tools to perform work, user research must try to describe and nuance user experiences that help people enjoy, play, keep in touch with loved ones. Such experiences do not need to have clear beginning and end that is observable to the observer and perceived by the end-users [1]. They could span hours or days, and can sometimes result from infrequent, subtle interactions with technology that occupies a person's peripheral reach [22]. For example, the HomeRadio project at Philips Research [11] explored the use of ambient displays, discrete abstract visualizations of activity at a remote household. Different types and levels of activity would be portrayed in the constitution of colored patterns that would be projected on a ceiling or a wall. An individual in the presence of such a system can hardly be called a user. There is no clear task to be performed and this individual will not always be aware of interacting with the technology. Usability testing, which is central to usability engineering practices developed in the 80's and 90's is practically irrelevant in this case.

To better understand user experiences such as the one just described, new concepts, methods and instruments need to be developed. For example, the traditional view on usability can be replaced by more targeted and refined concepts that characterize the nature of the human experience defined. In the later sections of this chapter we shall present some of these examples, such as social presence and affective benefits from communication.

Rather than the typical usability test that might last an hour, the experience described above cannot be assessed credibly without exposing individuals for a much longer period of time. Such an exposure will also be pointless if it takes place in a bland usability-testing laboratory divorced from the domestic context that makes it meaningful to the individual. Techniques for selecting and analyzing the usage of technology at the long term also need to be developed and refined [18].

Long term field-testing, in realistic surroundings and in realistic social contexts seems to be an ideal for conducting user research. However, where experimental technologies are concerned this ideal is not always reached, because installation of technology in the field would be disruptive for people's home life. For example, the FRIDGE prototype [21] developed at the technical university of Eindhoven relied on a noisy projector that was mounted on the ceiling. Such an installation would not be possible to install in people's homes and would be too disruptive to use. Further, field-testing can be costly for logistical reasons (installing and uninstalling technology at homes, finding homes with the right infrastructure, etc.) and it brings about a loss of control and a difficulty in obtaining reliable observational data. In 2000 first serious plans were launched to build an advanced laboratory that could be used to conduct feasibility and user studies in Ambient Intelligence. After two years of designing and building, HomeLab was eventually opened on April 24, 2002 by Gerard Kleisterlee, the president of Philips Electronics. On the occasion of the opening an international technology seminar was held and a booklet was published explaining the purpose and the ambition of HomeLab [3]. The opening event officially marks the start of the ambient intelligence research in HomeLab.

2 HomeLab

The HomeLab is built as a two-stock house with a living, a kitchen, two bedrooms, a bathroom and a study. At a first glance, the home does not show anything special (see Fig. 1), but a closer look reveals the black domes at the ceilings that are hiding cameras and microphones. Equipped with 34 cameras throughout the home, HomeLab provides behavioral researchers a perfect instrument for studying user human behavior inside HomeLab.

Fig. 1. The HomeLab living room

Adjacent to the Home there is an observation room. From this room you have a direct view into the home. The signals captured by the cameras installed inside HomeLab, can be monitored on any of the 4 observation stations. Through an observation leader post (see Fig. 2) signals can be routed to these observation stations. The observation leader modifies camera setups, routes video and audio signals, and monitors the capture stations. Each observation station is equipped with two monitors and one desktop computer to control the cameras and to mark observed events. The marked events are time-stamped and appended to the video data. All captured signals and marked events are recorded by means of the four capture stations.

Broadband Internet facilities enable various ways to connect parts of the HomeLab infrastructure to the Philips High Tech Campus network or even to the outside world. A wireless Local-Area Network (LAN) offers the possibility to connect people in HomeLab without running cables. However, if cables are required, double floors provide nice hiding places. Corridors, adjacent to the rooms in HomeLab, accommodate the equipment that researchers and developers need to realize and control their systems and to process and render audio and video signals for the large flat screens in HomeLab.

Fig. 2. The observation leader post showing Vic Teeven, technical manager of the facility

A power control system features remote controllable light settings and power switches. But it still leaves the possibility for participants to simply turn on and off the lights by using "ordinary" switches. Future intelligent systems that aim to enhance people's emotions and experiences by means of lighting will be able to interface with the HomeLab power control system.

When setting up an experiment in HomeLab, the researcher designs a coding scheme for the observation session. A coding scheme lists all prototypical behaviors that are expected to be observable during the session. The observers mark the occurrence of these behaviors during the HomeLab session. Further analysis of this data can consist of a simple frequency analysis up to a sophisticated data mining analysis for finding hidden patterns in the data set. For this HomeLab is equipped with a software tool capable of detecting repeated patterns that are hidden to observers and very hard or impossible to detect with other available methods. It is particularly suitable for analyzing behavioral data. This tool is able to detect patterns that are obscured by other events, and finds patterns that no form of frequency count, lag sequential or time series analysis can identify. As such, it is an effective way to detect patterns in user-system interaction and to identify the precursors or consequences of specific behavioral events. This tool has been used extensively in studies of human communication, spoken dialogues, gestures, protocol analysis, etc. [15]. The Philips HomeLab is the first corporate user of this data analysis tool in the domain of user-system interaction.

2.1 Case Studies

Since the opening of HomeLab many technologies have been brought into innovative applications that create a wide range of user experiences [9]. In the next sections we will present three examples of research conducted in HomeLab for creating user experiences enabled by developments in the area of connectivity and lighting.

2.1.1 The Feeling of Being Together

The introduction of advanced technologies such as interactive TV has not resulted in the expected behavioral change of consumers. We contend that one of the most important causes for this has been the absence of sufficient content to offer attractive user benefits. Consequently our research in the Philips HomeLab explores the potential user benefits from interconnected CE devices. Through this interconnection people and places can be connected to each other, thereby letting consumers create and enjoy their own content for social communication purposes.

One potential benefit from this interconnection is social presence, which refers to the sensation of "being together" [14] that may be experienced when people interact through a telecommunication medium. As connectivity permeates our daily lives we expect that network infrastructures will become enablers for social interactions. While communication media such as e-mail, telephony, text messaging services for mobile phones, etc., are common, there is more to system-mediated communication than exchanging information.

Research in social presence goes back 30 years, starting with the seminal work of Short, Williams and Christie 1976 at British Telecom. The impetus in that early research into the psychology of communication and in much of its offspring, has been to provide a way to describe and measure how does a computer mediated telecommunication get to emulating face to face interactions between people. Such research though covers only one half of the story. From an end-user (and consumer) perspective it is necessary to understand the benefits that social presence helps deliver and in what ways social presence can lead to a pleasurable user experience.

Our research on the feeling of being together has been focusing on the potential of attaining social presence by maintaining a peripheral awareness of a connected person or group of persons, when consuming content such as a broadcasted program [10]. This research assesses affective benefits that arise out of this interconnection and illustrate the positive impact of awareness on social interactions.

When developing the concept for creating this user experience, special attention was paid to: (i) safeguarding the privacy of the home environment, (ii) minimizing the shift of user attention away from the actual content being consumed and (iii) creating the feeling of being connected when consuming content over different locations.

This concept is embodied by means of presenting sketch like visualizations of the physical activities in the remote location.

2.2 The User Study

In total 33 participants, all Dutch males, participated in the experiment. They were recruited as groups of friends who enjoy watching soccer games. The groups were split (2–1) and the participants were placed in two different rooms. During the experiment all participants watched the same soccer game. Although they knew they were watching the same soccer game they did not know that they were in the same building.

The amount of visual information the subjects received about their friend(s) was varied over the different conditions and would range from: no visual information, a sketch like visualization representing physical movements of the people at the remote location and a full motion video of the remote location. Social presence and Group Attraction were measured after each condition by use of standardized psychometric instruments available in literature.

2.3 Results

The results from this study indicate that a low bandwidth visualization of the physical activities from remote locations is capable of establishing a sense of social presence. Furthermore, the feeling of being part of a group (i.e. group attraction) was increased.

When compared with the full motion video, the sketch visualization gave participants not so much the feeling of being observed by the remote location. This latter aspect of the sketchy visualization could be of great importance to create social presence enabling systems for the home environment: earlier research has shown that privacy considerations are a major obstacle for the acceptance of video communication at the home.

Test participants indicated that they would prefer different levels of social presence for different kind of programs. People prefer to watch sports and movies in presence of others, whereas they prefer to watch news and documentaries alone. They do not want to be disturbed while concentrating on more serious programs. For entertaining programs, viewers enjoy making a cozy atmosphere and to experience other person's reactions.

2.3.1 The Affective Benefits of Being Connected

As network technology is becoming part of our daily life, applications supporting close family members living apart to keep in touch with each other, become important. Such applications are typically described as Computer Mediated Communication (CMC) or System Mediated Communication in general. A special class of CMC systems are awareness systems. These systems support individuals to maintain, with low effort, a peripheral awareness of each other's activities. Pioneering awareness systems have focused on the workplace, as for example Media Spaces [5]. Early concepts of awareness systems for the home are the digital family portrait [18] and the design concepts proposed by the Casablanca project [12]. To date, only the interLiving project [13] has attempted to develop and deploy a functional awareness system for a home environment. Likewise, there has been little attempt to understand the concept of awareness itself.

Researchers approach the concept of awareness from several angles. Works like the digital family portrait focus on attaining awareness with minimal cognitive effort through pre-attentive processes of the user. More important for the design of awareness systems, is to study the benefits that users experience through interacting with awareness systems and the role they can play in the gamut of social interactions of an individual. This wider perspective is adopted by [16] who define *affective awareness* as the general sense of being in touch with one's friends and family. However, the exact nature of this feeling, how to achieve it and how to assess it, remain unexplored to date.

In the ASTRA project [17] we addressed this limitation in the context of the design of a novel service for maintaining awareness between households and mobile individuals. This service allows informal, social communication to be enjoyed at moments of leisure and relaxation at home. Access from home is combined with spontaneous, lightweight sharing of experiences on the move.

With the ASTRA system, an individual can capture daily experiences using a mobile device that supports picture taking, freehand drawing and handwriting. For the field tests, we used existing functionality of the Sony Ericsson P800 mobile phone, which contains a camera, to create and send e-mail messages to a server. With a homebound device, related family members at home can access these annotated pictures created with the mobile device. When the homebound device is not used it shows an overview of such messages in a spiral visualisation (see Fig. 3). Users operate the ASTRA home system through a Philips desXcape wireless LCD monitor, which affords touch-screen interaction. The touch-screen interface can be used to browse similar messages and check the availability of others.

The information displayed in the spiral overview consists of pictures or drawings and text notes made on the mobile user. The spiral is divided in 3 areas: The main area of the spiral holds 6 pictures, whose size and position portrays recency. The centre of the spiral gives access to earlier items, the outer end to more recent ones.

2.4 The User Study

A formative evaluation of the prototype was conducted in the HomeLab. Four families (parents with children) and 1 group of close friends were recruited

Fig. 3. The User Interface of the homebound ASTRA system

through email adverts. In total 19 people (11 males and 8 females) took part, 11 were adults and 8 children.

From each group 1 person stayed in the HomeLab, while the others visited a nearby open-air museum. They were asked to spend 2 hours there and to send at least 6 messages to the person in the HomeLab, using the P800 mobile phone. HomeLab participants took part in a usability test and were then requested to occasionally monitor the ASTRA device. When the other participants returned from the museum, the HomeLab participants explained the system to them, verbalising in this way their conception of ASTRA. Finally, a group interview was conducted.

2.5 Results

Participants found the interface easy to understand and to use, though they helped identify some points for improvement. On average, participants sent 12 pictures. Though interviewed participants liked to share their experiences, they did not feel a particular need to use an awareness system like ASTRA. This was attributed to the short time they experienced the system and to that they used the system to perform test-tasks rather than to satisfy an actual communication need. The HomeLab study identified several areas for improvement both in the underlying service and the interface.

2.6 Conclusion

From the results of the HomeLab study it was concluded that a more longitudinal field study was required. For giving users the opportunity to experience the interaction concept and to collect early user feedback, the HomeLab proves to be a valuable instrument. To gain more insight into the role such technology might have in daily life, a field test is required.

This field test involved 2 pairs of households: In total 7 adults, 4 teenagers and 2 children were involved. The families participated out of interest but received a financial reward as well.

We used a within-subjects design consisting of 2 phases lasting a week each. In the first week the current communication between the 2 households was observed. In the second week ASTRA was introduced to them. Each household received 1 homebound device and 1 mobile device. Communication was observed by means of a diary, by logging ASTRA usage and by group interviews. The affective benefits of using ASTRA were assessed using the ABC questionnaire [20].

The field test confirmed the hypothesis that an awareness system such as the ASTRA system would enrich communication between people. More specific, the test indicated that the affective benefits of being connected were significantly more positive when an awareness system is introduced into people's daily life. For example, test-participants experienced strong affective benefits, like feeling closer to the other person, feeling in touch, and reported feeling higher levels of group attraction when compared to not using the designed appliance.

2.6.1 The Feeling of Being Immersed

A favorite leisure activity of many is watching movies, either in the cinema, or at home. The way we watch movies at home has changed much since the first television was introduced: from black-and-white to full color, from mono to stereo to surround sound. In their research on creating immersive experiences, Elmo Diederiks and Jettie Hoonhout developed and tested the Living Light concept in the Philips HomeLab. The Philips Living Light system can be considered as the next step in the Home Cinema Experience. It offers light ambiances and light effects for film and music (see Fig. 4).

The system comprises four LightSpeakers (left-right front-back), a CenterLight and a SubLight, which is situated underneath the couch. Light scripts for selected pieces of film and music have been developed in conjunction with light designers, theatre lighting experts, filmmakers and musicians. This system has been installed in the living room of HomeLab, to provide the right context for experiencing this concept.

2.7 The User Study

The concept has been evaluated with specified user groups, primarily to determine its appeal in terms of acceptance, usability, and excitement.

The secondary purpose was to collect additional information on possible improvements of the concept from a user point of view. Also, participants'

Fig. 4. The living light concept

opinions were collected regarding the appropriateness of the light settings that were presented. The 32 participants were representative for the target group: they had a keen interest in watching movies or listening to music at home. Half of the participants were shown film and the other half were asked to listen to music, both were enhanced by light ambiance settings and light effects provided by the Living Light system.

2.8 Results

The evaluation of the concept points out that Living Light is a potential winner concept as participants highly appreciated the concept. The concept truly appealed to the participants, especially in case of film. Participants indicated that the lighting made watching movies or listening to music a very enjoyable and a more immersive experience.

In addition, participants indicated that they would like use such a system to create the right Ambiance in the Home, e.g., when having friends over for dinner, or for enjoying an evening at home with the family. Light is seen as a key factor in creating the right setting and ambiance at home. Current technologies do not provide sufficient and easy to use means to create the desired light settings. The fact that the Living Light system offers possibilities to manipulate for example color temperature and intensity was therefore highly appreciated. These findings provided additional directions for future research: lighting concepts that offer solutions that take away the hassles of setting up appropriate lighting, and in addition provide extra means to enhance activities and create enjoyable ambiances.

2.9 Conclusion

The above three case studies approach three different technical innovations that entertain different user experiences. These experiences were relating to watching a match *together* with close friends, maintaining with low effort a peripheral awareness of another person and finally enjoying the immersion of in a rich audiovisual experience while watching a film.

In all the cases discussed the innovation and the main experience studied do not refer to the primary activity of a person but rather to the periphery of their attention, following the call of Mark Weiser for calm computing. Interestingly, by not requiring the full attention of the user and not aiming to cognitively engineer interaction tasks so that they are more effectively and efficiently performed, the studies above impact the type of research method that can be applied to study them.

For example, techniques like paper prototyping, usability testing and heuristic evaluations, are of marginal interest, helping to study the more mundane interactions with the system. The designed experiences have to be created and experienced in a realistic context, requiring realistic and serious deployment, see also [7].

Such deployment is rarely feasible in the intended context of use, e.g., the test-participants' home, as technical infrastructure and servicing are quite demanding for the experimental nature of Ambient Intelligence technologies. In this light, the HomeLab facility constitutes an invaluable tool for understanding a future that does not yet exist.

The studies described in this chapter, present some first steps into the direction of studying the interaction of humans with technology in an Ambient Intelligence world. Our research and this discussion have focused only on the home environment and on leisure activities, but a similar mindset should apply to testing workspaces, e.g., investigating the office or the hospital clinic of the future. Aiming at realistic prototypes and experiencing realistic deployments in controlled infrastructure, rich insights can be obtained. The research in the HomeLab brings together technological innovation, design creativity and scientific inquiry and can help both orientate a human centric development of Ambient Intelligence technologies and teach us more about human nature.

Acknowledgments

Conceptualizing the Ambient Intelligence vision has involved many people, realizing this vision involves even more people. Special acknowledgements to the many researchers from Philips Research and TU/e involved in HomeLab projects.

References

1. Abowd, G.D. & Mynatt, E.D. (2000). *Charting Past, Present, and Future Research in Ubiquitous Computing*, ACM ToCHI, 7(2), 29–58. ACM.

2. Aarts, E.H.L., Harwig, R., & Schuurmans, M. (2001), *Ambient Intelligence*, in: P. Denning, The Invisible Future, McGraw Hill, New York, pp. 235–250.
3. Aarts, E.H.L., & Eggen, B. (Eds.) (2002), *Ambient Intelligence in HomeLab*, Neroc, Eindhoven, The Netherlands.
4. Aarts, E.H.L. & Marzano, S. (Eds.) (2003), *The New Everyday: Vision on Ambient Intelligence*, 010 Publishers, Rotterdam, The Netherlands.
5. Bly, S.A., Harrison, S.R. & Irwin, S. (1993). *Media Spaces: Bringing People Together in a Video, Audio and Computing Environment.* Communications of the ACM, 36, 1, 28–47.
6. Branaghan, R.J. (2001). *Design by people for people: essays on usability*, Usability Professional's Association.
7. Davies, N. & Gellersen, H.W. (2002). *Beyond Prototypes: Challenges in Deploying Ubiquitous Computing Systems,* Pervasive Computing, 1(1), 26–35.
8. De Ruyter, B. (2003). *User Centred Design*, In (Eds.): Aarts, Emile, and Stefano Marzano (editors) (2003), The New Everyday: Vision on Ambient Intelligence, 010 Publishers, Rotterdam, The Netherlands.
9. De Ruyter, B. (2003). 365 Ambient Intelligence Research in HomeLab, Neroc, Eindhoven, The Netherlands.
10. De Ruyter, B., Huijnen, C., Markopoulos, P., Ijsselstein, W., (2003). *Creating social presence through peripheral awareness*, HCI International 2003, Crete June 22–27, Greece.
11. Eggen, B., Hollemans, G., & van de Sluis, R. (2003). *Exploring and enhancing the home experience.* Cognition Technology and Work, 5, 44–54.
12. Hindus, D., Mainwaring, S.D., Leduc, N., Hagström, A.E., & Bayley, O. (2001) *Casablanca: Designing social communication devices for the home.* Proceedings CHI 2001, 325–332.
13. Hutchinson, H., Mackay, W., Westerlund, B., Bederson, B.B., Druin, A., Plaisant, C., Beaudoin-Lafon, M., Conversy, S., Evans, H., Hansen, H., Roussel, N., Eiderbäck, B., Lindquist, S. & Sundblad, Y. (2003) *Technology Probes: Inspiring Design for and with Families*, Proceedings CHI 2003, CHI Letters 5 (1), 17–24.
14. IJsselsteijn, W.A., de Ridder, H., Freeman, J., & Avons, S.E. (2000). *Presence: Concept, determinants and measurement.* Proceedings of the SPIE, 3959: 520–529.
15. Jonsson, G.K., Bjarkadottir, S.H., Gislason, B., Borrie, A. & Magnusson, M.S. (2003), *Detection of real-time patterns in sports: interactions in football.* L'éthologie appliquée aujourd'hui. (C. Baudoin, ed), Volume 3 - Ethologie humaine. Levallois-Perret, France: Editions ED. ISBN 2-7237-0025-9.
16. Liechti O. & Ichikawa T. (2000). *A Digital Photography Framework Supporting Social Interaction and Affective Awareness.* Personal and Ubiquitous Computing, 4(1).
17. Markopoulos, P., Romero, N, van Baren, J., IJsselsteijn, W.A., de Ruyter, B. & Farschian, B. (2004). *Keeping in touch with the Family: Home and Away with the ASTRA awareness system*, CHI2004, Vienna, Austria, 24–29 April 2004.
18. Mynatt E.D., Rowan J., Jacobs A. & Craighill S. (2001). Digital family portraits supporting peace of mind for extended family members, Proceedings CHI 2001, CHI Letters 3(1), 333–340.
19. van Vugt, H. & Markopoulos, P. (2003). *Evaluating technologies in domestic con-texts: extending diary techniques with field-testing of prototypes.*

Proceedings HCI International, vol. III, (pp. 1039–1044). Lawrence Erlbaum and Associates.

20. van Baren, J., IJsselsteijn, W.A., Romero, N., Markopoulos, P., de Ruyter, B. (2003). *Affective Benefits in Communication: The development and field-testing of a new questionnaire measure*, PRESENCE 2003, Aalborg, Denmark, October 2003.

21. Vroubel, M., Markopoulos, P., & Bekker, M.M. (2001). *FRIDGE: exploring intuitive interaction styles for home information appliances.* In Jacko, J., & Sears, A., (Eds.), CHI 2001, Extended Abstracts, (pp. 207–209). ACM Press.

22. Weiser, M. (1991), *The computer for the twenty-first century*, Scientific American 265(3), 94–104.

How Ambient Intelligence will Improve Habitability and Energy Efficiency in Buildings

E. Arens, C.C. Federspiel, D. Wang, and C. Huizenga

Abstract. Ambient intelligence has the potential to profoundly affect future building operations. Recent breakthroughs in wireless sensor network technology will permit, (1) highly flexible location of sensors and actuators, (2) increased numbers and types of sensors informing more highly distributed control systems, (3) occupants' involvement in control loops, (4) demand responsive electricity management, (5) integration among now-separate building systems, and (6) the adoption of mixed-mode and other new types of air conditioning systems that require more sensor information to operate efficiently. This chapter describes the issues with current building automation technology, assesses how some applications of wireless sensor technology can increase the quality of control and improve energy efficiency, and suggests opportunities for future development.

1 Introduction

Buildings are primarily constructed to produce indoor environments in which their occupants are comfortable, healthy, safe, and productive. A complex mixture of systems (heating, ventilating, air-conditioning (HVAC), lighting, life safety equipment, the architecture itself, and the building's occupants) is used to achieve this purpose. Since buildings tend to be designed and built individually, the mixture of systems is virtually unique for each building. Most buildings are essentially prototype designs, but rather than being used for testing, they are put directly into operation. Designers and operators rarely have the chance to evaluate systematically how *effectively* their buildings produce desirable environments, or how *energy-efficiently* they do so. There is a great shortage of such information throughout buildings' lives – they are delivered to the operators without instructions, and once in operation, operators often cannot determine how they perform because there are insufficient channels for collecting physical data and occupant feedback. As a result, they tend to be operated in rather ad-hoc ways – often whatever works to cause the least complaints. It would help if more information were available.

In the past two decades, the adoption of computer control systems in commercial buildings has greatly improved the access to and management of physical data. However, these systems still communicate with relatively few sensors and actuators, so their information is not detailed or reliable enough to truly operate the building effectively or efficiently. In addition, few of them integrate HVAC with related but independently marketed systems like lighting, security, fire, or occupant information. Residential buildings tend to be intrinsically much simpler than commercial ones, but even here the amount of sensing and the information provided to systems and to occupants is less than optimal – usually all contained within a single thermostat.

In the US, 38% of all primary energy is used to condition buildings, divided evenly between commercial and residential buildings. This is the largest single energy use sector, exceeding transportation and industry. In commercial buildings, heating, ventilating, and air-conditioning (HVAC) consumes approximately 28% of total energy consumption, followed by interior lighting at 25%. In residential buildings, space heating and cooling have the highest energy consumption at 43%, followed by miscellaneous use at 16%, and water heating at 14%. The Department of Energy [5] estimates that in both building types, roughly half the total energy use could be economically avoided. Reducing energy use in buildings is both important and feasible.

There have been many approaches to achieve this objective. For example, buildings may be designed using passive temperature control, natural ventilation, solar control, and daylighting to reduce the energy used for HVAC and electric lighting. New air-conditioning systems such as underfloor air distribution, displacement ventilation, and chilled/heated ceilings can reduce operational costs. Old HVAC equipment, lighting, and windows can be replaced by newer versions which are generally more energy-efficient.

This chapter discusses how expanding the *ambient intelligence in building controls* might also reduce energy consumed in building operation. In some cases, it could be the fastest and most cost-effective way to obtain a given level of energy saving. In others, expanded intelligence may be necessary for some of the more efficient new building design techniques to become feasible in practice.

Increased ambient intelligence should also help produce more habitable indoor environments. In commercial buildings, our surveys consistently show thermal complaints (too hot and too cold) are the highest sources of dissatisfaction, with air quality, acoustics and lighting also high. The percentage of occupants voting dissatisfied typically exceeds 20%. For manufactured objects, this level of dissatisfaction would be totally unacceptable, but for current buildings it is clearly very hard to do better. We will argue that in order to do better, occupants need to be informed about and involved in the control of their indoor environment.

2 Current Building Controls: Problems and Needs

Ideally, building control systems maintain occupant comfort at a low energy cost. The state-of-the-art in building control has greatly advanced in recent years. In commercial buildings digital controls are replacing pneumatic controls [13], and energy management and control systems (EMCS) now are increasingly used to monitor and manage the HVAC systems in large commercial buildings. Some of these are web-enabled and most allow for remote monitoring and control. However, while the communication and hardware technology of building controls has changed, the control functions are still rudimentary, with very little use of supervisory control or embedded intelligence. The sensing is far more complete on the HVAC machinery than in the building and its interior spaces. Lighting control technology still consists primarily of switching large banks of fixtures based on a time clock. The intelligence employed in these controls is low because with limited numbers of sensors and actuators one cannot practically do much more.

Sensors and actuators have historically been so expensive that keeping their numbers minimal has been taken for granted. The cost of installing a single sensor or unit controller in a commercial building can be as high as $1000. As much as 90% of that cost is in running the wires needed to power the sensors and communicate with them. Installing wire usually requires making openings in walls and ceilings and then having to refinish them. In some cases the most appropriate sensor position (say on an office worker's desk or chair) is unavailable to a wired sensor, which must be on one of the building's surfaces. So compromises are made such that the sensor is positioned where it is most convenient and inexpensive. This leads to a situation where buildings are "sensory starved". The building is run on a small amount of sensor data whose accuracy cannot be cross-checked, and whose measurement locations may not represent the environments that the occupants actually experience. Because such sensory shortcomings are taken for granted by designers, the whole approach to building design is essentially distorted. Buildings must be conceived as simplified mechanisms appropriate for this level of control–large indoor spaces are considered as a single nodes, mechanical systems are designed to mix the air in such spaces uniformly even when this imposes an energy and air-quality penalty, and lights are arrayed in uniform banks even when the need for light varies across the space.

Occupant complaints decrease occupants' work productivity and increase maintenance cost by millions of dollars annually. For example, Federspiel [7] reported that the most common action taken in response to thermal sensation (hot/cold) complaints is to adjust a control system setting, and that automating these actions could reduce HVAC maintenance costs by 20%. Additional sensors would make it easier to determine when problems reported by occupants can be resolved automatically, and when it is necessary to dispatch maintenance personnel to solve the problem. In addition, thermal comfort depends on multiple factors besides temperature. If a space is controlled with a single temperature sensor, the temperature needs to be tightly controlled within a narrow range to avoid potential discomfort caused by other variables such as air movement or radiation that the thermostat cannot detect. Such tight control requires extra energy consumption by the HVAC system. If the environment were more completely sensed, it could be possible to tune it to provide comfort and ventilation as efficiently as possible.

Occupants' comfort is now never considered directly in building operation. Controls that could obtain information about the comfort of individual occupants have been proposed [6], but have not yet been put into use in buildings. Occupancy and predetermined preferences could be identified by sensors in the chair, as is now done in some automobiles. A person's thermal state could also be predicted from measured skin temperatures sensed through contact or remotely by infrared radiation. None of these things is readily possible if sensors must be mounted on building surfaces, such as walls or ceilings. The workstation furniture is the closest to, indeed in contact with, the occupants. But the difficulty of making hard-wired connections to furniture systems makes such placement traditionally impossible.

The heating and cooling of relatively small local body parts like the hands, feet, or face have a disproportionately strong effect on comfort and satisfaction. If these could be comfortably conditioned with a relatively tiny energy input, the overall ambient space temperature could be allowed to float in a relatively wide range, generating great energy savings. Workstation furniture within a building provides promising sites for occupant sensing and comfort control, perhaps using a parallel local HVAC system allowing individual control independent of the central building HVAC system. The localized actuation of heating and cooling panels and jets within the furniture would probably be best controlled by wireless means, as with a television remote.

3 Wireless Sensor-Networks: An Enabling Technology

There are at least four attributes of emerging wireless sensor network technology that could be significant for building applications: small size, low power, and self-organization. These attributes will enable a number of new applications that will improve habitability and improve energy efficiency.

Although buildings are large systems, the small size that is achievable with MEMS technology is desirable for buildings because it allows sensors to be embedded in building materials and furnishings without causing aesthetic problems. For example, Hill [9] describes the development of a single-chip wireless sensor node of just five square millimeters. Small size is also expected to help reduce the per-unit cost of wireless sensors.

In the past, the need for wired power was one of the key attributes of wireless sensor technology that prevented its widespread use in buildings. Low-power radios such as those described by Rabaey et al. [15] combined with ambient energy harvesting systems such as those described by Roundy et al. [16] and firmware designed to conserve energy stored in batteries or capacitors will allow wireless sensors to operate without wired power for years. This will enable the placement of sensors in locations that have been desirable but impractical in the past. It will also enable mobile sensors.

Self-organizing embedded software will allow large networks to configure themselves so that the labor associated with system installation, operation, and maintenance will be lower than it is today. It will enable data from mobile sensors to get where it needs to go.

There are a number of emerging techniques for automatically determining the location of sensors. This is very important even with today's wired sensors. In today's buildings, the CAD drawings that should describe the location of sensors is often inaccurate either because the building was not constructed exactly as planned or because it has been renovated without adequate documentation. Sensors that can self-locate will make it much easier to maintain buildings, and will reduced the need for detailed documentation every time a portion of the building is renovated. In addition, where sensors are embedded within office partitions or furnishings, continued locationing would help operators keep track of the network as office cubicles are moved or relocated.

4 Opportunities for Improvements

From the building's perspective, here are some of the opportunities enabled by wireless sensor networks.

4.1 Improved Sensor Locations

Sensors are essential components in control systems. For thermal comfort, a thermostat sensor should sense how a building occupant feels about the environment. The thermostat should therefore be placed near the occupant. However, thermostats in closed offices are usually mounted close to the door for convenience of wiring. In open plan offices, thermostats have to be mounted on an external wall, an internal wall or on a column. Thermostats mounted on external walls can easily be affected by nearby sunlight or thermal transfer through the wall. In the interior, air circulation patterns cause local differences between the thermostat and occupant locations. Poorly located sensors therefore misrepresent the room conditions that the occupants experience and produce sensing delays or inaccurate information.

Wireless sensors can make it much easier to sense variables of interest directly within the occupied zone. Sensors on a desk, within chairs, on phones, or computer keyboard or mouse could measure air temperature and air motion within the occupant's local microclimate. Sensors at various levels on furniture, partitions, and ceiling tiles could detect vertical stratification in the environment. In addition, the increased sensor densities that we envision will allow measurement errors and sensor faults to be more easily spotted and corrected than is possible at present.

Figure 1 shows sensors and actuators that might be found in an office in the near future. Sensors on walls, windows, lights and blinds, furniture,

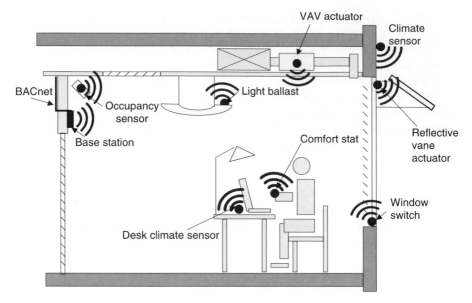

Fig. 1. Diagram showing the use of sensor networks indoors

exterior, HVAC system, even *on* the occupant (a temperature-sensing ring, or a voting device) might prove useful. The sensor information is ported via the open-source building-automation protocol BACnet to the building's energy management system from a base station functioning as a gateway. At the room scale, some of the control and actuation could take place within the room itself. The comfort-stat (a kind of remote controller) could control the lights or the variable-air-volume (VAV) diffuser in the ceiling.

Going beyond this, occupants feel comfortable in, and often prefer, a wider range of environmental conditions if they have *control* over their local conditions [4]. This range could be controlled in ways that are energy efficient if there were sensors providing information on the air movement, thermal radiation, and temperature gradients within the space.

In both tenant-occupied and owner-occupied commercial buildings, churn (renovations resulting from organizational changes) is a significant operating cost. Today wired sensors must often be re-wired and re-located when a space is renovated. According to IFMA [11], the average move in the government costs $1340 (per person). If new walls, new or additional wiring, new telecommunications systems, or other construction is needed to complete the move, the average cost in a government setting is $3640. Wireless sensor networks could significantly reduce the churn costs associated with re-wiring because sensors can be freely located and easily moved.

4.2 More Sensor Types

MEMS technology makes it possible to add sensing modalities to a sensor node without significantly increasing the cost. Sensors that would be useful for building applications include light spectrum (in three bands to differentiate daylight, sunlight, and artificial light), sound spectrum (to differentiate noise sources and control masking systems), averaging air velocity sensors across a deep column of air, carbon dioxide concentration, and perhaps a pollutant or tracer-gas-specific concentration sensor.

4.3 Occupants' Involvement in Control Loops

Networks of physical sensors and actuators will make it possible for occupants to become more involved in controlling their local space. This is a good thing. "Smart buildings" controlled without occupant intelligence or involvement are often highly unpopular. Occupant control helps satisfy occupants by widening their comfort zone, so that such buildings do not need to be as rigorously conditioned as buildings with totally automated control. The "adaptive model" quantifies this widening effect [4]. It is currently being incorporated in ASHRAE Standard 55, one of the two major standards for indoor thermal environmental control [1].

In operation, occupants could operate the room's overhead lights without leaving their desks. Programmable switches and ballasts could give occupants more flexibility to adjust ambient light levels according to their preferences. Similarly, the VAV damper positions could be independently overridden while the central control system is informed of the action via BACnet.

At a broader level than just providing individual controllers, we view occupants as a useful resource to control the environment. By providing occupants with information that allows them to play a more effective role in the environmental control of their buildings they will be more satisfied, and arguably more productive, than is possible today [19]. They can also save energy expense: 3M corporate headquarters in Minnesota uses their public address system two or three times per year to control demand during peak price periods. They broadcast a message asking workers to close fume hoods, shut off lab equipment not in use, shut off lights, shut off office equipment not needed, close blinds, etc. The net result of one such recent use was that the building's electrical demand dropped from 15 MW to 13 MW in 15 minutes, and then to 11 MW over 2 hours. This type of information could take place continuously in a less obtrusive way.

A two-way communication infrastructure could be constructed to manage large commercial buildings. For the buildings with wideband communication infrastructure, it would be easy to provide occupants access to facility management through an intranet. They could then report problems and track facility management's response more conveniently. Such an advanced facility management could also allow occupants to receive messages. For example, an occupant in a perimeter zone might receive messages that ask him to close the blinds at a certain time to reduce use of peak-rate electric power.

It would be desirable to design devices that provide critical information to the occupants. Such devices could be wireless motes that only receive price signals. For example, in residential buildings, a lighting mote could be red when the electricity price is high, yellow when medium, green when low. It could also flash these colors to indicate an upcoming price. Thus informed, occupants could decide how to operate their appliances, such as to postpone washing clothes when the red light is on or flashing, or to precool the house with the air conditioner when high prices are foreseen. Some of this could be automated but ideally the occupants should have access to the system's control strategy, and also to be able to override it at any given time.

5 Applications

5.1 HVAC: Optimizing Energy and Comfort in Multiple Rooms

In a air conditioning system in commercial buildings, it is common to use one sensor to control multiple spaces or rooms, while these multiple rooms could be experiencing different load profiles and occupancy patterns and therefore have widely varying temperatures. The potential energy benefit of increasing sensing resolution in office buildings has been investigated by Lin et al. [12] using computer simulations. They showed that by increasing the number of sensors they could simultaneously reduce discomfort and energy consumption. Table 1 shows the results compared to the standard case where a set of rooms on a perimeter exposure of a building is controlled by a single sensor. In the multi-sensor case, each room has a sensor. A strategy optimized for comfort reduced energy consumption by four percent and reduced the predicted percent dissatisfied (PPD, a measure of the fraction of people who are uncomfortable) by 9.5%. A strategy optimized for energy consumption reduced energy use by 17.3% while still reducing PPD by 5.7%. Even simple, ad-hoc strategies worked well. Controlling the average temperature reduced energy consumption by 7.1% and reduced PPD by 9.1% while controlling the average of the highest and lowest temperature reduced the two metrics by 7.1% and 6.5% respectively. For the case of controlling the average, the authors could show that the improvements arise when one or more rooms requires cooling while one or more other rooms in the same zone requires heating at the same time.

5.2 HVAC: Optimizing Energy and Comfort Within A Single Room

Wang et al. [18] studied the energy needed to condition an office room with air stratification present in the space, and quantified the effects of

Table 1. Predicted percent reductions in energy and discomfort from increasing sensing density

	Cooling	Heating	Total	PPD
Comfort opt	2.1	7.3	4.0	9.5
Energy opt	8.3	32.9	17.3	5.7
Average	3.5	13.4	7.1	9.1
Span	2.8	14.6	7.1	6.5

adding temperature sensors at foot level to the traditional sensors at chest height – a scenario that would be readily accomplished with a net of wireless temperature sensors. The stratification was produced by air supplied from an underfloor plenum, a relatively new technology with a number of attractive features. The additional foot-level sensors enable a more sophisticated variable-temperature-and-volume (VTV) system for controlling the air supply than could be possible with a conventional single thermostat. The traditional single-thermostat systems use variable-air-volume (VAV) and constant-air-volume (CAV). The three systems were compared.

The two-sensor (VTV) system used the least energy among three cases with savings of 8% compared with the VAV system, and 24% when compared with the CAV system. The major energy difference came from fan energy consumption: the VTV consumed 14% less fan energy than that of VAV and 37% less than that of CAV. All three of these systems had the same average temperature in the occupied zone, and all three had vertical temperature gradients within acceptable limits.

A similar situation occurs when controlling indoor air quality. Ventilation air is wastefully distributed in fully mixed spaces. 10 l/s may be supplied per person; only 0.1 l/s or 1% is actually inhaled [a person doing moderate work consumes 16 ml/s oxygen]. Efficiency could be gained by supplying fresh air directly to the occupants' breathing zone. Velocity and temperature sensors near the occupants' head would make it possible for the system to view the air movement around the occupants to achieve a desirable airflow pattern.

5.3 HVAC: Fault Tolerance, Fault Detection and Diagnosis

It is often difficult to quantify the benefit of fault detection and diagnosis capabilities in building control systems because not enough is understood about the frequency and severity of even the most common faults. This fact combined with the high cost of wired sensors and the relatively low cost of energy makes it uncommon to have redundancy that would be useful for fault-tolerant control, fault detection, or diagnostics. Braun and Li [3] reported that 75% of the labor spent on preventative maintenance could have been avoided by using existing fault detection and diagnosis technology. Furthermore, the energy performance of 15 of the 21 units (70%) they inspected were negatively impacted by faults, causing the efficiencies to be reduced by 20–30%.

We expect that the energy benefits of increasing sensor density will create sensor redundancies that can be exploited for fault tolerance, fault detection, and diagnosis. For example, the study by Lin et al. [12] involved adding more space temperature sensors and changing the control software to reduce energy consumption and thermal discomfort. In such a system, a failure of a single space temperature sensor would not necessarily cause the control system to fail because it could revert to an alternative strategy that made use of the available, properly functioning sensors.

5.4 Distributed Control of Data Centers

With the advent of increased power densities in today's computer chips and the explosion of demand for centralized computing services, data centers have grown hotter and larger, respectively. The power densities per unit area have risen in recent years from $500 \, \text{W/m}^2$ to $3000 \, \text{W/m}^2$ [14]. These energy rates yield computer racks that each dissipate 10–15 kW [17]. Therefore, a cooling control system that could modestly increase cooling efficiency could save a significant amount of money and justify the equipment costs required for such a system.

Boucher [2] describes a control system that has the sophistication to optimize the cooling of each rack in a data center, to maintain proper thermal conditions for all computers. This system requires:

1. A distributed sensor network to indicate the local subfloor conditions of the data center.
2. The ability to vary cooling resources locally (in this case by a moving nozzle that directed cooling supply air at hot spots, like a fire hose).
3. Knowledge of how local cooling variation affects the overall conditions of the data center.

Using these three things, a "smart" dynamic cooling controller was developed and implemented to automatically optimize the computer room air conditioner (CRAC) settings with respect to minimum energy usage. To achieve (1), wireless sensor technology makes the addition of a distributed sensor network much more viable and flexible from a client service point of view.

Table 2 shows the energy performance achieved by the smart dynamic cooling controller implemented in the data center shown in Fig. 2. The experiment was conducted by isolating the section of the data center in the upper right region of the figure and using false compute loads in the two rows of racks. The compute load was 52 kW. The two CRAC units were controlled in parallel.

The table illustrates that simply by replacing the single sensor at the return point with a sensor network at the rack inlet locations the power consumption could be reduced by almost 50% from the base case. Explicitly

Table 2. Energy performance of data center cooling controls

Configuration	Fan Speed, %	Supply Temperature, °C	Power Consumption, kW
Return temperature control (base case)	95	13.8	45.8
Rack inlet control (uses sensor network)	95	20.4	31.2
Rack inlet control with fan optimization	45	20.0	13.5

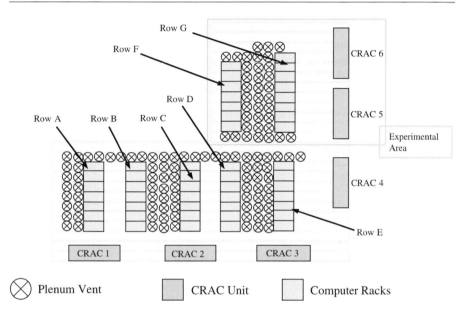

Fig. 2. Schematic diagram of data center used for experiment

regulating the rack inlet conditions enabled us to use a fan optimization strategy that reduced energy consumption by 70% from the base case.

5.5 Lighting Control

Lighting in commercial buildings uses almost the same energy as HVAC (28% of total energy use and over 50% of electricity consumption). Like HVAC controls, lighting controls are also 'starved' for sensors (daylight, occupancy) and actuation (too few switches, unnecessary lights often on when few would suffice). Wireless sensors and actuators would permit programmable switches for each occupant, switching and dimming in response to daylight sensors, occupancy sensors, or commands from occupants. At UCB we are developing a system of wireless sensors, wireless switches, and wireless controls that can be easily integrated with existing wired systems. This system promises to greatly increase energy efficiency while simultaneously improving controllability and lighting quality for occupants.

Wireless lighting control networks can provide control at the fixture or ballast level, as illustrated in Fig. 3. Ballast level control can be implemented using stand-alone relay devices that switch power to the ballast or by ballasts with integrated wireless capability (currently under development). Ballast-level control provides greatly enhanced flexibility for the occupants and can lead to significant energy savings. With traditional wired switches, occupants in open plan offices usually have limited control over ambient lighting. One switch normally controls lights for many different workspaces and "ownership" of the switch is unclear. As a result, occupants leave lights on unnecessarily. A programmable wireless network enables individuals to control only the lights affecting their workspace, greatly increasing the likelihood that lights will be turned off when unneeded.

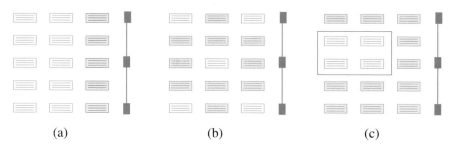

(a)   (b)   (c)

Fig. 3. Wireless lighting network control allows many control patterns for a given set of light fixtures. Shown here in plan view (**a**) perimeter daylight control where fixtures near the windows are turned off when sufficient daylight is available, (**b**) a demand reduction control scheme where 2/3 of the lights are turned off to shed load; this could also function as an emergency lighting scheme, and (**c**) lights for an individual workspace are on while adjoining workspaces lights are off

Light level sensors can be integrated into the network to take advantage of available daylight. In the most advanced systems, dimming ballasts can be used to maintain the desired light levels. Even without dimming ballasts, fixtures with standard ballasts can be turned off in areas that have adequate daylight. Many fixtures provide bi-level control by using two ballasts and provide an intermediate level of electric light in response to daylight levels. Motion sensors can be added to the network to turn lights on or off as needed in response to occupancy.

Wireless networks have great potential to lower lighting electric demand in response to requests from the electric utility. A fraction of fixtures can be designated as non-critical and can turn off in response to a broadcast request.

Since lighting is the largest electric end-use in most office buildings, it has significant potential for demand reduction programs.

UC Berkeley's prototype wireless lighting control system can easily be applied as a retrofit in existing buildings. As shown in Fig. 4, the system includes relays, light sensors, motion sensors and control switches that communicate via a wireless network.

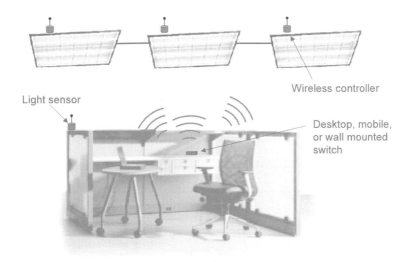

Fig. 4. Arrangement of wireless control hardware

5.6 Demand-Responsive Electricity Management

Wireless technology provides an important opportunity in residential and small-commercial buildings as well. Such buildings typically do not have the complex environmental systems and central control of large commercial buildings. Wireless networks of sensors and controllers could enable residential-scale systems of considerable sophistication that are now only possible in commercial buildings.

In California and other states, demand-responsive electricity management is being proposed to solve the problem of energy demand and supply. It is intended to increase the efficiency and improve the control of the electricity-supply infrastructure in urban or regional levels. For instance, hourly pricing gives the users the option to reduce their usage during expensive periods and increase their usage during inexpensive periods; thus demand response in a connected market reduces load levels at high retail prices, and reduces pressure in the wholesale power market, allowing prices to fall. In aggregate, this allows energy generation resources to be managed more efficiently.

For demand response to be implemented in practice, a network combining distributed time-sensitive metering and smart energy-consuming appliances with cost-setting mechanisms is needed [10]. Figure 5 illustrates some of the components of a residential demand-response system. Within each house, the electricity meters should be capable of receiving real-time electricity tariffs and automatically initiating responses that reduce overall energy cost, while being responsive to the occupants' preferences. They must be flexible enough to respond to changing pricing plans and billing intervals, and to signal the consumer's usage to the utility. The meters should also be capable of acting as a platform to support other sensors and actuators, and have a user interface that is clear and intuitive to typical residential users. Finally, to be widely adopted, they must be more inexpensive to purchase and install than current solutions. For this, the sensors/actuators must be combined on one or two chips with wireless communications and power scavenging. Federspiel et al. [8] describe some of the design concepts for such a wireless demand-response system.

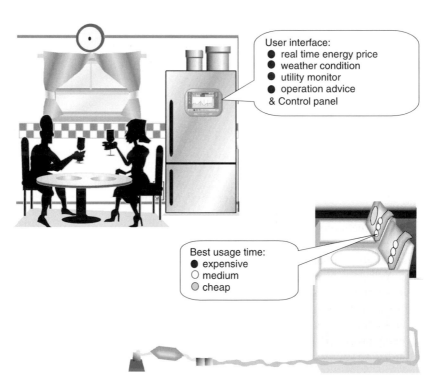

Fig. 5. Involving residential occupants in electricity use decisions

6 Integration of Systems

Wireless ambient intelligence may also help solve an environmental control problem that is unrelated to wiring: the general lack of integration among the different environmental control functions. Figure 6 shows how many of the control functions in buildings might be integrated. In commercial buildings, the environmental controls include the following:

- Heating, ventilating, and air conditioning (HVAC) systems control thermal conditions and air quality through mechanical heating, cooling, and ventilation equipment. In some cases, primarily overseas, such systems are interlinked with building envelope components, such as operable windows.
- Interior lighting is controlled by electric lighting systems and, in some buildings, by automated daylighting and solar controls.
- Acoustical systems mask, dampen, or eliminate unwanted sound.
- Fire safety systems detect combustion and activate fire suppression, smoke control, and evacuation systems.

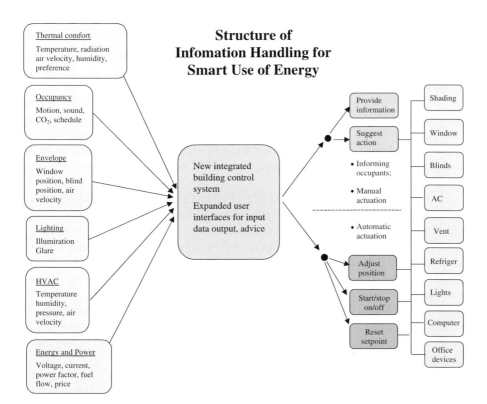

Fig. 6. Structure for an integrated wireless building control system

- Energy management systems are used in conjunction with some of the above systems to improve their efficiency; these typically sense both environmental and system state variables.
- Security, lighting, HVAC, and energy management systems all may monitor human occupancy through thermal, chemical, and acoustic means.

The controls associated with any one of these functions rarely respond to the operation of one of the other functions. For example, temperature controls do not respond pro-actively to the operation of lighting or automated window-shading devices. This is because distinct systems are dedicated to each of these functions. Consequently, it is difficult and expensive for the temperature control system to get information about the operation of lights or windows, and vice-versa. This problem could be solved if the sensory information used for each function were derived from the same source.

Ambient intelligence will involve multiple sensing modalities being embedded in a single sensing node. This will ensure that the sensory information for each of the functions described above is readily available for use by any control application. Ambient intelligence will also involve developing control algorithms designed to take advantage of a data-rich system. These algorithms will result in functionally integrated environmental controls.

7 Encouraging More Advanced Building Design

Although sensors are normally seen as devices used for operations, they can have an enormous influence on what is possible in building design. We see opportunities for wireless sensor networks to enable new more efficient architectural and HVAC design concepts that until now have not been accepted for a variety of reasons.

It has become routine in the US and elsewhere to rely on mechanical air conditioning to create the interior environment in buildings. In many less-developed parts of the world, air conditioning has become a synonym for "modernization", and their buildings follow western models which may be inappropriate for their economic context. In both developed and developing parts of the world, there are energy and environmental consequences to this mechanical approach, and occupants who inhabit the artificially controlled environments may suffer health symptoms and comfort dissatisfaction. It would be desirable to design buildings that were climate-adaptable, which could take advantage of natural climate whenever possible without degraded energy or environmental performance.

Recent buildings designed with operable windows, dynamic lighting and solar control, and hybrid/mixed-mode ventilation systems, have the potential to provide higher levels of occupant satisfaction, direct natural ventilation, and decreased AC energy use. When uncoordinated, such systems could waste energy (e.g., AC-cooled air escaping out the window, or unwanted humidity leaking in). With window sensors and temperature sensors in/near the window, the air movement in/out of the window can be determined and used to control for the "desired" direction of airflow.

The success of such new building systems may depend on wireless network technology, and the higher level of ambient intelligence it enables. Wired sensors for window-position, temperature, humidity, sunlight, and airflow have typically been too expensive to install at the sensor density needed. With continuous monitoring of enough of the important variables, such systems could be operated efficiently and lessons about their performance understood by both their operators and by designers of future advanced building systems.

8 Summary

This chapter discussed how wireless sensor network technology may affect future building design and operation.

Flexible location of sensors and increased sensing density, as well as increased variety of sensor types, can make significant improvements to building energy efficiency and the well-being of their occupants.

The technology will make the following changes in the near future: to include building occupants in control loops via information and distributed interfaces, to achieve demand responsive electricity management in residential buildings, to integrate now-separate building mechanical, electrical, security, and fire/safety systems in commercial buildings. Challenges for researchers and design practitioners are to develop exploitable applications.

In the long term, the interaction between wireless technology and applications would encourage the adoption of sophisticated climate-adapting building designs, new types of air conditioning systems that provide individual control at the workstation level, and sophisticated energy-cost and comfort-management for small residential-scale buildings.

References

1. American Society of Heating, Refrigerating, and Air-Conditioning Engineers. 2004. "ANSI/ASHRAE Standard 55, Thermal Environmental Conditions for Human Occupancy". Atlanta.
2. Boucher, T., 2004, "Self-Optimizing Cooling Control for Data Centers," MS Thesis, Department of Mechanical Engineering, UC Berkeley.
3. Braun, J. and H. Li, 2003, "Fault Detection and Diagnostics for Rooftop Air Conditioners," Attachment 1, California Energy Commission report 500-03-096.
4. de Dear, R.J. and Brager, G.S. 2001, "The Adaptive Model of Thermal Comfort and Energy Conservation in the Built Environment", *International Journal of Biometeorology*, Vol. 45, No. 2, 100–108.

5. DOE. 2000. "Scenarios for a Clean Energy Future (Oak Ridge, TN; Oak Ridge National Laboratory and Berkeley, CA; Lawrence Berkeley National Laboratory)"; ORNL/CON-476 and LBNL-44029, November.
6. Federspiel, C. C. and Asada, H. 1994, "User-Adaptable Comfort Control for HVAC Systems," *Journal of Dynamic Systems, Measurement and Control,* 116(3), 474–486.
7. Federspiel, C. C., 1998, "Statistical Analysis of Thermal Sensation Complaints," ASHRAE Transactions, 104(1), 912-923.
8. Federspiel, C. C., E. Arens, T. Peffer, and D. M. Auslander, 2004 "Design Concepts for Residential Demand Response Systems," submitted to *2004 ACEEE Summer Study on Energy Efficiency in Buildings.*
9. Hill, J. L., 2003, *System Architecture for Wireless Sensor Networks,* Ph.D. dissertation, UC Berkeley.
10. Hirst, E., 2002, "Barriers to Price-Responsive Demand in Wholesale Electricity Market", June 2002, Prepared for Edison Electric Institute, Washington, DC.
11. IFMA, 1997, *Benchmarks III,* Research Report #18, International Facility Management Association, Houston, Texas.
12. Lin, C., C. C. Federspiel, and D. M. Auslander, 2002, "Multi-Sensor Single-Actuator Control of HVAC Systems," *International Conference for Enhanced Building Operations,* Austin, TX, October 14–18, 2002.
13. Moult, R., 2000, "Fundamentals of DDC." ASHRAE Journal. November.
14. Patel, C. D., 2003, "A vision of energy aware computing – from chips to data centers", *The International Symposium on Micro-Mechanical Engineering,* ISMME2003-K15, December.
15. Rabaey, J. M., M. J. Ammer, J. L. da Silva, D. Patel, and S. Roundy, 2000, "PicoRadio supports ad hoc ultra-low power wireless networking," *IEEE Computer,* 33(7), 42–48.
16. Roundy, S., P. K. Wright, and J. Rabaey, 2003, "A study of low level vibrations as a power source for wireless sensor nodes," *Computer Communications,* 26(11), 1131–1144.
17. Sharma, R. K., C.E. Bash, C.D. Patel, 2002, "Dimensionless parameters for evaluation of thermal design and performance of large-scale data centers", *American Institute of Aeronautics and Astronautics,* AIAA-2002-3091.
18. Wang, D., Arens, E., Webster, T., Shi, M., 2002, "How the Number and Placement of Sensors Controlling Room Air Distribution Systems Affect Energy Use and Comfort", International Conference for Enhanced Building Operations, Oct, 16–17, 2002. Richardson, Texas.
19. Wyon, D. 1997, "Individual Control at Each Workplace for Health, Comfort and Productivity", Creating the Productive Workplace Conference, October, London.

Part II

System Design and Architecture

Networked Infomechanical Systems (NIMS) for Ambient Intelligence

W.J. Kaiser, G.J. Pottie, M. Srivastava, G.S. Sukhatme, J. Villasenor, and D. Estrin

1 Introduction

Networked embedded sensor and actuator technology has developed over the last decade to now enable the vision of Ambient Intelligence. This will fundamentally advance our ability to monitor and control the physical world with applications for consumers, healthcare, the commercial enterprise, security, and for science and engineering in the natural environment. Significant progress has been made in the development of algorithms and complete systems for scalable, energy-aware networking, sensing, signal processing, and embedded computing. Now, new information technology, microelectronics, and sensor systems are being integrated and deployed in some of the first applications in critical environmental monitoring. This progress, however, reveals a new set of challenges. Specifically, distributed sensor networks have not yet acquired the essential capability to monitor and report their own spatiotemporally-dependent sensing uncertainty. Thus, while sensor networks may acquire information on events in the environment, these systems are not yet able to determine the probability that events may be undetected or determine how the combination of calibration error and unknown signal propagation characteristics may degrade the ability to fuse data across a distribution of sensors.

For example, in virtually all important application areas, static sensor nodes are confronted with unknown and evolving obstacles to vision or acoustic signal propagation that severely limit the ability to characterize features of interest and introduce uncertainty. Most importantly, self-awareness of sensing uncertainty will be required, for in many applications it is only the sensor network that may be present in an environment and must be depended upon to report its true performance. It is important to note that since it is *physical* phenomena and evolving environmental structures that induce uncertainty, then *physical* adaptation of a sensor network (for example, through robotic mobility) may provide the only practical method for detection and reduction of uncertainty.

This chapter describes a broad new research thrust, *Networked Infomechanical Systems (NIMS)*, that provides networked nodes exploiting infrastructure-supported mobility for autonomous operations and physical reconfiguration. As shown in Fig. 1, NIMS infrastructure and mobility allow nodes

Fig. 1. Networked Infomechanical Systems (NIMS) introduces a hierarchy of fixed and mobile sensing nodes and infrastructure enabling access to complex, three dimensional environments. NIMS mobility provides novel methods for establishing self-awareness of sensing uncertainty. Further, examples of new NIMS distributed services include node transport, physical sample acquisition, energy harvesting and delivery, wireless network relay functions, and many others

to explore complex, full three-dimensional environments. This also enables active reduction of uncertainty through physical reconfiguration of sensing nodes and infrastructures. NIMS adds a unique capability for acquisition and transport of physical samples (for example of water or atmosphere) thereby providing methods for detection and analysis of trace components that are not detectable by conventional in situ sensors. System operating lifetime is extended by NIMS infrastructure that provides energy harvesting (for example of solar energy) and energy distribution. Finally, NIMS mobility and aerial deployment provides networking resources that may be located and oriented to optimize wireless links for mobile and fixed node systems. The remainder of this chapter begins in Sect. 2 with a description of the challenge problem of sensing uncertainty that inevitably appears in complex environments. The NIMS sensor diversity capability is discussed next with its benefits for reducing sensing uncertainty, enabling adaptive sensor fusion, and extending rate-distortion, bandwidth and energy limits in distributed sensor networks. NIMS applications are also described for natural environmental science and civil (built environment) monitoring. Section 3 introduces sensing diversity and its information theoretic foundations. Sensing diversity reduces sensing uncertainty by exploiting the ability to introduce new sensor systems and to reconfigure sensor networks through robotic mobility. Section 3 then continues with description of the fusion-based detection and localization enabled by NIMS.

The development of NIMS introduces essential new tiers in the distributed sensing architecture. These new tiers permit sensing, sampling, and logistics for transport of nodes, physical samples, energy, and data. The NIMS system hierarchy combines static and mobile sensor nodes, and physically reconfigurable infrastructure that provide sustainable mobility in large, complex three-dimensional spaces. This System Ecology and its attributes are described in Sect. 4 along with the methods of Coordinate Mobility that exploit the System Ecology for self-aware sensing and sampling.

Finally, this chapter concludes with a description of a NIMS Ambient Intelligence application with a system deployment in natural environment monitoring.

2 Self-Awareness for Sensing Networks

2.1 The Sensing Uncertainty Problem

Early work in the development of distributed sensor networks has demonstrated feasibility for low power, compact, sensor nodes and wireless sensor networks [1–5]. Scalable and energy-aware networking for densely distributed sensor nodes has been developed [6, 7]. In addition, cooperative signal processing methods have been demonstrated [8, 9]. Now, a multidisciplinary, international research community is addressing the broad spectrum of information theory, information technology, and fundamental sensing principles, to enable Ambient Intelligence for many applications.

Together, the requirements for sensing fidelity and autonomous operation obviate the need for a new distributed sensor attribute, self-awareness. Self-awareness provides a sensor network with the means to autonomously determine its sensing-uncertainty. The autonomous nature of this self-aware operation is essential. Specifically, many emerging applications for distributed sensor networks require that sensor networks acquire and return data that are critical to users and society. For example, the sensor network may supply information required to guide natural environment protection or physical security. Of course, human operators and other system may not be present at all locations and times in order to provide assurance of proper information acquisition. Thus, scalable, reliable operation demands that the distributed sensor network be self-aware and autonomously probe, report, and optimize its own uncertainty.

Distributed sensor networks provide the critical data source for Ambient Intelligence. Past research has demonstrated feasible operation of sensor networks. However, the value of this data source for Ambient Intelligence depends on ensuring its fidelity for acquiring information on physical phenomena. There are many limitations contributing to degraded measurement *fidelity* in sensor networks; some examples can be provided to illustrate. First, since phenomena under investigation are, in typical applications, inherently unpredictable, then the required density of measurement sampling and sampling rate required to achieve low distortion measurement is temporally and spatially variable and may be unknown. Further, since the development of phenomena and the evolution of the environment are unpredictable, then the propagation of sensor signals is also unpredictable. To illustrate, normal urban traffic patterns, or changes in natural environments, may introduce unexpected obstacles to vision sensors, sharply reducing sensing fidelity. Similarly, changes in foliage patterns or atmospheric conditions affect acoustic propagation.

The unpredictability of arrival of events and the appearance of environmental obstacles to sensing limit sensing fidelity. This, in turn, also limits the capability of sensor data fusion methods that rely on many sensor inputs to test a hypothesis regarding the presence and behavior of phenomena and signal sources in the environment. It is most important to note, however, that it is a fundamental goal for distributed sensor networks to enable *autonomous* monitoring of the physical environment and to acquire information about the evolution of events.

To illustrate these principles of *sensing self-awareness*, consider Fig. 2. Here a mobile source moves through an environment, producing an acoustic

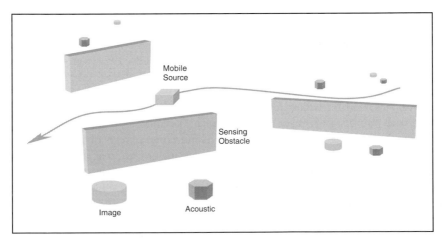

Fig. 2. Environments inherently present obstacles to sensing, as shown here for distributed imaging and acoustic sensors. Unpredictable source motion leads to obscuration of the source from sensors

signal detectable by acoustic sensors and also presenting a visible signature for image sensors. However, obstacles in the environment (natural foliage in a natural environment, mobile or static objects or structures in an urban environment) lead to fundamental *distortion* in measurements. Specifically, the motion of the mobile source may not be observed by occluded image sensors and signals emitted by mobile acoustic sources may be attenuated and distorted with an unpredictable nature as a result of propagation through distributed obstacles.

It is most important to note that the sensor systems do not presently possess the means to rapidly and accurately determine the presence of obstacles. An obstacle to sensing may not be detectable (for example by an acoustic sensor) and may not be interpreted properly by local analysis of images. Of course, a sensor network may hypothesize the presence of obstacles based on an observation of few or no detectable sources. However, verification of this hypothesis itself requires that a reliable model exist for the expected arrival of sources and that these sources arrive so frequently that a model for obstacle presence may be rapidly derived. This is clearly not a reliable and general solution to this most important problem.

2.2 NIMS Infrastructure-Enabled Mobility for Sensing Self-Awareness

It is clear that sensing uncertainty is perhaps the primary concern in distributed sensor networks since it limits the acquisition of high-fidelity environmental information and it is this, after all, that motivates distributed sensor deployment. It is also clear that physical reconfiguration achieved through proper forms of mobility may be required to circumvent sensing obstacles (as will be discussed further in Sect. 4.1). However, here are conditions on the form of mobility that can enhance the full set of distributed sensing operational capabilities. For example, in addition to providing diverse location and perspective and providing navigation through complex environments, it is also essential that mobility methods be predictable and precise. Specifically, the mobility mechanism must *reduce* system-wide spatio-temporal uncertainty as opposed to increasing uncertainty as a result of errors or limitations in motion or navigation. As will be seen, this generally requires the introduction of an *infrastructure*.

The requirements for sensor mobility control for applications in environment monitoring are as follows: (1) Sensor mobility must permit a wide range of location and viewing perspectives. This requires the ability to change separation between sources and sensors over a wide range and choose a wide range of viewing or sensing perspectives. In the natural environment, this will require overhead viewing perspective. (2) Sensor mobility must be precise so that sensor location uncertainty does not degrade sensing uncertainty yet further. (3) Sensor mobility must accommodate complex terrain and surfaces that may incompatible with surface vehicle navigation (or may themselves be disturbed by vehicle passage). (4) Sensor mobility must also be sustainable in that energy requirements and the rate of system degradation must be low. At the same time, the impact of mobility on the environment (for example acoustic noise or power plant exhaust emissions) must be minimized. (5) Finally, the sensor mobility system must also permit logistics for motion and delivery of components that may include physical samples, energy sources, replacement nodes, and other subsystems.

The addition of an infrastructure immediately addresses the above requirements in a way that would not be possible with other robotic forms. While many infrastructure types are anticipated, the "cableway" infrastructure discussed previously and discussed further below provides an example that meets these requirements, and is compatible with a broad range of environmental science applications. Further, it requires small logistics cost for deployment (that is a deployment cost no greater than deploying fixed sensors at elevation). The cableway infrastructure will be discussed with reference to the above requirements. (1) First, the cableway permits a wide range of location and viewing perspectives by allowing aerial suspension of nodes that may themselves probe a three-dimensional volume, as shown in Fig. 3a. (2) The cableway provides precise sensor mobility. (3) Also, the cableway system allows sensor nodes to negotiate complex terrain. (4) The cableway system also enables sustainable operation. Energy requirements for mobility

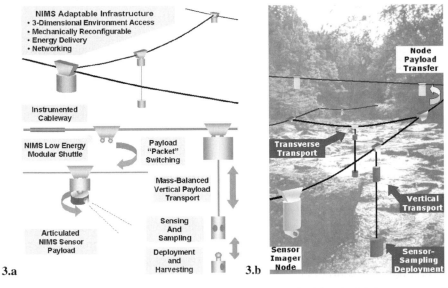

Fig. 3. (a) NIMS Systems include fixed and mobile nodes along with instrumented and adaptable infrastructure. NIMS nodes may be fixed to the infrastructure, may move on the infrastructure, or be delivered to locations and recovered by other nodes. (b) A schematic view of a NIMS deployment in a riparian stream environment with distributed sensing, sampling, and node transport

are modest and may be made vanishingly small when transport velocity is low and mass-balancing is employed to reduce gravity-work. (5) Finally, the cableway system provides a means to acquire physical samples and deploy sampling systems. It also permits low energy transport of massive payloads (if required) and permits the implementation of logistics for energy, node, and sample transport.

2.3 NIMS Sensing, Sampling, and Applications

The NIMS architecture of fixed and mobile devices and infrastructure enables an expanded set of new applications for distributed sensing and monitoring that were beyond the scope of fixed sensors alone. These exploit the capabilities summarized in Table 1 of Sensing Diversity, Fusion Based Identification and Localization, and Distributed Physical Sampling. These methods will be further described in Sect. 3.

Table 1. Networked Infomechanical Systems, with its precise and sustainable aerial mobility and reconfiguration capability, enables a series of new capabilities for distributed sensing and sampling

NIMS Sensing and Sampling Methods	
Sensing Diversity	Sensing Diversity methods exploit mobility to select and distribute available sensing resources to both map sensing uncertainty in space and time and then adaptively reduce this uncertainty.
Fusion Based Identification and Localization	NIMS mobility and the diverse resources available to infrastructure-supported nodes enable a large, multi-dimensional solution space for optimizing the cooperative identification and localization or sources.
Distributed Physical Sampling	NIMS mobility enables the acquisition of *in situ physical samples* for identification and localization of phenomena by sensitive and accurate *ex situ* methods based on, for example, laboratory analysis.

2.3.1 Natural Environmental Monitoring

Ambient Intelligence has been extended to the natural environment with distributed sensing deployments for fundamental science investigations of phenomena including global change, and for providing the data required for environmental stewardship. The application of NIMS to natural environment monitoring provides a means to reach the full three-dimensional region where the ecosystem exists. This application exploits sensing diversity that addresses the biocomplexity of the natural environment and its obstacles to sensing. Further this application relies on effective identification and localization. Finally, in addition to sensing, NIMS also provides essential sampling capability for the many investigations relying on laboratory analysis of chemical and biological phenomena for which no in situ sensors are available.

NIMS applications also include monitoring of environmental resources with example applications to efficient and safe use of agricultural land, harvesting of coastal resources, management of effluent, and collection of consumer water resources. All of these applications require monitoring by sensing and sampling of complex, dynamic terrestrial and marine environments. NIMS sensing and sampling will provide the unique capability of precise deployment and recovery of sensor systems in harsh aquatic environments. Also, NIMS physical sampling, driven by algorithms based on regular, triggered, or model-based sampling trajectories will allow for acquisition and processing of samples containing critical dissolved and suspended agents for which compact sensing systems do not yet exist. Sample-based measurements of aquatic resources may include monitoring of nutrients (nitrates) and biological pathogens. This also includes monitoring the effect of ecosystem dynamics (estuary flow, currents, tides, wave action, UV radiation) on the origin and fate of these agents.

2.3.2 Public Safety and Emergency Response

Physical safety and security includes a vast range of applications that have been supported by distributed sensors. With new concerns regarding public health and safety, monitoring in urban environments is now critical. Monitoring methods are required that provide high fidelity sensing in complex environments and that may rapidly adapt to emergency. For example, in the event of fire or structural collapse, highly mobile and sustainable sensing systems are required for accurately assessing damage and directing assistance where required. The urban environment presents a high spatial density of obstacles to imaging and sensing. In the event of structural collapse, new obstacles will appear and may create environments that are unsafe for emergency responders.

The NIMS capability for self-awareness of sensing uncertainty brings substantial new value to this application area. Specifically, as remote systems are deployed in environments, it may be that no manual observation of the environment is otherwise available (since no personnel may be present or because the environment may become unsafe for personnel presence). Thus, if environmental events cause sensing systems to be degraded (for example an environmental change introduces a new obstruction to an image sensor) then the NIMS principles of sensor diversity and coordinated mobility will be essential to recover system reliability.

NIMS capability may be deployed in place, integrated with structures, or may be rapidly deployed in response to events. For example, unconstrained robotic systems (ground-based or aerial vehicles) may deploy NIMS infrastructure to provide a means for sustainable, intensive monitoring of a disaster environment with a diverse array of sensing and sampling devices.

3 Networked Infomechanical Systems (NIMS): Enabling Self-Awareness

3.1 Information Theoretic Foundations

Fixed sensor networks inevitably confront sensing uncertainty due to inherent and evolving environmental evolution and the presence of distortion-inducing obstacles. This is manifested as an uncertainty in the support of a hypothesis derived from distributed sensor data. For example, this may result in a reduced detection probability, an identification fault, a tracking error, or a misestimate of the population of individual sources.

We can first consider the problem of detection, identification, and localization of sources by observation of an environment with a distributed sensor network. First, consider N types of individual sensors, $\mathbf{s}_k$, in the environment. Their location will be described by a manifold, $M(t)$, with locations $\mathbf{x}_k$ and time, t. In typical applications of fixed, distributed sensors, these locations will be on the surface in the environment, or perhaps attached to natural or artificial structures that may or may not be under investigation themselves. Now, the set of sources (passive or active objects of interest) appear at locations $\mathbf{y}$ in a volume V, with location distribution $p(y(t))$ at time t. Sensors will yield an observation set, Z, from one or more sensors. This set will generally form a time series or sequence of images.

The nature of propagation from a source to a sensor will, clearly, determine the limits to sensing fidelity. Of course, it is this propagation, not the properties of sensors elements or sampling characteristics that set the strictest limits on sensing fidelity. Further, propagation characteristics may include frequency and phase dependent transfer functions as well as interference and noise. For imaging sensors, obstacles in the line-of-vision as well as confusion in background images combine to complicate propagation. The coupling between a sensor and its environment lends another important propagation consideration. For example, the coupling between a ground-deployed seismic sensor and the surface introduces an additional transfer function that must be included in source characterization.

Observations, therefore, depend on propagation gains $G(\mathbf{x}, \mathbf{y}, t)$ between sensors at $\mathbf{x}$ and sources $\mathbf{y}$ at a time t. Models for propagation gains may be either deterministic or based on the propagation loss statistics of the inhomogeneous medium with respect to different sensing modes. For example, in imaging a 3-dimensional obstruction model is required. It is clearly elevation dependent and thus different parts of the sensor deployment manifold have different loss values, and these values themselves will be time-dependent as the environment evolves.

Together, these contributions to sensing uncertainty have been present in distributed sensing and are well-known in specific applications. Normally, complex site survey, preparation of the environment (for example the creation of a massive pier for a seismic sensor or the clearing of foliage for imaging sensors), and manual effort are devoted to each sensor. However, distributed sensor networks are planned for rapid deployment directly in unprepared, complex environments and will confront sensing uncertainty to a degree not reached for previous, isolated sensor deployments. As will be discussed, there is a new pathway, based on NIMS, for addressing these fundamental problems.

3.2 NIMS Sensor Diversity

NIMS *sensor diversity* methods exploit NIMS mobility and physical reconfiguration to combine diverse sensing types, diverse sensor locations, and perspectives for applications including (1) Reducing fundamental sensing uncertainty, (2) Enabling an actively optimized form of sensor data fusion, (3) Extension of rate-distortion limits, and, (4) Extension of energy and bandwidth constraints.

3.2.1 Reducing Sensing Uncertainty with Sensor Diversity

Sensing uncertainty in a conventional fixed sensor network arises due to the unknown and unpredictable characteristics of $G(\mathbf{x},\mathbf{y},t)$. As was noted previously, since the arrival of events are unpredictable, and since obstacles to sensing may themselves be passive (and not detectable by sensors) then the fixed sensor network may generally never *determine* or *reduce* its uncertainty. However, self-awareness of sensing uncertainty can be obtained through sensor diversity. To illustrate, consider Fig. 4. Note that for this example, increasing the density of sensors deployed on the surface has negligible impact on sensing uncertainty if, as is often the case, the density of obstacles is similar or even greater than that of sources. An example is that of imaging sensors deployed at a low level (the understory) of a forest environment. Here, experimental observations show that obstacle densities limit line-of-site viewing segments to distances of only one to several meters. Thus, high probability detection of sources via imaging (should this be required to support a scientific investigation) requires an extremely high sensor deployment density.

Sensor diversity, however, introduces methods for determination and reduction of sensing uncertainty through, deployment, operation, and redeployment of sensors that provide diverse detection methods and perspectives.

An illustrative example is shown in Fig. 4. Here we observe that sensing obstacles obscure the mobile source from the view of fixed sensors. However,

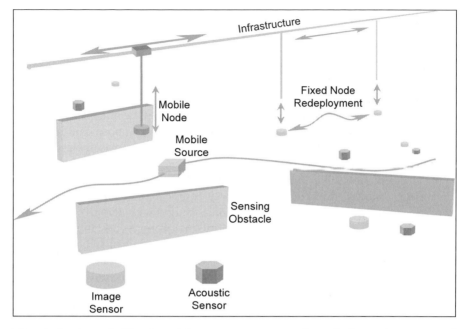

Fig. 4. As shown in Fig. 2, mobile sources propagate through the environment and are generally not observable by fixed distributed sensors deployed at the surface. However, the introduction of Networked Infomechanical Systems (NIMS) mobile devices permits nodes to be physically relocated at optimized locations and with optimal viewing perspectives. In addition, the presence of many source events may also lead to the physical redeployment of a fixed node to an optimized location and viewing perspective

the introduction of sensor diversity through use of sensor nodes that are mobile and supported by (in this example) an overhead infrastructure offers a reduction in sensing uncertainty by placing image sensors in optimal locations and affording optimal viewing perspectives.

Sensor diversity itself depends on NIMS for providing viewing perspective and enabling precise mobility. NIMS, in this example, itself depends on infrastructure for both the support of nodes and the ability to move and replace nodes. Note that in contrast to conventional mobility systems, the NIMS infrastructure provides a high degree of certainty associated with motion and orientation, as required to reduce uncertainty.

3.2.2 Exploiting Sensor Diversity by Active Fusion

The advantages of sensor diversity and an example of its implementation are provided in an *active sensor fusion* method that seeks to optimize sensing uncertainty for each component contributing to the derived hypothesis. This can be applied to the source identification problem. Here an optimization criterion is to select the largest likelihood $p(\mathbf{s}_k|Z(\mathbf{x},\mathbf{y},t))$ according to a fusion rule over the set of sensor locations $\mathbf{x}$ (each within $\mathbf{M}$), over the ensemble $\mathbf{y}$ of locally relevant source positions (within $\mathbf{V}$), at some time t, subject to a global cost constraint for placing sensors at the best locations, and the propagation model, G.

Note that again, in contrast to conventional approaches, the location of sensors, $\mathbf{x}$, may be actively adjusted in optimization. Of course, there is a cost for this sensor re-location. However, the cost constraint above can further include the energy cost of re-location as well as the cost processing data and communicating decisions to the end user. When a fidelity constraint (e.g., identification probability and latency) is added, this is then a network rate-distortion problem, with distributed lossy source coding and possibly cooperative communication. In other variations, sources may actually follow some trajectory, each node may follow a distinct trajectory (possibly responsive to the source), and the fusion rules/optimality criteria can be varied. While not always the best approach, the Bayes criterion is particularly interesting, since optimal fusion among multiple sensors for detection problems amounts to a maximization of mutual information [10, 11]. It thus applies to sensors of widely diverse types (e.g., imagers and acoustic), with the main problems being determination of prior probabilities and the specific signal conditioning that leads to conversion of observations into sufficiently good approximations of the likelihoods with reasonable complexity. More generally one can form weighted sums of (quantized) log-likelihoods for other optimization criteria.

While many versions of these problems are being studied in the context of static sensor networks [8,9], NIMS dramatically transforms the available solution space. Rather than sensors remaining at static locations constraining the network to the sensing performance first observed at deployment time, the manifold of permitted locations is much larger and more topologically complex. In addition, the cost of moving nodes among positions is relatively low. Therefore, nodes can be redeployed to take advantage of better knowledge of $p(y)$ or G acquired through sustained observations, or with actual changes in the source and obstruction distributions (e.g., with season). Further, the larger manifold of permitted positions enables choice of positions with lower obstruction, thereby allowing improved detection probabilities or lower communication energy costs between the static nodes and the infrastructure. The ability to reposition static nodes and also have nodes that sense while moving provides greater scope for investigation of adaptive algorithms. This allows, for example, a direct implementation of iterative optimization algorithms.

3.2.3 Extending Rate-Distortion Limits

Another example illustrating the benefits of sensor diversity achieved by NIMS mobility is the network rate-distortion problem. For the Gaussian n-helper problem, a main sensor X_0 and n helpers X_1, X_2, ... X_n observe a Gaussian source and then fuse their information while minimizing the set of transmission rates $\{R_0, R_1, \ldots, R_n\}$ subject to a distortion (fidelity) constraint D. The resulting rate-distortion region is bounded by:

$$R_0(D) \geq \frac{1}{2} \log^+ \left\{ \frac{\sigma_0^2}{D} \left[\prod_{k=1}^{n} \left(1 - \rho_{0k}^2 + \rho_{0k}^2 2^{-2R_k} \right) \right] \right\} \qquad (3.1)$$

where ρ_{0k} is the correlation between observations at sensors 0 and k. As usual, $\log^+(x) = \max\{\log(x), 0\}$. For static networks, the sensor positions are fixed and the adaptation choices are limited to determining which nodes will participate in fusing information and at what rates. In the context of NIMS, on a local scale one can additionally ask for a given node deployment, manifold M, propagation model G and source distribution $p(y)$ where the next node should be placed to minimize the expected rate, over the ensemble of source positions (or trajectories). Similar questions can be posed in terms of communication cost, rate savings that would be realized by repositioning some or all of the sensors X_n, or some combination of these quantities. The solution may serve for example as the iteration in a greedy deployment algorithm, possibly supplemented by occasional redeployment steps. Alternatively, this question may be asked in terms of specific estimates of source locations for tracking of a mobile source as it traverses the volume V, and based on the model of source motion and the constraints on node mobility how the nodes should be marshaled in a neighboring region. A global planning problem is to consider detection probabilities for nodes with given location distributions within a manifold M, and assess detection probabilities according to some fusion rule given G and $p(y)$ (e.g., over the ensemble of locations). Then variations in M and node density may be considered to determine whether the average detection probability improves to meet the desired performance level.

3.2.4 Extending Energy and Bandwidth Limits

The NIMS infrastructure also dramatically changes the energy and bandwidth constraints for these optimizations so as to radically transform the solutions. Thus, while in static sensor networks energy constraints dictate intense processing at sources with careful management of observation duty cycles [12], in NIMS, the nodes connected to the cable network may have no such constraints, enabling a rich set of design trades that exploit the asymmetry between different classes of observers. The nodes and infrastructure may also be laid out to allow for lower energy ground to air links (single hop) or with a small number of hops between any given node and the infrastructure to mitigate scalability [13–16] and energy consumption issues. NIMS also allows local signal processing and sensing to be augmented with new resources, leading to a far less uniform distribution of resources throughout the sensing volume V (slowly varying with time), to achieve a given level of observation fidelity. That is, resources can be adaptively matched to actual conditions.

3.3 NIMS Fusion-Based, Detection, Identification and Localization

Fusion-based detection, identification, and localization of sources through cooperative algorithms operating over distributed nodes are fundamentally limited in the event of sensing uncertainty. Typical distributed sensor applications expose sensing elements to a variable, uncertain sensing environment, and with potentially uncertain sensor calibration. This fundamentally degrades the performance of essential cooperative algorithms that must be relied upon for the essential function of fusion-based information acquisition regarding sources. NIMS self-awareness and sensor diversity through physical reconfiguration directly addresses this most important and long-standing problem.

For distributed sensor networks, source identification involves an interaction of layered suites of signal processing algorithms, networking algorithms, and distributed database access. The generic optimization problem is to maximize the mutual information subject to resource constraints (for example, energy reserves and number of nodes in the volume, V). In static networks, the energy constraints dictate layered processing, with low energy operations at the bottom level operating with constant vigilance, and higher levels operating episodically as detection thresholds are met or activation signals from other nodes are received. Similarly, nodes whenever possible process information to avoid communication; such as cooperative fusion of (approximate) likelihoods among a subnetwork of nodes, and then, only if necessary, exchange raw data for coherent combining. Data is queued based on its likelihood of being needed in a later query from a neighbor or remote observer. Node density can be adjusted to reduce the likelihood of more than one target being in the regard of a sensor so that expensive cooperative source separation algorithms can be avoided, with some balance against node cost. These problems individually and collectively are the subject of many interesting research efforts [18–21].

Note however that distributed fixed sensors are constrained to the limited locations and orientations X, and fixed energy resources that are provided at time of deployment – both planned based on knowledge of $p(y)$ and G at deployment time. However, the environment provides unscheduled, surprising events that are distributed in space and time and may not be compatible with the mix of deployed sensors or other critical aspects of their deployment. While the group of nodes that participates in fusion in response to an event may adapt, the performance may not be adequate unless the initial deployment will greatly overprovision resources in the environment; that is, node densities and energy reserves corresponding to worst case conditions throughout the entire volume, V.

All of the design considerations for fixed networks play a role in NIMS, but mobility and the far greater resources available to nodes connected to or serviced by the infrastructure allow for a far broader source identification solution space.

This is illustrated by the example in Fig. 4. Here, the path of a mobile source is not detectable by sensors that are obstructed from viewing this source. The obstruction may be an obstacle to viewing by imaging sensors (for example foliage in an natural environment), an obstacle to acoustic propagation, or may also be a source of interference for a chemical sensor and or an ecosystem event that separates a chemical sensor from its medium to be sensed (e.g., change in water level). Sensor diversity addresses these problems: if optimal fusion of information from the resources deployed in a set of positions X is not sufficient to meet detection or identification performance criteria, then NIMS will allow resources to be re-deployed automatically. Self-aware operation results from this approach through algorithms that use diverse sensor types and perspectives to continuously perform measurements of sensing performance to ensure that adequate sensing coverage exists for establishing a high probability for detection of events.

The (Bayes) data fusion problem in this context is to adapt the fusion rules to maximize the probability of selecting the most likely hypothesis based on the prior information and the set of observations. Consider the recursive estimation of log-likelihood functions for a single sensor:

$$\ln p_S(s|Z^r) = \ln p_S(s|Z^{r-1}) + \ln\left[\frac{p_Z(z(r)|s)}{p_Z(z(r)|Z^{r-1})}\right] \qquad (3.2)$$

where S is the set of hypotheses, $z(r)$ is the observation taken at time r and Z^i is the set of observations up to time i. Taking expectations on both sides, this equation may be interpreted as stating that the posterior information is equal to the prior information (to time $r - 1$) plus the information obtained from the current observation. Data fusion is obtained by replacing the last term by a sum of the log-likelihoods over the set of sensors [11]. A variety of weighting strategies are possible, resulting in a broad set of fusion algorithms. Unfortunately, there may be considerable initial uncertainties in the propagation environment, the hypothesis priors, and the calibration of the sensors, all of which make the choice of effective fusion rules difficult.

With NIMS, the ability to deploy a wide variety of devices makes the fusion problem (and subsequent identification problems) both richer and paradoxically more tractable through the ability to reduce these uncertainties. Consider for example the problem of autonomous in situ calibration. Calibration is described by the above equation, where now the hypotheses are known with near-certainty, and the objective remains to determine the log-likelihood function given a test observation set (a standard) whose priors will generally not match those of the environment to be sensed. In the fusion context, the reliability of the measurements of individual sensors can be gauged according to how the likelihoods they report compare to the known hypotheses. With NIMS, it is possible to obtain the standards in several new ways.

The shuttle network can actually create events: broadcast sound patterns, present a visible target for detection by imaging sensors, and introduce a seismic signal to calibrate acoustic, imaging, or seismic sensors, respectively. This enables the distributed determination of G. It may also transport an instrument that has been calibrated off-line into a region and measure the test events to precisely determine G and also measurement errors of instruments in its vicinity. Alternatively, samples can be collected and analyzed off-line, with the results compared to elements in the field. Based upon the model for instrument drift, an interpolation function can be applied to adjust measurements made in between calibration events.

Note further that there is no essential difference between determination of G and the basic update required for adaptive data fusion. In both instances, observations reduce the uncertainties and can serve as one step in a recursive update of the log-likelihood function. However, when observing natural phenomena, there will be decision uncertainty so that less weight will be assigned in updating according to the degree of that uncertainty. Further, in dealing with heterogeneous sensing modes or instruments with different accuracies (or calibration confidence) again not every observation will be accorded the same level of reliability. Indeed, unreliable sensors can be detected based on the extent their reported likelihoods match the weighted group consensus. New instruments can be brought into a region in which there is insufficient consensus or progress in reducing measurement uncertainties. Consequently, within the same simple mathematical framework it is possible with NIMS to explore in situ automated calibration, reliability, and adaptive data fusion.

Consider for example the following localization algorithm. Nodes with arrays can estimate direction of arrival (DOA) for a source. It is desired to use the minimum number of such nodes to achieve a given accuracy. One way to proceed is to incrementally add nodes that lead to maximum reduction in the uncertainty following fusion [22, 23]. It may be shown that given our current estimate of $p(y)$ based on the sensors making observations, the potential for reduction of uncertainty is the entropy of the predicted DOA minus the entropy of the estimator. Our initial experimental results show great promise for static networks. This basic approach can be extended in NIMS: if the available nodes cannot produce sufficient accuracy, then additional resources can be brought to bear, or nodes moved to more advantageous positions, e.g., using a gradient search algorithm guided by uncertainty reduction. Further, the principle extends to other data fusion problems, provided the appropriate pre-processing can be accomplished to produce an estimate of the marginal reduction in uncertainty.

3.4 NIMS Distributed Physical Sampling

A limitation of the contribution of distributed sensors and even sensing diversity to information acquisition includes the limitations of fundamental sensing elements. A primary goal of enabling scalable deployment of distributed sensors has been that individual elements be compact and low in mass (to reduce the logistics cost of deployment) and to present low energy demands. Of course, it is also required that sensing elements provide reliability with adequate sensitivity (noise-equivalent signal spectral density) in the environment of interest. However, many environmental characterization problems involving chemical sensing (solid, liquid, or gas phase) confront the need for detection of trace elements within interfering media. In addition, these sensor systems may require subsystems for management of media flow and filtering. Also, in the event that trace element detection or isotopic analysis is required for an investigation, then compact sensors may not be available and laboratory-scale spectrometers may be needed. Taken together, these fundamental measurement requirements may limit the capability of conventional distributed sensor networks since the fundamental measurement may not be possible with distributed, compact sensors. However, again NIMS sensor diversity may be applied, but, now with physical sampling capability.

NIMS infrastructure enabled mobility provides another high precision method with the ability to acquire physical samples (solid, liquid, or gas phase) from the environment for transport to centralized assets for analysis. As shown in Fig. 5, this includes the ability to acquire a compact sample and in addition to re-provision sensors that may require entire replacement or replacement of materiel required for operations. In addition, the NIMS

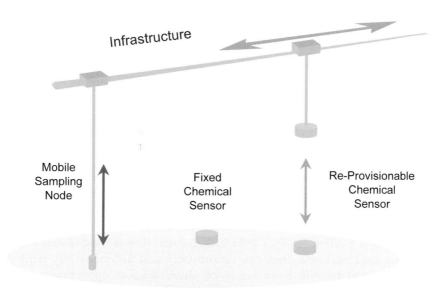

Fig. 5. In addition to physical sensing, NIMS enables physical sampling where mobile devices may acquire samples according to an event-driven or scheduled algorithm and convey samples to centralized sample analysis facilities (that may include remote laboratory analysis). NIMS also permits re-provisioning of in situ sensors that may be otherwise limited by a short operating lifetime in the medium

infrastructure can enable the accurate recover, re-calibration and replacement of sensor systems. The NIMS infra structure effectively enables a distributed sensor to consist of two components, a remote forward area sampler, and a fixed base and possibly centralized, analysis system.

4 NIMS System Ecology

4.1 NIMS System Ecology

The central requirements for self-awareness has motivated the development of sensor diversity and coordinated mobility. It is clear that physical reconfiguration (or a very high three-dimensional, volume density of deployed static sensor nodes) is required for enabling autonomous measurement and active reduction of sensing uncertainty in complex environments. Further, it is also clear that to achieve the sustainable, precise, and capable sensing and sampling needed, infrastructure-enabled mobility is required. However, to achieve the ability to adapt to varying environments, and to scale to large deployments, an architecture is required that properly combines the advantages of fixed and mobile nodes and infrastructure. In particular, it is important to introduce hierarchy to enable scalability with the assets requiring the largest resource costs being sparsely distributed and yet supporting a high spatial density of less capable nodes. Further, this hierarchy of node architecture tiers must include standardized interfaces and methods for cooperation between tiers in order to exploit hierarchy in favor of scalable, sustainable, robust and high performance operations. Specific applications may favor a larger distribution of elements in a specific tier and self-aware, self-adapting systems will adjust their own distribution to optimize application-specific resource costs and benefits.

The hierarchy of fixed and mobile nodes tiers along with interaction among tiers, forming a System Ecology, as shown in Table 2. Here, the resources exchanged between tiers along with system architecture define the System Ecology. These resources include data, samples, nodes assets, and energy. In some cases, resources are extracted from the environment (for example, sensor data, physical material samples, and solar energy) and in other cases these are supplied at the time of deployment.

The lowest System Ecology level includes untethered fixed nodes, such as wireless sensor networks that can be precisely and autonomously deployed and maintained by NIMS for study of phenomena at appropriate spatial scales. The next level consists of tethered fixed assets such as wired suspension networks, mobility drive mechanisms, gateways (for energy and communications), position beacons, storage depots, and chemical analysis engines. Together, the three levels in this info-mechanical network provide a means for generating and transporting energy and information, where information may be in the form of bits or physical samples.

Table 2. The NIMS System Ecology includes Fixed and Mobile Node and Infrastructure Tiers to enable the adaptation required to optimize the Dimensions of sensing fidelity, energy efficiency, and reach the largest spatio-temporal coverage. In this Table, the benefits contributed by each tier to these sensing dimensions are listed

	Sensing Fidelity Dimension	Energy Efficiency Dimension	Spatial Coverage Dimension	Temporal Coverage Dimension
Mobile Node Tier	Adaptive Topology and Perspective	Enable Low Energy Transport and Communications	Enable Both Sensing and Sampling in 3-D	Enable Long Term Sustainability
Connected Fixed Node Tier	Optimal, Precise Deployment of Nodes	Enable Energy Production and Delivery Logistics	Enable Optimized Node Location and Sensing Perspective in 3-D	Continuous, In Situ Sensing Sampling
Untethered Fixed Nodes Tiers	Localized Sensing and Sampling Capability	Event Detection and Guidance for Mobile Assets	Access to Non-Navigable Areas	Continuous Low Energy Vigilance

NIMS operation algorithms confront the challenges of rapid spatio-temporal formation of teams (linking multiple tiers) that enhance sensing and sampling capability by autonomously allocating appropriate tasks and roles. This is related to previous progress in homogeneous [33] and to a lesser extent, heterogeneous teams [34] for agents and robotics. NIMS, however, departs from previous development by including a System Ecology, organized hierarchically, with a diversity of communication pathways and sensing assets. NIMS operation also depended on a multi-objective optimization, (engaging all ecology dimensions), spatially distributed, and operates over a wide range of temporal scales (for example, defined by the speed of data transport and the speed of mechanical transport).

Figure 7 shows an image of a NIMS prototype system developed for forest environment monitoring. Its objective is the monitoring of critical parameters, including complex microclimate dynamics and also the spatiotemporally dynamic light environment that affect plant physiology and in particular, photosynthetic production by plants. The NIMS node also includes capability for imaging of the forest ecosystem. The NIMS node and its cable may be suspended between trees (or other structure). In addition to horizontal transport, vertical node transport is included as well. Thus, the NIMS system may access nearly the entire volume of a transect defined by a plane between two trees.

This prototype system includes an embedded processing platform (Linux operating system) and horizontal motion drive in a horizontally mobile Class II node. This node also includes a two-axis articulated image sensor. The NIMS node also carries a vertical transport mechanism for a vertically-suspended Class III NIMS node. This second node includes atmospheric temperature and relative humidity meteorological sensors along with an optical sensor for detection of downwelling photosynthetically active radiation (PAR). Wireless networking supports links between the Class II and Class III NIMS nodes, fixed nodes, and gateway access points to the Internet that are distributed in the environment. While developed for forest monitoring, it is clear that this NIMS system is applicable in many other environments and is also one application-specific example of a very large configuration space of NIMS architecture choices.

NIMS has recently been deployed in both test environments for fundamental algorithm and system research as well as in a natural environment, the Wind River Canopy Crane Research Facility in the Wind River Experimental Forest in Washington. A view of the NIMS node suspended in the forest environment is shown in Figs. 7 and 8 with both detail and panoramic views.

This system includes an embedded processing platform (Linux operating system) and horizontal motion drive in a horizontally mobile Class II node. This node also includes a two-axis articulated image sensor. The NIMS node also carries a vertical transport mechanism for a vertically-suspended Class III NIMS node. This second node includes atmospheric temperature and relative humidity meteorological sensors along with an optical sensor for detection of downwelling photosynthetically active radiation (PAR). Wireless networking supports links between the Class II and Class III NIMS nodes, fixed nodes, and gateway access points to the Internet that are distributed in the environment. While developed for forest monitoring, it is clear that this NIMS system is applicable in many other environments and is also one application-specific example of a very large configuration space of NIMS architecture choices.

The System Ecology opens a complex design space that enables adaptation to application demands. For example, the relative demands of spatial sampling density and physical configuration latency both contribute to determining the required rate-distortion operating point. By exploiting the System Ecology, both at design-time and run-time, therefore, the distribution of static and mobile sensors with varying operation range may be selected to match evolving environmental and application demands. For example, at the cost of increased measurement latency, a slowly moving mobile sensor node may explore a region of space with a high sampling point density and at the cost of only few mobile assets. Alternatively, at the cost of node resources, static or mobile nodes may be relocated and remain resident at locations that best benefit the sensing task. Such adaptations may evolve in time and space.

Finally, the System Ecology may include both infrastructure-supported nodes of primary focus in this Chapter, as well as unsupported and freely moving surface-bound or aerial robotic systems that further augment monitoring capability.

4.2 Reactive and Proactive Coordinated Mobility

Sensor diversity enables a method for determining and reducing sensing uncertainty. Now, since sensing uncertainty arises from limitations associated with *physical* configuration of sensor network nodes, then *physical reconfiguration* in the form of articulation, mobility, and the distribution of new sensing assets is *required* for reducing uncertainty. However, this then creates the requirements for systems that combine sensor diversity based self-awareness to enable *coordinated mobility* for measurement of sensing uncertainty and methods for effecting its reduction.

The relocation of sensing assets may be in rapid response to a triggering event that results from physical phenomena directly or model-based analysis of phenomena. This exploits progress in multi-robot operations, [24, 25] however, with the new features of NIMS constrained and precise mobility. This is enabled by *reactive coordinated mobility*. However, the NIMS system many also *proactively* probe the sensor network environment to determine the spatio-temporal regions where sensing uncertainty is expected to be large. This forms a *proactive coordinated mobility operating regime*.

A domain specific application applies to the problem of detection of mobile objects (sources) in natural environments. For example, acoustic sensors may typically be deployed in environments where acoustic propagation is highly variable with source-sensor range, terrain foliage, and meteorological conditions. Yet, it is at the same time required that detection of sources remain effective throughout these variations. Figure 6 illustrates an example where acoustic sensors are able to detect that sources have moved through their area, however, due to obstacles to sensing, these acoustic sensors are not able to support detection of an important large aggregation of sources. A combination of both event detection and an awareness of sensing uncertainty level produce a trigger for *reactive coordinated mobility* of mobile sensors and redeployment of nodes. Figure 6 illustrates that coordinated mobility enables a potentially drastic advance in performance by optimizing sensor population and position with both mobile nodes (imaging devices with powerful viewing perspective) and redeployed sensors. In this example, a static node acts as a trigger and the system is able to physically relocate sensing assets to acquire data at higher resolution and diversity at the trigger location.

Examples of *proactive coordinated mobility* include those where mobile nodes may analyze historical data (obtained via sensor diversity algorithms) and realize that particular areas are mapped with less certainty, causing them to revisit those areas at higher frequency until they are better mapped. Another reason for opportunistic motion is exploration, where in the absence

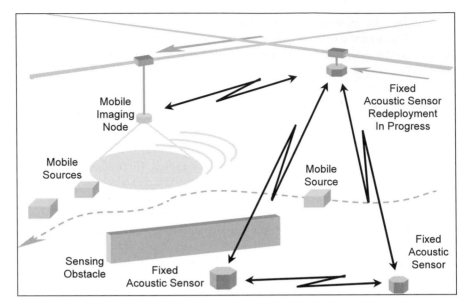

Fig. 6. An application example illustrates the aggregation of mobile objects (sources) at a location. In this typical example, fixed acoustic sensors are able to detect motion of some sources, but due to obstacles, are not able to detect the aggregated source population. However, their combination of event detection and awareness of their sensing uncertainty enables a trigger of coordinated mobility where the fixed and mobile elements collaborate on both detection of souces and redeployment of nodes. Coordination includes redeployment of nodes on more than one infrastructure element

of triggers from the static sensors on the ground, the mobile nodes proactively explore their configuration space, to detect phenomena of interest. For example, it is a general occurrence that in situ sensors may not successfully detect features of interest (i.e. acoustic sensors may not detect sound sources or chemical sensors may not detect compounds for which they are sensitive.) However, of course, it cannot be concluded by the system that the lack of sensor signals means that no sources are present – they may simply be occluded by current environment conditions or be unexpected with respect to initial sensor deployment. Thus, proactive exploration of the sensing space is *essential* for establishing performance. Here, *proactive coordinated mobility* provides a constant background probing of system performance and at the same time surveying for unanticipated events and sources. Without this, the distributed sensor system may detect, at best, only those sources that were expected to deployment.

The underlying problem in coordinated mobility for any distributed actuated system is the selection of actions at the individual node level such that the entire system performance is optimized, or at least improved. Typically performance is measured using a task-specific objective function (give example from one of our applications here). This underlying action selection problem is widely studied in the mobile robotics community. Approaches fall broadly under one of two sets of algorithms: those that minimize spatial interference [26] (e.g., avoid collisions between nodes at junctions), and those that focus on task allocation [27–29] which dynamically assign nodes to tasks. Each of the two problems (interference and task allocation) exist for both regimes (proactive and reactive) in which NIMS operates.

4.3 Coordinated Mobility Examples: Sampling

An example illustrating NIMS reactive coordinated mobility is referred to as *sampling*. An example of proactive coordinated mobility is referred to as *exploration*. The exploration operating regime enables the detection and mapping of obstacles to sensing.

Consider the sampling problem where samples are acquired at k source locations s_1, s_2, ... s_k. The mobile nodes are at locations $x(t)$. Assume there are N mobile nodes, with locations $x_1(t)$, $x_2(t)$, ... $x_N(t)$ and Q nodes on the ground with locations $h_1(t)$, $h_2(t)$, ... $h_Q(t)$. While the NIMS nodes are autonomously mobile, the sensors on the ground are not (however, they may be deployed up by the mobile nodes and relocated as well). The sensors generate observations $Z(X, H, Y, t)$, where Y is the set of source locations.

In a simple version of sampling, the sensors on the ground are pre-deployed at fixed locations, and we can solve for X such that $p(\mathbf{s}_k|Z)$ is maximized subject to visibility constraints, G. These constraints express knowledge about the map of the environment. Informally put, this asks the following question – "What are good vantage points for the mobile sensors in response to a trigger from a sensor on the ground?" For small values of k (few sources), the corresponding task-allocation problem (assigning action(s) to a few mobile nodes), and the interference problem (avoiding mobile node contention for use of infrastructure) by central planning. For small values of k ($k \ll N$) both of these problems can be successfully solved using a central planner, and re-planning is relatively low in computational burden. At medium values of k ($k \sim N$), a hybrid technique may solve the problem. The NIMS system may be decomposed into a collection of clusters. Each cluster will have some mobile nodes and some fixed nodes as members. $p(\mathbf{s}_k|Z)$ is then approximated by the product of terms depending on the configurations of the clusters instead of individual nodes. A central planner will handle the task-allocation and interference problems across clusters, whereas within each cluster, reactive techniques will be used to address these problems. At values of k larger than N, synoptic sampling by re-positioning mobile nodes alone, is impossible – a somewhat more complex version of sampling is needed where the sensors on the ground are repositioned by the mobile nodes, followed by a repositioning of the mobile nodes themselves.

This latter version of the sampling problem requires us to solve for H *and* X such that $p(\mathbf{s}_k|Z)$ is maximized subject to the visibility constraints G. Informally put – "where should the ground sensors be deployed *and* where should the mobile sensors position themselves?" In this regime the conditional probability of sampling at locations $\mathbf{s}_k$ can be factorized into the probability of sampling conditioned on the positions of the mobile nodes, and the probability of sampling conditioned on the positions of the nodes on the ground. This enables a solution for the configurations of the mobile and ground nodes separately. The task allocation problem for assigning mobile nodes to fixed ground nodes using reactive techniques will follow a greedy assignment, and planner-based techniques will optimize the sum of the distances traveled by the mobile nodes.

4.4 Coordinated Mobility Examples: Exploration

We now consider a second problem, exploration: an example of the proactive regime. In particular, consider the case where the mobile nodes explore the environment opportunistically to build a map of the environment. The mobile nodes identify the locations of obstacles in the environment to update the visibility constraints G. Formally, this is the problem of minimizing uncertainty in G given the observations Z. This problem has been studied in vision [30] (given the camera pose, estimate the poses of feature points) and robotics [31] (build a map of the environment given the location of the robot(s)). A completely distributed, greedy solution to the problem would be for each mobile node to update G individually. This is likely to lead to suboptimal task allocation (node visits to each area would overlap) and possibly high interference. A hybrid, cluster-based approach could mitigate the suboptimal allocation to a certain extent, but interference problems would still have to be solved reactively. Another approach will be to embed the 'rules of the road' into the NIMS infrastructure itself, so that spatial contention between nodes would be reduced. However, given that this regime does not necessarily require as fast a response as the pull regime, it may be possible to rely on centralized planning to a large extent. We observe that this version of the problem where nodes explore opportunistically to build a map of the environment is made significantly harder if all the nodes positions are not known precisely. This, more general, problem is that of simultaneously localizing nodes and mapping the environment, and is known to be very difficult to solve in large part due to issues in data-association [32]. Incremental but approximate solutions exist [32], which interleave the estimation of node locations with map estimations. These approaches interleave the two estimation problems by maximizing the probability of the node locations and map conditioned on the sensor readings. The most likely map that maximizes this conditional is used to estimate locations. This process is repeated until convergence. We propose to focus primarily on the mapping problem.

We will solve the localization problem to a high degree of accuracy for mobile nodes [33] by imaging known GPS locations on the ground. Once locations (and hence visibility constraints) are known, maps may be constructed accurately.

5 NIMS for Environmental Ambient Intelligence

Figure 7 shows an image of a NIMS prototype system developed for forest environment monitoring. Its objective is the monitoring of critical parameters, including complex microclimate dynamics and also the spatiotemporally dynamic light environment that affect plant physiology and in particular, photosynthetic production by plants. The NIMS node also includes capability for imaging of the forest ecosystem. The NIMS node and its cable may be suspended between trees (or other structure). In addition to horizontal transport, vertical node transport is included as well. Thus, the NIMS system may access nearly the entire volume of a transect defined by a plane between two trees.

Fig. 7. (*Left panel*) A Class II NIMS Node system deployed in a forest environment. This node includes embedded computing, wireless networking, horizontal transport, images, sensing. This node also supports a vertically suspended meteorological sensing Class III Node carrying atmospheric temperature, relative humidity, and photosyntbetically active radiation (PAR) sensor devices. Wireless links provide access between the nodes and conventional wide area networks

This prototype system includes an embedded processing platform (Linux operating system) and horizontal motion drive in a horizontally mobile Class II node. This node also includes a two-axis articulated image sensor. The NIMS node also carries a vertical transport mechanism for a vertically-suspended Class III NIMS node. This second node includes atmospheric temperature and relative humidity meteorological sensors along with an optical sensor for detection of downwelling photosynthetically active radiation (PAR). Wireless networking supports links between the Class II and Class III NIMS nodes, fixed nodes, and gateway access points to the Internet that are distributed in the environment. While developed for forest monitoring, it is clear that this NIMS system is applicable in many other environments and is also one application-specific example of a very large configuration space of NIMS architecture choices.

NIMS Node Classes	
NIMS node and infrastructure appear in four classes defined by the nature of infrastructure and nodes reconfigurability and mobility. These are listed below.	
NIMS Class I	Systems composed of fixed infrastructure with fixed nodes supported by the infrastructure
NIMS Class II	Systems composed of fixed infrastructure (this may be fixed cableways or fixed rigid infrastructure) and mobile nodes that propagate on the infrastructure.
NIMS Class III	Systems composed of both mobile infrastructure (this may include moving cableways to which nodes are attached).
NIMS Class IV	Systems composed of both mobile infrastructure and mobile nodes. An example of class IV system is shown in Fig. 2a where two parallel cableways support NIMS nodes that themselves support a cable transverse relative to these parallel cables.

NIMS has recently been deployed in both test environments for fundamental algorithm and system research as well as in a natural environment, the Wind River Canopy Crane Research Facility in the Wind River Experimental Forest in Washington. A view of the NIMS node suspended in the forest environment is shown in Figs. 7 and 8 with both detail and panoramic views.

This system includes an embedded processing platform (Linux operating system) and horizontal motion drive in a horizontally mobile Class II node. This node also includes a two-axis articulated image sensor. The NIMS node also carries a vertical transport mechanism for a vertically-suspended Class III NIMS node. This second node includes atmospheric temperature and relative humidity meteorological sensors along with an optical sensor for detection of downwelling photosynthetically active radiation (PAR). Wireless networking supports links between the Class II and Class III NIMS nodes, fixed nodes, and gateway access points to the Internet that are distributed in the environment. While developed for forest monitoring, it is clear that this NIMS system is applicable in many other environments and is also one application-specific example of a very large configuration space of NIMS architecture choices.

Fig. 8. The NIMS node system is shown in a panoramic view. This system was deployed at the Wind River Canopy Crane Research natural forest facility

6 Summary

The realization of Ambient Intelligence in many environments will require methods for mapping and reducing spatio-temporally varying sensing uncertainty. Sensing uncertainty results from the presence of unknown path loss in sensor signal propagation and unknown sensor system calibration. Obstacles to sensing, often the structures in the environment of interest, may occlude light or sound propagation or introduce. Uncertainty may enter detection of dissolved chemical agents in water, for example, due to changing currents. Clearly, sensing uncertainty is a general problem that will limit distributed sensor system performance in many applications by reducing the probability of detection and introducing distortion in fusion-based detection, identification, and localization.

After the deployment of a fixed sensor network, distortion or occlusion in the physical sensing channel will be ultimately manifested as sensing uncertainty. Since the sensing channel depends on physical properties of the environment and sensor elements, then in general, only a physical reconfiguration can change this distortion or occlusion. Thus, mobility and articulation of perspective are required to be present in some elements of distributed sensors networks. This mobility, however, must be precise (so as to not introduce further uncertainty relative to location), must probe 3-D spaces, and must also operate with sustainable characteristics to enable long-term operations.

The Networked Infomechanical Systems (NIMS) architecture has been introduced to provide these characteristics of precise navigation in complex 3-D environments with a low energy transport method. Now, to match the spatiotemporal variation of environments, the distributed sensor system must incorporate fixed and mobile devices along with systems that provide services, including transport of computation, communication, time synchronization, and transport of node systems and energy resources. The systematic implementation of architectures that match these requirements invokes the need for a complete System Ecology hierarchy of nodes and infrastructure. Now, development of Ambient Intelligence can access a vastly expanded design space to match environmental monitoring goals.

For autonomous Ambient Intelligence operations, the NIMS systems must itself autonomously explore the environment in an adaptive fashion to produce the required spatiotemporal map of sensing uncertainty. Sensor diversity algorithms are introduced to exploit many sensor types, sensing perspectives, and locations. Through coordinated mobility algorithms, fixed and mobile nodes may cooperate to proactively probe the environment to establish the uncertainty map and then adaptively adjust available sensing resources to reduce sensing uncertainty. In addition, coordinated mobility may respond also to events on demand. NIMS introduces a further mobility-enabled approach for environmental characterization with the ability for autonomous physical sampling of material in the environment – relaxing demands on sensing requirements and creating opportunities for sensitive trace analysis of components.

NIMS capabilities are matched to a broad range of Ambient Intelligence applications in natural, indoor, and urban environments. Applications for environmental science and public health and safety have been described. Another set of Ambient Intelligence NIMS applications may focus on indoor "built" environments with functionality intended to support individuals and groups to promote collaboration, productivity, and safety. These benefits may apply to healthcare clinics, to the workplace, to architectural and artistic and to entertainment environments. For example, the lighting, displays devices, and acoustics within a space may be dynamically modified to best benefit productivity or artistic goals. Personnel may exploit the electromechanical reconfigurability of NIMS technology also to support architectural features that modify structures in response to needs or environmental changes.

NIMS technology adds new dimensions to Ambient Intelligence by introducing an entire System ecology of distributed devices, infrastructures and autonomous systems. NIMS research is anticipated to enable many new applications with the ability to explore environments, actively optimize system performance, and also adapt environments to benefit users.

References

1. Pottie GJ, and Kaiser WJ, "Wireless Integrated Network Sensors," Communications of the ACM, vol 43, pp. 51–58, 2000.
2. K Bult, et al, "Low Power Systems for Wireless Microsensors," 1996 International Symposium on Low Power Electronics and Design, Digest of Technical Papers (IEEE/ACM), 1996.
3. Frazer Bennett, David Clarke, Joseph B. Evans, Andy Hopper, Alan Jones, and David Leask, "Piconet: Embedded Mobile Networking", IEEE Personal Communications Magazine, Vol 4, pp. 8–15, 1997.

4. Jason Hill, Robert Szewczyk, Alec Woo, Seth Hollar, David Culler, Kristofer Pister. "System Architecture Directions for Network Sensors". Proceedings of ACM ASPLOS 2000, Cambridge, November 2000.
5. D Estrin, GJ Pottie, M Srivastava, "Instrumenting the world with wireless sensor networks," ICASSP 2001, Salt Lake City, May 7–11, 2001.
6. Ya Xu, John Heidemann, Deborah Estrin, "Geography-informed Energy Conservation for Ad-hoc Routing," Proceedings of the Seventh Annual ACM/IEEE International Conference on Mobile Computing and Networking (ACM Mobi-Com), Rome, Italy, July 16–21, 2001.
7. C Schurgers, V Tsiatsis, S Ganeriwal, and M.B. Srivastava, "Optimizing Sensor Networks in the Energy-Density-Latency Design Space," IEEE Transactions on Mobile Computing, vol 1, pp. 70–80, 2002.
8. M Ahmed, Y-S Tu, and G Pottie, "Cooperative Detection and Communication in Wireless Sensor Networks," 38th Allerton Conf On Comm Control, and Computing, Oct 4–6, 2000, pp. 755–764, 2000.
9. Feng Zhao, Jaewon Shin, James Reich. "Information-Driven Dynamic Sensor Collaboration for Target Tracking", IEEE Signal Processing Magazine, vol 19, 2002.
10. JW Fisher III and AS Willsky, "Information Theoretic Feature Extraction for ATR," Proc 34th Asilomar Conf On Signals, Systems and Computers, Pacific Grove CA, Oct 1999.
11. A Pandya, A Kansal, G Pottie, and M Srivastava, "Bounds on the Rate Distortion of Multiple Cooperative Gaussian Sources", Center for Embedded Networked Sensing (CENS) Technical Report 0027, Sept 2003.
12. MJ Dong, KG Yung, WJ Kaiser, "Low power signal processing architectures for network microsensors", Proceedings 1997 International Symposium on Low Power Electronics and Design, Monterey, CA, USA, 18–20 Aug 1997. New York, NY, USA: ACM, pp. 173–7, 1997.
13. P Gupta and PR Kumar, "The Capacity of Wireless Networks," IEEE Trans Inform Theory, vol 46, pp. 388–404, March 2000.
14. M Gastpar and M Vetterli, "On the Capacity of Wireless Networks: the Relay Case," IEEE Infocom 2002, New York, June 2002.
15. SD Servetto, "On the Feasibility of Large Scale Wireless Sensor Networks," Proc 40th Allerton Conf On Comm, Control and Computing, Oct 2002.
16. J Li, C Blake, D De Couto, HI Lee, and R Morris, "Capacity of Ad Hoc Wireless Networks," Proceedings of the 7th Annual International Conference on Mobile Computing and Networking (MobiCom'01), Rome, July, 2001.
17. J Gao, "Energy Efficient Routing for Wireless Sensor Networks", PhD Dissertation, UCLA EE Dept, 2000.
18. T Clouqueur, V Phipatanasuphorn, P Ramanathan, and Kewal Saluja, "Sensor Deployment Strategy for Target Detection", Proceedings of the First ACM International Workshop on Wireless Sensor Networks and Applications (WSNA-2002), Atlanta, GA, USA, September, 2002.
19. S Meguerdichian, F Koushanfar, M Potkonjak, and MB Srivastava, "Coverage Problems in Wireless Ad-Hoc Sensor Networks". Proceedings of the IEEE Conference on Computer Communications (Infocom), April 2001.
20. S Meguerdichian, F Koushanfar, G Qu, and M Potkonjak. "Exposure In Wireless Ad Hoc Sensor Networks". Proceedings of the 7th Annual International Conference on Mobile Computing and Networking (MobiCom '01), pp. 139–150, 2001.

21. Yi Zou, Krishnendu Chakrabarty. "Sensor Deployment and Target Localization Based on Virtual Forces". Proceedings of the 22nd Annual Joint Conference of the IEEE Computer and Communications Societies (InfoCom 2003), April 2003.
22. M Chu, H Haussecker, and F Zhao, "Scalable Information-Driven Sensor Querying and Routing for Ad Hoc Heterogeneous Sensor Networks", International Journal of High Performance Computing Applications, vol 16, pp. 293–314, 2002.
23. F Zhao, J Shin, J Reich, "Information-Driven Dynamic Sensor Collaboration for Tracking Applications." IEEE Signal Processing Magazine, vol 19, pp. 61–72, March 2002.
24. MJ Mataric, GS Sukhatme, and D Ostergaard, "Multi-robot Task Allocation in Uncertain Environments," Autonomous Robots (to appear).
25. GS Sukhatme, JF Montgomery, and MJ Mataric, "Design and Implementation of a Mechanically Heterogeneous Robot Group," Proceedings of SPIE: Sensor Fusion and Decentralized Control in Robotic Systems II Vol 3839, Boston, September 1999, pp. 122–133, 1999.
26. D Goldberg and MJ Matariæ, "Interference as a Tool for Designing and Evaluating Multi-Robot Controllers," Proceedings, AAAI-97, Providence, Rhode Island, July 27-31, 1997.
27. B Gerkey and MJ Matariæ, "Multi-Robot Task Allocation: Analyzing the Complexity and Optimality of Key Architectures", to appear in Proceedings of the IEEE International Conference on Robotics and Automation (ICRA 2003), Taipei, Taiwan, May 12-17, 2003.
28. Lynne E Parker, "ALLIANCE: An Architecture for Fault Tolerant Multi-Robot Cooperation", IEEE Transactions on Robotics and Automation, vol 14, 1998.
29. Cao Yu Uny Fukunaga, Alex S, and Kahng Andrew B, "Cooperative Mobile Robotics: Antecedents and Directions", Autonomous Robots, vol 4, pp. 7–27, 1997. Symposium on Robot Navigation, March, pp. 1–5, 1989.
30. E Menegatti and E Pagello, "Omnidirectional distributed vision for multi-robot mapping" Proc International Symposium on Distributed Autonomous Robotic Systems (DARS02), June 2002.
31. S Thrun, "Robotic Mapping: A Survey", Exploring Artificial Intelligence in the New Millenium, Morgan Kaufmann, Lakemeyer, G. and Nebel, B. (eds), 2002.
32. S Thrun, D Fox and W Burgard, "A Probabilistic Approach to Concurrent Mapping and Localization for Mobile Robots", Machine Learning, vol 31, pp. 29–53, 1998.
33. GS Sukhatme, J Montgomery, and RT Vaughan, "Experiments with Aerial-Ground Robots", in Robot Teams: From Diversity to Polymorphism, Eds T Balch and LE Parker, AK Peters, 2001.
34. MJ Matariæ, GS Sukhatme, and D Ostergaard, "Multi-robot Task Allocation in Uncertain Environments," Autonomous Robots (to appear).
35. GS Sukhatme, JF Montgomery, and MJ Matariæ, "Design and Implementation of a Mechanically Heterogeneous Robot Group," Proceedings of SPIE: Sensor Fusion and Decentralized Control in Robotic Systems II Vol 3839, Boston, September 1999, pp. 122–133, 1999.
36. Cerpa A, J Elson, D Estrin, L Girod, M Hamilton and J Zhao, "Habitat Monitoring: Application Driver for Wireless Communications Technology", ACM SIGCOMM Workshop on Data Communications in Latin America, San Jose, Costa Rica, 2001.

37. Mainwaring A, J Polastre, R Szewczyk and D Culler, "Wireless Sensor Networks for Habitat Monitoring", ACM International Workshop on Wireless Sensor Networks and Applications, Atlanta, Georgia, pp. 88–97, 2002.
38. Bucheli TD, Muller SR, Heberle S, Schwarzenbach RP, "Occurrence and Behavior of Pesticides in Rainwater, Roof Runoff, and Artificial Stormwater Infiltration", Environ Sci Technol, vol 32, pp. 3457–3464. 1998.
39. T Cohen, SS Que Hee, and RF Ambrose. "Comparison of Trace Metal Concentrations in Fish and Invertebrates in Three Southern California Wetlands", Marine Pollution Bulletin vol 42, pp. 224–232, 2002.
40. Choe JS, Bang KW, and Lee JH Characterization of Surface Runoff in Urban Areas. Water Sci Technol, vol 45, pp. 249–254, 2002.
41. P Fischer, MS Gustin, "Influence of Natural Sources on Mercury in Water, Sediment and Aquatic Biota in Seven Tributary Streams of the East Fork of the Upper Carson River, California." Water Air Soil Poll, vol 133, pp. 283–295, 2002.
42. Boehm AB, Grant SB, Kim JH, Mowbray SL, McGee CD, Clark CD, Foley DM, Wellman DE, "Decadal and Shorter Period Variability of Surf Zone Water Quality at Huntington Beach, California". Environ. Sci Technol, vol 36, pp. 3885–3892, 2002.

TinyOS: An Operating System for Sensor Networks

P. Levis, S. Madden, J. Polastre, R. Szewczyk, K. Whitehouse, A. Woo,
D. Gay, J. Hill, M. Welsh, E. Brewer, and D. Culler

Abstract. We present TinyOS, a flexible, application-specific operating system
for sensor networks, which form a core component of ambient intelligence systems.
Sensor networks consist of (potentially) thousands of tiny, low-power nodes, each of
which execute concurrent, reactive programs that must operate with severe memory and power constraints. The sensor network challenges of limited resources,
event-centric concurrent applications, and low-power operation drive the design of
TinyOS. Our solution combines flexible, fine-grain components with an execution
model that supports complex yet safe concurrent operations. TinyOS meets these
challenges well and has become the platform of choice for sensor network research;
it is in use by over a hundred groups worldwide, and supports a broad range of applications and research topics. We provide a qualitative and quantitative evaluation
of the system, showing that it supports complex, concurrent programs with very
low memory requirements (many applications fit within 16KB of memory, and the
core OS is 400 bytes) and efficient, low-power operation. We present our experiences
with TinyOS as a platform for sensor network innovation and applications.

1 Introduction

Advances in networking and integration have enabled small, flexible, low-
cost nodes that interact with their environment and with each other through
sensors, actuators and communication. Single-chip systems are now emerging
that integrate a low-power CPU and memory, radio or optical communication
[75], and MEMS-based on-chip sensors. The low cost of these systems enables
embedded networks of thousands of nodes [18] for applications ranging from
environmental and habitat monitoring [11,51], seismic analysis of structures
[10], and object localization and tracking [68].

Sensor networks are a very active research space, with ongoing work on
networking [22, 38, 83], application support [25, 27, 49], radio management
[8,84], and security [9,45,61,81], as a partial list. A primary goal of TinyOS
is to enable and accelerate this innovation.

Four broad requirements motivate the design of TinyOS:

1. **Limited resources:** Motes have very limited physical resources, due to
   the goals of small size, low cost, and low power consumption. Current
   motes consist of about a 1-MIPS processor and tens of kilobytes of storage.

We do not expect new technology to remove these limitations: the benefits of Moore's Law will be applied to reduce size and cost, rather than increase capability. Although our current motes are measured in square centimeters, a version is in fabrication that measures less than $5\,\text{mm}^2$.

2. **Reactive Concurrency:** In a typical sensor network application, a node is responsible for sampling aspects of its environment through sensors, perhaps manipulating it through actuators, performing local data processing, transmitting data, routing data for others, and participating in various distributed processing tasks, such as statistical aggregation or feature recognition. Many of these events, such as radio management, require real-time responses. This requires an approach to concurrency management that reduces potential bugs while respecting resource and timing constraints.

3. **Flexibility:** The variation in hardware and applications and the rate of innovation require a flexible OS that is both application-specific to reduce space and power, and independent of the boundary between hardware and software. In addition, the OS should support fine-grain modularity and interpositioning to simplify reuse and innovation.

4. **Low Power:** Demands of size and cost, as well as untethered operation make low-power operation a key goal of motedesign. Battery density doubles roughly every 50 years, which makes power an ongoing challenge. Although energy harvesting offers many promising solutions, at the very small scale of moteswe can harvest only microwatts of power. This is insufficient for continuous operation of even the most energy-efficient designs. Given the broad range of applications for sensor networks, TinyOSmust not only address extremely low-power operation, but also provide a great deal of flexibility in power-management and duty-cycle strategies.

In our approach to these requirements we focus on two broad principles:

- *Event Centric:* Like the applications, the solution must be event centric. The normal operation is the reactive execution of concurrent events.
- *Platform for Innovation:* The space of networked sensors is novel and complex: we therefore focus on flexibility and enabling innovation, rather then the "right" OS from the beginning.

TinyOS is a tiny (fewer than 400 bytes), flexible operating system built from a set of reusable components that are assembled into an application-specific system. TinyOS supports an event-driven concurrency model based on split-phase interfaces, asynchronous *events*, and deferred computation called *tasks*. TinyOS is implemented in the NesC language [24], which supports the TinyOS component and concurrency model as well as extensive cross-component optimizations and compile-time race detection. TinyOS has enabled both innovations in sensor network systems and a wide variety of applications. TinyOS has been under development for several years and is currently in its third generation involving several iterations of hardware, radio stacks, and programming tools. Over one hundred groups worldwide use it, including several companies within their products.

This chapter details the design and motivation of TinyOS, including its novel approaches to components and concurrency, a qualitative and quantitative evaluation of the operating system, and the presentation of our experience with it as a platform for innovation and real applications. This paper makes the following contributions. First, we present the design and programming model of TinyOS, including support for concurrency and flexible composition. Second, we evaluate TinyOS in terms of its performance, small size, lightweight concurrency, flexibility, and support for low power operation. Third, we discuss our experience with TinyOS, illustrating its design through three applications: environmental monitoring, object tracking, and a declarative query processor. Our previous work on TinyOS discussed an early system architecture [30] and language design issues [24], but did not present the operating system design in detail, provide an in-depth evaluation, or discuss our extensive experience with the system over the last several years.

Section 2 presents an overview of TinyOS, including the component and execution models, and the support for concurrency. Section 3 shows how the design meets our four requirements. Sections 4 and 5 cover some of the enabled innovations and applications, while Sect. 6 covers related work. Section 7 presents our conclusions.

2 TinyOS

TinyOS has a component-based programming model, codified by the NesC language [24], a dialect of C. TinyOS is not an OS in the traditional sense; it is a programming framework for embedded systems and set of components that enable building an application-specific OS into each application. A typical application is about 15K in size, of which the base OS is about 400 bytes; the largest application, a database-like query system, is about 64 K bytes.

2.1 Overview

A TinyOS program is a graph of components, each of which is an independent computational entity that exposes one or more *interfaces*. Components have three computational abstractions: *commands*, *events*, and *tasks*. Commands and events are mechanisms for inter-component communication, while tasks are used to express intra-component concurrency.

A *command* is typically a request to a component to perform some service, such as initiating a sensor reading, while an *event* signals the completion of that service. Events may also be signaled asynchronously, for example, due to hardware interrupts or message arrival. From a traditional OS perspective, commands are analogous to downcalls and events to upcalls. Commands and events cannot block: rather, a request for service is *split phase* in that the request for service (the command) and the completion signal (the corresponding event) are decoupled. The command returns immediately and the event signals completion at a later time.

Rather than performing a computation immediately, commands and event handlers may post a *task*, a function executed by the TinyOS scheduler at a later time. This allows commands and events to be responsive, returning immediately while deferring extensive computation to tasks. While tasks may perform significant computation, their basic execution model is run-to-completion, rather than to run indefinitely; this allows tasks to be much lighter-weight than threads. Tasks represent internal concurrency within a component and may only access state within that component. The standard TinyOS task scheduler uses a non-preemptive, FIFO scheduling policy; Sect. 2.3 presents the TinyOS execution model in detail.

TinyOS abstracts all hardware resources as components. For example, calling the `getData()` command on a sensor component will cause it to later signal a `dataReady()` event when the hardware interrupt fires. While many components are entirely software-based, the combination of split-phase operations and tasks makes this distinction transparent to the programmer. For example, consider a component that encrypts a buffer of data. In a hardware implementation, the command would instruct the encryption hardware to perform the operation, while a software implementation would post a task to encrypt the data on the CPU. In both cases an event signals that the encryption operation is complete.

The current version of TinyOS provides a large number of components to application developers, including abstractions for sensors, single-hop networking, ad-hoc routing, power management, timers, and non-volatile storage. A developer composes an application by writing components and wiring them to TinyOS components that provide implementations of the required services. Section 2.2 describes how developers write components and wire them in NesC. Figure 1 lists a number of core interfaces that are available to application developers. Many different components may implement a given interface.

2.2 Component Model

TinyOS's programming model, provided by the NesC language, centers around the notion of *components* that encapsulate a specific set of services, specified by *interfaces*. TinyOS itself simply consists of a set of reusable system components along with a task scheduler. An application connects components using a *wiring specification* that is independent of component implementations. This wiring specification defines the complete set of components that the application uses.

The compiler eliminates the penalty of small, fine-grained components by whole-program (application plus operating system) analysis and in-lining.

Interface	Description
Clock	Hardware clock
EEPROMRead/Write	EEPROM read and write
HardwareId	Hardware ID access
I2C	Interface to I2C bus
Leds	Red/yellow/green LEDs
MAC	Radio MAC layer
Mic	Microphone interface
Pot	Hardware potentiometer for transmit power
Random	Random number generator
ReceiveMsg	Receive Active Message
SendMsg	Send Active Message
StdControl	Init, start, and stop components
Time	Get current time
TinySec	Lightweight encryption/decryption
WatchDog	Watchdog timer control

Fig. 1. Core interfaces provided by TinyOS

Unused components and functionality are not included in the application binary. In-lining occurs across component boundaries and improves both size and efficiency; Sect. 3.1 evaluates these optimizations.

A component has two classes of interfaces: those it *provides* and those it *uses*. These interfaces define how the component directly interacts with other components. An interface generally models some service (e.g., sending a message) and is specified by an *interface type*. Figure 2 shows a simplified form of the TimerM component, part of the TinyOStimer service, that provides the StdControl and Timer interfaces and uses a Clock interface (all shown in Fig. 3). A component can provide or use the same interface type several times as long as it gives each instance a separate name.

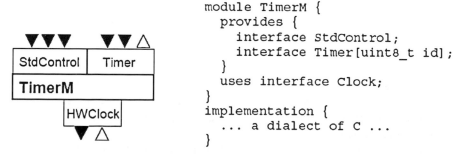

```
module TimerM {
  provides {
    interface StdControl;
    interface Timer[uint8_t id];
  }
  uses interface Clock;
}
implementation {
  ... a dialect of C ...
}
```

Fig. 2. Specification and graphical depiction of the TimerM component. Provided interfaces are shown above the TimerM component and used interfaces are below. Downward arrows depict commands and upward arrows depict events

```
interface StdControl {
  command result_t init();
  command result_t start();
  command result_t stop();
}

interface Timer {
  command result_t start(char type, uint32_t interval);
  command result_t stop();
  event result_t fired();
}

interface Clock {
  command result_t setRate(char interval, char scale);
  event result_t fire();
}

interface SendMsg {
  command result_t send(uint16_t address,
                        uint8_t length,
                        TOS_MsgPtr msg);
  event result_t sendDone(TOS_MsgPtr msg,
                          result_t success);
}
```

Fig. 3. Sample TinyOS interface types

Interfaces are *bidirectional* and contain both *commands* and *events*. A command is a function that is implemented by the providers of an interface, an event is a function that is implemented by its users. For instance, the Timer interface (Fig. 3) defines start and stop commands and a fired event. Although the interaction between the timer and its client could have been provided via two separate interfaces (one for its commands and another for its events), grouping them in the same interface makes the specification much clearer and helps prevent bugs when wiring components together.

NesC has two types of components: *modules* and *configurations*. Modules provide code and are written in a dialect of C with extensions for calling and implementing commands and events. A module declares private state variables and data buffers, which only it can reference. Configurations are used to wire other components together, connecting interfaces used by components to interfaces provided by others. Figure 4 illustrates the TinyOS timer service, which is a configuration (TimerC) that wires the timer module (TimerM) to the hardware clock component (HWClock). Configurations allow multiple components to be aggregated together into a single "supercomponent" that exposes a single set of interfaces. For example, the TinyOS networking stack is a configuration wiring together 21 separate modules and 10 sub-configurations.

Each component has its own interface namespace, which it uses to refer to the commands and events that it uses. When wiring interfaces together, a configuration makes the connection between the local name of an interface used by one component to the local name of the interface provided by another. That is, a component invokes an interface without referring explicitly

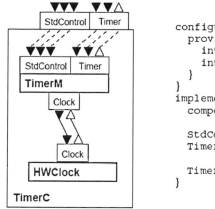

```
configuration TimerC {
  provides {
    interface StdControl;
    interface Timer[uint8_t id];
  }
}
implementation {
  components TimerM, HWClock;

  StdControl = TimerM.StdControl;
  Timer = TimerM.Timer;

  TimerM.Clk -> HWClock.Clock;
}
```

Fig. 4. TinyOS's Timer Service: the `TimerC` configuration

to its implementation. This makes it easy to perform inter-positioning by introducing a new component in the component graph that uses and provides the same interface.

Interfaces can be wired multiple times; for example, in Fig. 5 the `Std-Control` interface of `Main` is wired to `Photo`, `TimerC`, and `Multihop`. This fan-out is transparent to the caller. NesC allows fan-out as long as the return type has a function for combining the results of all the calls. For example, for `result_t`, this is a logical-AND; a fan-out returns failure if any sub-call fails.

A component can provide a *parameterized interface* that exports many instances of the same interface, parameterized by some identifier (typically a

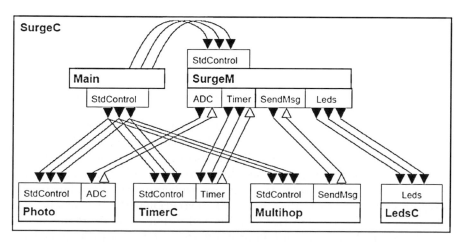

Fig. 5. The top-level configuration for the Surge application

small integer). For example, the the `Timer` interface in Fig. 2 is parameterized with an 8-bit `id`, which is passed to the commands and events of that interface as an extra parameter. In this case, the parameterized interface allows the single `Timer` component to implement multiple separate timer interfaces, one for each client component. A client of a parameterized interface must specify the ID as a constant in the wiring configuration; to avoid conflicts in ID selection, NesC provides a special `unique` keyword that selects a unique identifier for each client.

Every TinyOS application is described by a *top-level configuration* that wires together the components used. An example is shown graphically in Fig. 5: `SurgeC` is a simple application that periodically (`TimerC`) acquires light sensor readings (`Photo`) and sends them back to a base station using multi-hop routing (`Multihop`).

NesC imposes some limitations on C to improve code efficiency and robustness. First, the language prohibits function pointers, allowing the compiler to know the precise call graph of a program. This enables cross-component optimizations for entire call paths, which can remove the overhead of cross-module calls as well as inline code for small components into its callers. Section 3.1 evaluates these optimizations on boundary crossing overheads. Second, the language does not support dynamic memory allocation; components statically declare all of a program's state, which prevents memory fragmentation as well as runtime allocation failures. The restriction sounds more onerous than it is in practice; the component abstraction eliminates many of the needs for dynamic allocation. In the few rare instances that it is truly needed (e.g., TinyDB, discussed in Sect. 5.3), a memory pool component can be shared by a set of cooperating components.

2.3 Execution Model and Concurrency

The event-centric domain of sensor networks requires fine-grain concurrency; events can arrive at any time and must interact cleanly with the ongoing computation. This is a classic systems problem that has two broad approaches: (1) atomically queuing work on arrival to run later, as in Click [41] and most message-passing systems, and (2) executing a handler immediately in the style of active messages [74]. Because some of these events are time critical, such as start-symbol detection, we chose the latter approach. NesC can detect data races statically, which eliminates a large class of complex bugs.

The core of the execution model consists of run-to-completion *tasks* that represent the ongoing computation, and interrupt handlers that are signaled asynchronously by hardware. Tasks are an explicit entity in the language; a program submits a task to the scheduler for execution with the `post` operator. The scheduler can execute tasks in any order, but must obey the run-to-completion rule. The standard TinyOS scheduler follows a FIFO policy, but we have implemented other policies including earliest-deadline first.

Because tasks are not preempted and run to completion, they are atomic with respect to each other. However, tasks are not atomic with respect to interrupt handlers or to commands and events they invoke. To facilitate the detection of race conditions, we distinguish synchronous and asynchronous code:

- **Synchronous Code (SC):** code that is only reachable from tasks.
- **Asynchronous Code (AC):** code that is reachable from at least one interrupt handler.

The traditional OS approach toward AC is to minimize it and prevent user-level code from being AC. This would be too restrictive for TinyOS. Component writers need to interact with a wide range of real-time hardware, which is not possible in general with the approach of queuing work for later. For example, in the networking stack there are components that interface with the radio at the bit level, the byte level, and via hardware signal-strength indicators. A primary goal is to allow developers to build responsive concurrent data structures that can safely share data between AC and SC; components often have a mix of SC and AC code.

Although non-preemption eliminates races among tasks, there are still potential races between SC and AC, as well as between AC and AC. In general, any update to shared state that is *reachable from AC* is a potential data race. To reinstate atomicity in such cases, the programmer has two options: convert all of the conflicting code to tasks (SC only), or use *atomic sections* to update the shared state. An atomic section is a small code sequence that NesC ensures will run atomically. The current implementation turns off interrupts during the atomic section and ensures that it has no loops. Section 3.2 covers an example use of an atomic section to remove a data race. The basic invariant NesC must enforce this as follows:

Race-Free Invariant: *Any update to shared state is either SC-only or occurs in an atomic section.*

The NesC compiler enforces this invariant at compile time, preventing nearly all data races. It is possible to introduce a race condition that the compiler cannot detect, but it must span multiple atomic sections or tasks and use storage in intermediate variables.

The practical impact of data race prevention is substantial. First, it eliminates a class of very painful non-deterministic bugs. Second, it means that composition can essentially ignore concurrency. It does not matter which components generate concurrency or how they are wired together: the compiler will catch any sharing violations at compile time. Strong compile-time analysis enables a wide variety of concurrent data structures and synchronization primitives. We have several variations of concurrent queues and state machines. In turn, this makes it easy to handle time-critical actions directly in an event handler, even when they update shared state. For example, radio events are always dealt with in the interrupt handler until a whole packet has arrived, at which point the handler posts a task. Section 3.2 contains an evaluation of the concurrency checking and its ability to catch data races.

2.4 Active Messages

A critical aspect of TinyOS's design is its networking architecture, which we detail here. The core TinyOS communication abstraction is based on Active Messages (AM) [74], which are small (36-byte) packets associated with a 1-byte handler ID. Upon reception of an Active Message, a node dispatches the message (using an event) to one or more handlers that are registered to receive messages of that type. Handler registration is accomplished using static wiring and a parameterized interface, as described above.

AM provides an unreliable, single-hop datagram protocol, and provides a unified communication interface to both the radio and the built-in serial port (for wired nodes such as basestations). Higher-level protocols providing multihop communication, larger ADUs, or other features are readily built on top of the AM interface. Variants of the basic AM stack exist that incorporate lightweight, link-level security (see Sect. 4.1). AM's event-driven nature and tight coupling of computation and communication make the abstraction well suited to the sensor network domain.

2.5 Implementation Status

TinyOS supports a wide range of hardware platforms and has been used on several generations of sensor motes. Supported processors include the Atmel AT90L-series, Atmel ATmega-series, and Texas Instruments MSP-series processors. TinyOS includes hardware support for the RFM TR1000 and Chipcon CC1000 radios, as well as several custom radio chipsets. TinyOS applications may be compiled to run on any of these platforms without modification. Work is underway (by others) to port TinyOS to ARM, Intel 8051 and Hitachi processors and to support Bluetooth radios.

TinyOS supports an extensive development environment that incorporates visualization, debugging, and support tools as well as a fine-grained simulation environment. Desktops, laptops, and palmtops can serve as proxies between sensor networks and wired networks, allowing integration with server side tools implemented in Java, C, or MATLAB, as well as interfaces to database engines such as PostgreSQL. NesC includes a tool that generates code to marshal between Active Message packet formats and Java classes.

TinyOS includes TOSSIM, a high-fidelity mote simulator that compiles directly from TinyOS NesC code, scaling to thousands of simulated nodes. TOSSIM gives the programmer an omniscient view of the network and greater debugging capabilities. Server-side applications can connect to a TOSSIM proxy just as if it were a real sensor network, easing the transition between the simulation environment and actual deployments. TinyOS also provides JTAG support integrated with gdb for debugging applications directly on the mote.

3 Meeting the Four Key Requirements

In this section, we show how the design of TinyOS, particularly its component model and execution model, addresses our four key requirements: limited resources, reactive concurrency, flexibility and low power. This section quantifies basic aspects of resource usage and performance, including storage usage, execution overhead, observed concurrency, and effectiveness of whole-system optimization.

3.1 Limited Resources

We look at three metrics to evaluate whether TinyOS applications are lightweight in space and time: (1) the footprint of real applications should be small, (2) the compiler should reduce code size through optimization, and (3) the overhead for fine-grain modules should be low.

(1) Absolute Size: A TinyOS program's component graph defines which components it needs to work. Because components are resolved at compile time, compiling an application builds an application-specific version of TinyOS: the resulting image contains exactly the required OS services.

As shown in Fig. 6, TinyOS and its applications are small. The base TinyOS operating system is less than 400 bytes and associated C runtime primitives (including floating-point libraries) fit in just over 1 KB. `Blink` represents the footprint for a minimal application using the base OS and a primitive hardware timer. `CntToLeds` incorporates a more sophisticated timer service which requires additional memory. `GenericBase` captures the footprint of the radio stack while `CntToRfm` incorporates both the radio stack and the generic timer, which is the case for many real applications. Most applications fit in less than 16 KB, while the largest TinyOSapplication, TinyDB, fits in about 64 KB.

(2) Footprint Optimization: TinyOS goes beyond standard techniques to reduce code size (e.g., stripping the symbol table). It uses whole-program compilation to prune dead code, and cross-component optimizations remove redundant operations and module-crossing overhead. Figure 6 shows the reduction in size achieved by these optimizations on a range of applications. Size improvements range from 8% for Maté, to 40% for habitat monitoring, to over 60% for simple applications.

Component Overhead: To be efficient, TinyOS must minimize the overhead for module crossings. Since there are no virtual functions or address-space crossings, the basic boundary crossing is at most a regular procedure call. On Atmel-based platforms, this costs about eight clock cycles.

Using whole-program analysis, NesC removes many of these boundary crossings and optimizes entire call paths by applying extensive cross-compo nent optimizations, including constant propagation and common sub-expression elimination. For example, NesC can typically inline an entire component into its caller.

Application	Size			Structure		
	Optimized	Unoptimized	Reduction	Tasks	Events	Modules
Blink	683	1791	61%	0	2	8
Blink LEDs						
GenericBase	4278	6208	31%	3	21	19
Radio-to-UART packet router						
CntToLeds	6121	9449	35%	1	7	13
Display counter on LEDs						
CntToRfm	9859	13969	29%	4	31	27
Send counter as radio packet						
Habitat monitoring	11415	19181	40%	9	38	32
Periodic environmental sampling						
Surge	14794	20645	22%	9	40	34
Ad-hoc multihop routing demo						
Maté	23741	25907	8%	15	51	39
Small virtual machine						
Object tracking	23525	37195	36%	15	39	32
Track object in sensor field						
TinyDB	63726	71269	10%	18	193	91
SQL-like query interface						

Fig. 6. Size and structure of selected TinyOS applications

In the TinyOS timer component, triggering a timer event crosses seven component boundaries. Figure 7 shows cycle counts for this event chain with and without cross-component optimizations. The optimization saves not only 57% of the boundary overhead, but also 29% of the work, for a total savings of 38%. The increase in the crossing overhead for the interrupt occurs because the in-lining requires the handler to save more registers; however, the total time spent in the handler goes down. The only remaining boundary crossing is the one for posting the task at the end of the handler.

Cycles	Optimized	Unoptimized	Reduction
Work	371	520	29%
Boundary crossing	109	258	57%
Non-interrupt	*8*	*194*	*95%*
Interrupt	*101*	*64*	*-36%*
Total	**480**	**778**	**38%**

Fig. 7. Optimization effects on clock event handling. *This figure shows the breakdown, in CPU cycles, for both work and boundary crossing for clock event handling, which requires 7 module crossings. Optimization reduces the overall cycle count by 38%*

Anecdotally, the code produced via whole-program optimization is smaller and faster than not only unoptimized code, but also the original hand-written C code that predates the NesC language.

3.2 Reactive Concurrency

We evaluate TinyOS's support for concurrency by looking at four metrics: (1) the concurrency exhibited by applications, (2) our support for race detection at compile time, (3) context switching times, and (4) the handling of concurrent events with real-time constraints.

(1) **Exhibited Concurrency:** TinyOS's component model makes it simple to express the complex concurrent actions in sensor network applications. The sample applications in Fig. 6 have an average of 8 tasks and 47 events, each of which represents a potentially concurrent activity. Moreover, these applications exhibit an average of 43% of the code (measured in bytes) reachable from an interrupt context.

As an example of a high-concurrency application, we consider TinyDB, covered in Sect. 5.3, an in-network query processing engine that allows users to pose queries that collect, combine and filter data from a network of sensors. TinyDB supports multiple concurrent queries, each of which collects data from sensors, applies some number of transformations, and sends it up a multi-hop routing tree to a basestation where the user receives results. The 18 tasks and 193 events within TinyDB perform several concurrent operations, such as maintenance of the routing tables, multi-hop routing, time synchronization, sensor recalibration, in addition to the core functionality of sampling and processing sensor data.

(2) **Race Detection:** The NesC compiler reports errors if shared variables may be involved in a data race. To evaluate race detection, we examine the reported errors for accuracy.

Initially, TinyOS included neither an explicit `atomic` statement nor the analysis to detect potential race conditions; both TinyOS and its applications had many data races. Once race detection was implemented, we applied detection to every application in the TinyOS source tree, finding 156 variables that potentially had a race condition. Of these, 53 were false positives (discussed below) and 103 were genuine data races, a frequency of about six per thousand code statements. We fixed each of these bugs by moving code into tasks or by using `atomic` statements. We then tested each application and verified that the presence of atomic sections did not interfere with correct operation.

Figure 8 shows the locations of data races in the TinyOS tree. Half of the races existed in system-level components used by many applications, while the other half was application specific. `MultihopM`, `eepromM`, and `TinyAlloc` had a disproportionate number of races due to the amount of internal state they maintain through complex concurrent operations. `IdentC` tracks node

Component	Type	Data-race variables
RandomLFSR	System	1
UARTM	System	1
AMStandard	System	2
AMPromiscious	System	2
BAPBaseM	Application	2
ChirpM	Application	2
MicaHighSpeedRadioM	System	2
TestTimerM	Application	2
ChannelMonC	System	3
NoCrcPacket	System	3
OscilloscopeM	Application	3
QueuedSend	System	3
SurgeM	Application	3
SenseLightToLogM	Application	3
TestTemp	Application	3
MultihopM	System	10
eepromM	System	17
TinyAlloc	System	18
IdentC	Application	23
Total		103

Fig. 8. Component locations of race condition variables

interactions, records them in flash, and periodically sends them to the basestation; it has complex concurrency, lots of state, and was written before most of the concurrency issues were well understood. The NesC version is race free.

The finite-state-machine style of decomposition in TinyOS led to the most common form of bug, a non-atomic state transition. State transitions are typically implemented using a read-modify-write of the state variable, which must be atomic. A canonical example of this race is shown in Fig. 9, along with the fix.

```
/* Contains a race: */    /* Fixed version: */
if (state == IDLE) {      uint8_t oldState;
  state = SENDING;        atomic {
  count++;                  oldState = state;
  // send a packet          if (state == IDLE) {
}                             state = SENDING;
                            }
                          }
                          if (oldState == IDLE) {
                            count++;
                            // send a packet
                          }
```

Fig. 9. Fixing a race condition in a state transition

The original versions of the communication, TinyAlloc and EEPROM components contained large numbers of variable accesses in asynchronous code. Rather than using large atomic sections, which might decrease overall responsiveness, we promoted many of the offending functions to synchronous code by posting a few additional tasks.

False positives fell into three major categories: state-based guards, buffer swaps, and causal relationships. The first class, state-based guards, occurred when access to a module variable is serialized at run time by a state variable. The above state transition example illustrates this; in this function, the variable count is safe due to the monitor created by state. Buffer swaps are a controlled kind of sharing in which ownership is passed between producer and consumer; it is merely by this convention that there are no races, so it is in fact useful that NesC requires the programmer to check them. The third class of false positives occurs when an event conflicts with the code that caused it to execute, but because the two never overlap in time there is no race. However, if there are other causes for the event, then there is a race, so these are also worth explicitly checking. In all cases, the norace type qualifier can be used to remove the warnings.

(3) Context Switches: In TinyOS, context switch overhead corresponds to both the cost of task scheduling and interrupt handler overhead. These costs are shown in Fig. 10 based on hand counts and empirical measurements. The interrupt overhead consists of both switching overhead and function overhead of the handler, which varies with the number of saved registers.

Overhead	Time (clock cycles)
Interrupt Switching	8
Interrupt Handler Cost	26-74
Task Switching	108

Fig. 10. TinyOS scheduling overhead

(4) Real-Time Constraints: The real-time requirements in the sensor network domain are quite different from those traditionally addressed in multimedia and control applications. Rather than sophisticated scheduling to shed load when many tasks are ongoing, sensor nodes exhibit bursts of activity and then go idle for lengthy intervals. Rather than delivering a constant bit rate to each of many flows, we must meet hard deadlines in servicing the radio channel while processing sensor data and routing traffic. Our initial platforms required that we modulate the radio channel bit-by-bit in software. This required tight timing on the transmitter to generate a clean waveform and on the receiver to sample each bit properly. More recent platforms provide greater hardware support for spooling bits, but start-symbol detection requires precise timing and encoding, decoding, and error-checking must keep pace with the data rate. Our approach of allowing sophisticated handlers has proven sufficient for meeting these requirements; typically the handler performs the time-critical work and posts a task for any remaining work. With a very simple scheduler, allowing the handler to execute snippets of processing up the chain of components allows applications to schedule around a set of deadlines directly, rather than trying to coerce a priority scheme to produce the correct ordering. More critical is the need to manage the contention between the sequence of events associated with communication (the handler) and the sampling interval of the application (the tasks). Applying whole-system analysis to verify that all such jitter bounds are met is an area for future work.

3.3 Flexibility

To evaluate the goal of flexibility, we primarily refer to anecdotal evidence. In addition to the quantitative goal of fine-grain components, we look at the qualitative goals of supporting concurrent components, hardware/software transparency, and interposition.

Fine-grained Components: TinyOS allows applications to be constructed from a large number of very fine-grained components. This approach is facilitated by cross-module in-lining, which avoids runtime overhead for component composition. The TinyOS code base consists of 401 components, of which 235 are modules and 166 are configurations. The 42 applications in the tree use an average of 74 components (modules and configurations) each. Modules are typically small, ranging from between 7 and 1898 lines of code (with an average of 134, median of 81).

Figure 11 shows a per-component breakdown of the data and code space used by each of the components in the TinyOS radio stack, both with and without in-lining applied. The figure shows the relatively small size of each of the components, as well as the large number of components involved in radio communication. Each of these components can be selectively replaced, or new components interposed within the stack, to implement new functionality.

Concurrent Components: As discussed in the previous section, any component can be the source of concurrency. Bidirectional interfaces and explicit support for events enable any component to generate events autonomously. In addition, the static race detection provided by NesC removes the need to worry about concurrency bugs during composition. Out of our current set of 235 modules, 18 (7.6%) contain at least one interrupt handler and are thereby sources of concurrency.

Hardware/Software Transparency: The TinyOS component model makes shifting the hardware/software boundary easy; components can generate events, which may be software upcalls or hardware interrupts. This feature is used in several ways in the TinyOS code base. Several hardware interfaces (such as analog-to-digital conversion) are implemented using software wrappers that abstract the complexity of initializing and collecting data from a given sensor hardware component. In other cases, software components (such as radio start-symbol detection) have been supplanted with specialized hardware modules. For example, each of the radios we support has a different hardware/software boundary, but the *same* component structure.

Interposition: One aspect of flexibility is the ability to *interpose* components between other components. Whenever a component provides and uses the same interface type, it can be inserted or removed transparently.

Component	Code Size		Data Size
(Sizes in bytes)	inlined	noninlined	
AM	456	654	9
Core Active Messages layer			
MicaHighSpeedRadioM	1162	1250	61
Radio hardware interface			
NoCRCPacket	370	484	50
Packet framing without CRC			
CrcFilter	–	34	0
CRC filtering			
ChannelMonC	454	486	9
Start symbol detection			
RadioTimingC	42	56	0
Timing for start symbol detection			
PotM	50	82	1
Transmit power control			
SecDedEncoding	662	684	3
Error correction/detection coding			
SpiByteFifoC	344	438	2
Low-level byte interface			
HPLPotC	–	66	0
Hardware potentiometer interface			

Fig. 11. Breakdown of code and data size by component in the TinyOS radio stack. A '-' in the inlined column indicates that the corresponding component was entirely in-lined. Dead code elimination has been applied in both cases

One example of this is seen in work at UVA [26], which interposes a component in the network stack at a fairly low level. Unknown to the applications, this component buffers the payload of each message and aggregates messages to the same destination into a single packet. On the receive side, the same component decomposes such packets and passes them up to the recipients individually. Although remaining completely transparent to the application, this scheme can actually *decrease* network latency by increasing overall bandwidth.

A similar type of interpositioning can be seen in the object tracking application described in Sect. 5.2. The routing stack allows the interpositioning of components that enable, for example, reliable transmission or duplicate message filtering. Similarly, the sensor stacks allow the interpositioning of components that implement weighted-time averaging or threshold detection.

3.4 Low Power

The application-specific nature of TinyOS ensures that no unnecessary functions consume energy, which is the most precious resource on the node.

132   P. Levis et al.

However, this aspect alone does not ensure low power operation. We examine three aspects of TinyOS low power operation support: application-transparent CPU power management, power management interfaces, and efficiency gains arising from hardware/software transparency.

CPU Power Usage: The use of split-phase operations and an event-driven execution model reduces power usage by avoiding spinlocks and heavyweight concurrency (e.g., threads). To minimize CPU usage, the TinyOS scheduler puts the processor into a low-power sleep mode whenever the task queue is empty. This decision can be made very quickly, thanks to run-to-completion semantics of tasks, which maximizes the time spent in the sleep mode. For example, when listening for incoming packets, the CPU handles 20000 interrupts per second. On the current sensor hardware, the CPU consumes 4.6 mA when active and 2.4 mA when idle, and the radio uses 3.9 mA when receiving. System measurements show the power consumption during both listening and receiving to be 7.5 mA. The scheduler, which needs to examine the task queue after every event, still manages to operate in idle mode 44% of the time.

Power-Management Interfaces: The scheduler alone cannot achieve the power levels required for long-term applications; the application needs to convey its runtime requirements to the system. TinyOS addresses this requirement through a programming convention which allows subsystems to be put in a low power idle state. Components expose a `StdControl` interface, which includes commands for initializing, starting, and stopping a component and the subcomponents it depends upon. Calling the `stop` command causes a component to attempt to minimize its power consumption, for example, by powering down hardware or disabling periodic tasks. The component saves its state in RAM or in nonvolatile memory for later resumption using the `start` command. It also informs the CPU about the change in the resources it uses; the system then uses this information to decide whether deep power saving modes should be used. This strategy works well: with all components stopped, the base system without the sensor board consumes less than 15 μA, which is comparable to self discharge rate of AA alkaline batteries. The node lifetime depends primarily on the duty cycle and the application requirements; a pair of AA batteries can power a constantly active node for up to 15 days or a permanently idle node for up to 5 years (battery shelf life). By exposing the start/stop interface at many levels, we enable a range of power management schemes to be implemented, for example, using power scheduling to disable the radio stack when no communication is expected, or powering down sensors when not in use.

Hardware/Software Transparency: The ability to replace software components with efficient hardware implementations has been exploited to yield significant improvements in energy consumption in our platform. Recent work [36] has demonstrated a single-chip mote that integrates the microcontroller, memory, radio transceiver, and radio acceleration logic into a $5\,mm^2$

silicon die. The standard software radio stack consumes 3.6 mA (involving about 2 million CPU instructions per second); The hardware implementation of these software components consumes less than 100 µA and allows for much more efficient use of microcontroller sleep modes while providing a 25-fold improvement in communication bit rate.

4 Enabled Innovations

A primary goal for TinyOS is to enable innovative solutions to the systems challenges presented by networks of resource constrained devices that interact with a changing physical world. The evaluation against this goal is inherently qualitative. We describe three subsystems where novel approaches have been adopted that can be directly related to the features of TinyOS. In particular, TinyOS makes several kinds of innovations simpler that appear in these examples: (1) cross-layer optimization and integrated-layer processing (ILP), (2) duty-cycle management for low power, and (3) a wide-range of implementation via fine-grain modularity.

4.1 Radio Stack

A mote's network device is often a simple, low-power radio transceiver that has little or no data buffering and exposes primitive control and raw bit interfaces. This requires handling many aspects of the radio in software, such as controlling the radio state, coding, modulating the channel, framing, input sampling, media access control, and checksum processing. Various kinds of hardware acceleration may be provided for each of the elements, depending on the specific platform. In addition, received signal strength can be obtained by sampling the baseband energy level at particular times. The ability to access these various aspects of the radio creates opportunities for unusual cross-layer optimization.

Integrated-Layer Processing: TinyOS enables ILP through its combination of fine-grain modularity, whole-program optimization, and application-specific handlers. One example is the support for link-layer acknowledgments (acks), which can only be generated after the checksum has been computed. TinyOS allows the radio stack to be augmented with addition error checking by simply interposing the checksum component between the component providing byte-by-byte radio spooling and the packet processing component. It is also important to be able to provide link-level acknowledgments so that higher levels can estimate loss rates or implement retransmission, however, these acks should be very efficient. The event protocol within the stack that was developed to avoid buffering at each level allows the checksum computation to interleave with the byte-level spooling. Thus, the ack can be generated immediately after receiving the last byte thus the underlying radio component can send the ack *synchronously*, i.e. reversing the channel direction without re-arbitration or reacquisition. Note that holding the channel is a real-time operation that is enabled by the use of sophisticated handlers that traverse multiple layers and components without data races. This collection of optimizations greatly reduces both latency and power, and in turn allows shorter timeouts at the sender. Clean modularity is preserved in the code since these time-critical paths span multiple components.

ILP and flexible modularity have been used in a similar manner to provide flexible security for confidentiality and authentication [2]. Although link-level security is important, it can degrade both power and latency. The ability to overlap computation via ILP helps with the latency, while interposition makes it easy add security transparently as needed. This work also showed that the mechanisms for avoiding copying or gather/scatter within the stack could be used to substantially modify packet headers and trailers without changing other components in the stack.

A TinyOS radio stack from Ye et al. [2] is an example that demonstrates ILP by combining 802.11-style media access with transmission scheduling. This allows a low-duty cycle (similar to TDMA) with flexible channel sharing.

Power Management: Listening on the radio is costly even when not receiving anything, so minimizing duty cycle is important. Traditional solutions utilize some form of TDMA to turn off the radio for long periods until a reception is likely. TinyOS allows a novel alternative by supporting fast fine-grain power management. By integrating fast power management with precise timing, we were able to periodically sample the radio for very short intervals at the physical layer, looking for a preamble. This yields the illusion of an always-on radio at a 10% duty cycle while listening, while avoiding a priori partitioning of the channel bandwidth. Coarse-grain duty cycling can still be implemented at higher levels, if needed.

TinyOS has also enabled an efficient solution to the epidemic wakeup problem. Since functionality can be placed at different levels within the radio stack, TinyOS can detect that a wakeup is likely by sampling the energy on the channel, rather than bring up the ability to actually receive packets. This low-level wake-up only requires 0.00125% duty cycle [29], a 400-fold improvement over a typical packet-level protocol. A similar approach has been used to derive network neighborhood and proximity information [73].

Hardware/Software Transparency: The existence of a variety of radio architectures poses a challenge for system designers due to the wide variation in hardware/software boundaries. There are at least three radio platforms that are supported in the TinyOS distribution: the 10 kbps first-generation RFM, the 40 kbps hardware-accelerated RFM, and the recent 40 kbps Chipcon. In addition, UART and I2C stacks are supported. The hardware-accelerated RFM platform exemplifies how a direct replacement of bit level processing with hardware achieves higher communication bandwidth [29]. In the extreme cases, the entire radio stack has been built in pure hardware in Spec(mote-on-a-chip) [36], as well as in pure software in

TOSSIM [44]. We have also transparently used hardware acceleration for encryption. Stack elements using a component remain unchanged, whether the component is a thin abstraction of a hardware element or a software implementation.

4.2 Time Synchronization and Ranging

Time and location are both critical in sensor networks due to the embodied nature of sensor nodes; each node has a real, physical relationship with the outside world. One challenge of network time synchronization is to eliminate sources of jitter such as media access delay introduced by the radio stack. Traditional layering often hides the details at the physical layer. Timing protocols often perform round-trip time estimation to account for these errors. TinyOS allows a component to be interposed deep within the radio stack to signal an event precisely when the first bit of data is transmitted; this eliminates media access delay from calculations. Similarly, receivers can take a timestamp when they hear the first data bit; comparing these fine-grain timestamps can reduce time synchronization error to less than a bit time ($< 25\,\mu$s). Although reference broadcast synchronization (RBS) [16] achieves synchronization accurate to within $4\,\mu$s without interposition by comparing time stamps of receivers, it does so at the cost of many packet transmissions and sophisticated analysis.

The ability to interact with the network stack at this low level also enabled precise time of flight (TOF) measurements for ranging in an ad-hoc localization system built on TinyOS [76]. A transmitter sends an acoustic pulse with a radio message. TinyOS's low context switching overhead enables receivers to check for the acoustic pulse and the radio message concurrently. Taking the difference between the timestamps of the two signals produces an acoustic TOF measurement. TinyOS can accurately measure both arrival times directly in their event handlers, since the handlers execute immediately; a solution based on queuing the work for later would forfeit precise timing, which is also true for the time-synchronization example above.

The newest version of the ranging application uses a co-processor to control the acoustic transducer and perform costly localization calculation. Controlling the acoustic transducer requires real time interactions between the two processors which is enabled by TinyOS's low overhead event handling. To exploit parallelism between the two processors, computation and communication must be overlapped; the split-phased nature of TinyOS's AM model makes this trivial.

4.3 Routing

The rigid, non-application specific communication stack found in industrial standards such as IEEE 802.11 [1] or Bluetooth [7] often limit the design space for routing protocols. TinyOS's component model and ease of interposition yield a very flexible communication stack. This opens up a platform for implementing many different routing protocols such as broadcast based routing [23], probabilistic routing, multi-path routing [37], geographical routing, reliability based routing [80,82], TDMA based routing [14], and directed diffusion [34].

The large number of routing protocols suggests that sensor network applications may need to use a diverse set within one communication stack. TinyOS's parameterized interfaces and extensible component model enable a coherent routing framework where an application can route by network address, geographic location, flooding, or along some application specific gradients [69].

4.4 Dynamic Composition and Virtual Machines

In our experience, most sensor network applications utilize a common set of services, combined in different ways. A system that allows these compositions to be concisely described could provide much of the flexibility of full reprogramming at a tremendous decrease in communication costs. Maté, a tiny byte-code interpreter that runs on TinyOS [43], meets this need. It is a single NesC module that sits on top of several system components, including sensors, the network stack, and non-volatile storage.

Maté presents a virtual stack architecture to the programmer. Instructions include sensing and radio communication, as well as arithmetic and stack manipulation. Maté has a set of user-definable instructions. These allow developers to use the VM as a framework for writing new VM variants, extending the set of TinyOS services that can be dynamically composed. The virtual architecture hides the split-phased operations of TinyOS behind synchronous instructions, simplifying the programming interface. This requires the VM to maintain a virtual execution context as a continuation across split-phase operations. The stack-based architecture makes virtual context switches trivial, and as contexts are only 78 bytes (statically allocated in a component), they consume few system resources. Contexts run in response to system events, such as timers or packet reception.

Programs virally propagate through a network; once a user introduces a single mote running a new program, the network rapidly and autonomously reprograms itself. Maté programs are extremely concise (orders of magnitude shorter than their binary equivalents), conserving communication energy. TinyOS' event-driven execution provides a clear set of program-triggering events, and the NesC's interfaces allow users to easily change subsystems (such as ad-hoc routing). Maté extends TinyOS by providing an inexpensive mechanism to dynamically compose programs. NesC's static nature allows it to produce highly optimized and efficient codes; Maté demonstrates that run-time flexibility can be re-introduced quite easily with low overhead. By eschewing aside the traditional user/kernel boundary, TinyOS allowed other possibilities to emerge. Maté suggests that the run-time/compile-time boundary in sensor networks might better be served by a lean byte-code interpreter that sits on top of a TinyOS substrate.

5 Applications

In this section, we describe three applications that have been built using the TinyOS platform: an environmental monitoring system, a declarative query processor, and magnetometer-based object tracking. Each of these applications represents a distinct set of design goals and exhibits different aspects of the TinyOS design.

5.1 Habitat Monitoring

Sensor networks enable data collection at a scale and resolution that was previously unattainable, opening up many new areas of study for scientists. These applications pose many challenges, including low-power operation and robustness, due to remote placement and extended operation.

One such application is a habitat monitoring system on Great Duck Island, off the coast of Maine. Researchers deployed a 35-node network on the island to monitor the presence of Leach's Storm Petrels in their underground burrows [51]. The network was designed to run unattended for at least one field season (7–9 months). Nodes, placed in burrows, monitored light, temperature, relative humidity, pressure, and passive infrared; the network relayed readings back to a base station with an Internet connection via satellite, to be uploaded to a database. Figure 12 illustrates the tiered system architecture for this application.

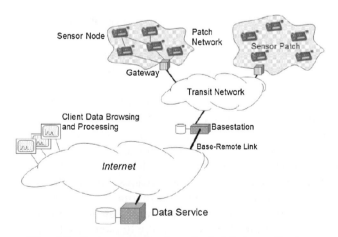

Fig. 12. System architecture for habitat monitoring

A simple TinyOS program ran on the motes. It periodically (every 68 s) sampled sensors and relayed data to the base-station. To achieve long network lifetimes, nodes used the power management facilities of TinyOS aggressively, consuming only 35 µA in low power state, compared to 18–20 mA when active. Nodes sampled sensors concurrently (using a split-phase data acquisition operation), rather than serially, resulting in further power reduction. During the 4 months of deployment, the network collected over 1.2 million sensor readings.

A specialized gateway node, built using a mote connected to a high-gain antenna, relayed data from the network to a wired base station. The gateway application was very small (3090 bytes) and extraordinarily robust: it ran continuously, without failing, for the entire 4 months of deployment. The gateway required just 2 Watt-hours of energy per day and was recharged with a 36 in^2 solar panel [63]. In comparison, an early prototype version of the gateway, an embedded Linux system, required over 60 Watt-hours of energy per day from a 924 in^2 solar panel. The Linux system failed every 2 to 4 days, while the gateway mote was still operating two months after researchers lost access to the island for the winter.

5.2 Object Tracking

The TinyOS object-tracking application (OTA) uses a sensor network to detect, localize and track an object moving through a sensor field; in the prototype, the object is a remote-controlled car. The object's movement through the field determines the actions and communication of the motes. Each mote periodically samples its magnetometer; if the reading has changed significantly since the last sample, it broadcasts the reading to its neighbors. The node with the largest reading change estimates the position of the target by computing the centroid of its neighbors' readings. Using geographic routing [38], the network routes the estimated position to the base-station, which controls a camera to point at the target. The operation of the tracking application is shown in Fig. 13.

OTA consists of several distributed services, such as routing, data sharing, time synchronization, localization, power management, and sensor filtering. Twelve different research groups are collaborating on both the architecture and individual subsystem implementation. TinyOS execution model enables running these services concurrently on limited hardware resources. The component model allows for easy replacement and comparative analysis of individual services. Currently, the reference implementation consists of 54 components. General purpose services, such as time synchronization or localization, have many competing implementations, enabled by different features of TinyOS. Replacement of low-level components used for sensing allowed OTA to be adapted to track using light values instead of magnetic fields.

Several research groups have successfully implemented application specific services within this framework. Hui et al. developed a sentry-based approach

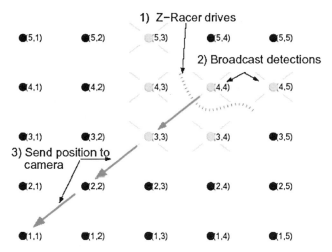

Fig. 13. Event-triggered activity in the object tracking application. (1) *The vehicle being tracked drives around position (4, 4)(dashed-line); (2) Six nodes broadcast readings (light-ened nodes); (3) node (4, 4) declares itself the leader, aggregates the readings, and routes them to the base station (dark arrows)*

[31] that addresses power management within an object tracking network. Their algorithm chooses a connected subset of sentry motes, which allows for degraded sensing; the non-sentry units are placed in a low power state. This service makes extensive use of the TinyOS power management interfaces, and is shown to reduce energy consumption by 30% with minimal degradation of tracking accuracy.

5.3 TinyDB

Many sensor network users prefer to interact with a network through a high-level, declarative interface rather than by low-level programming of individual nodes. TinyDB [50], a declarative query processor built on TinyOS, supports this view, and is our largest and most complex application to date. It poses significant challenges for concurrency control and limited resources.

In TinyDB, queries (expressed in an SQL-like syntax) propagate through the network and perform local data collection and in-network aggregation. Queries specify only what data the user is interested in and the data collection rate; the user does not specify any details of query propagation, data collection, or message routing. For example, the query:

```
SELECT AVG (light)
FROM sensors
WHERE temp >100°F
SAMPLE PERIOD 10s
```

tells the network to provide the average light value over all the nodes with temperature greater than 100°F once every 10 seconds. TinyDB uses in-network aggregation [42, 49] to greatly reduce network bandwidth requirements; this requires that nodes coordinate to produce the results.

TinyDB relies heavily on TinyOS' component-oriented design, concurrency primitives, and ability to perform cross-layer optimizations. TinyDB consists of components that perform query flooding, local data collection, formation of routing trees, aggregation of query data, and a catalog of available sensor devices and attributes (such as location) at each node. It uses the routing, data collection, and power management interfaces of TinyOS, and inter-operates with a variety of implementations of these services.

TinyOS's task model meshes well with the concurrency requirements of TinyDB, which supports multiple simultaneous queries by scheduling a timer for each query which fires when the next set of results for that query are due. Each timer event posts a task to collect and deliver results for the corresponding query. The non-preemptive nature of tasks and the support for safe concurrent handlers avoid data races despite extensive information sharing.

One example benefit of cross-layer optimization in TinyDB is message snooping, which is important for determining the state of neighboring nodes in the network. Snooping is used to enable query propagation: new nodes joining the network learn of ongoing queries by snooping for results broadcast by neighbors. This technique also enables message suppression; a node can avoid sending its local reading if it is superseded by a message from another node, as in the case of a query requesting the maximum sensor value in the network.

6 Related Work

Sensor networks have been the basis for work on *ad hoc* networking [34, 37, 38, 47], data aggregation [33, 49], distributed algorithms [25, 46, 59], and primitives such as localization [8, 76, 77], and time synchronization [16, 62]. In addition to our mote platform, a number of low-power sensor systems have been proposed and developed [3, 4, 12, 39, 55, 56, 64], though few of these systems have addressed flexible operating systems design. Several projects use more traditional embedded systems (such as PDAs [16]) or customized hardware [64].

A wide range of operating systems have been developed for embedded systems. These range from relatively large, general-purpose systems to more compact real-time executives. In [30] we discuss range of these embedded and real-time systems in detail. These systems are generally not suitable for extremely resource-constrained sensor nodes, which mandate very compact, specialized OS designs. Here, we focus our attention on a number of emerging systems that more closely match the resource budget and execution model of sensor networks.

Traditional embedded operating systems are typically large (requiring hundreds of KB or more of memory), general-purpose systems consisting of a binary kernel with a rich set of programming interfaces. Examples include WinCE [52], QNX [28], PalmOS [60], pSOSystem [79], Neutrino [65], OS-9 [54], LynxOS [48], Symbian [71], and uClinux [72]. Such OS's target systems with greater CPU and memory resources than sensor network nodes, and generally support features such as full multitasking, memory protection, TCP/IP networking, and POSIX-standard APIs that are undesirable (both in terms of overhead and generality) for sensor network nodes.

There is also a family of smaller real-time executives, such as CREEM [40], OSEKWorks [78], and Ariel [53], that are closer in size to TinyOS. These systems support a very restrictive programming model which is tailored for specialized application domains such as consumer devices and automotive control.

Several other small kernels have been developed that share some features in common with TinyOS. These systems do not support the degree of modularity or flexibility in TinyOS's design, nor have they been used for as wide a range of applications. EMERALDS [85] is a real-time microkernel, requiring about 13 KB of code, that supports multitasking using a hybrid EDF and rate-monotonic scheduler. Much of this work is concerned with reducing overheads for semaphores and IPC. AvrX [5] is a small kernel for the AVR processor, written in assembly, that provides multitasking, semaphores, and message queues in around 1.5 KB of memory. Nut/OS [15] and NESOS [58] are small kernels that provide non-preemptive multitasking, similar in vein to the TinyOS task model, but use somewhat more expensive mechanisms for inter-process communication than TinyOS's lean cross-module calls. The BTNode OS [39] consists mainly of library routines to interface to hardware and a Bluetooth communication stack, but supports an event-driven programming model akin to TinyOS. Modules can post a single-byte event to a dispatcher, which fires the (single) handler registered for that event type.

A number of operating systems have explored the use of component architectures. Click [41], Scout [57], and x-kernel [32] are classic examples of modular systems, but do not address the specific needs of low-power, low-resource embedded systems. The units [19] component model, supported by the Knit [67] language in OSKit [20], is similar to that in NesC. In Knit, components provide and use interfaces, and new components can be assembled out of existing ones. Unlike NesC, however, Knit lacks bidirectional interfaces and static analyses such as data race detection.

Several embedded systems have taken a component-oriented approach for application-specific configurability [21]. Many of these systems use heavyweight composition mechanisms, such as COM or CORBA, and several support runtime component instantiation or interpositioning. PURE [6], eCos [66], and icWORKSHOP [35] more closely match TinyOS's goal of lightweight, static composition. These systems consist of a set of components that are wired together (either manually or using a composition tool) to form an application. Components vary in size from fine-grained, specialized objects (as in icWORKSHOP) to larger classes and packages (PURE and eCos). VEST [70] is a proposed toolkit for building component-based embedded systems that performs extensive static analyses of the system, such as schedulability, resource dependencies, and interface type-checking.

7 Discussion, Future Work, and Conclusion

Sensor networks present a novel set of systems challenges, due to their need to react to the physical environment, to let nodes asynchronously communicate within austere resource constraints, and to operate under a very tight energy budget. Moreover, the hardware architectures in this new area are changing rapidly. When we began designing an operating system for sensor nets we believed that the layers and boundaries that have solidified over the years from mainframes to laptops were unlikely to be ideal. Thus, we focused on building a framework for experimenting with a variety of system designs so that the proper boundaries could emerge with time. The key elements being a rich component approach with bidirectional interfaces and encapsulated tasks, pervasive use of event-based concurrency, and whole-system analysis and optimization. It has been surprising just how varied those innovations are.

Reflecting on the experience to date, the TinyOS' component approach has worked well. Components see a great deal of re-use and are generally defined with narrow yet powerful interfaces. NesC's optimizations allow developers to use many fine-grained components with little penalty. This has facilitated experimentation, even with core subsystems, such as the networking stack. Some developers experience initial frustration with the overhead of building components with a closed namespace, rather than just calling library routines, but this is compensated by the ease of interpositioning, which allows them to introduce simple extensions with minimal overhead.

The resource-constrained event-driven concurrency model has been remarkably expressive and remains almost unchanged from the first version of the OS. We chose the task/event distinction because of its simplicity and modest storage demands, fully expecting that something more sophisticated might be needed in the future. Instead, it has been able to express the degree of concurrency required for a wide range of applications. However, the mechanics of the approach have evolved considerably. Earlier versions of TinyOS made no distinction between asynchronous and synchronous code and provided inadequate support for eliminating race conditions, many of which were exceedingly difficult to find experimentally. At one point, we tried introducing a hard boundary to AC, so all "user" processing would be in tasks. This made it impossible to meet the real-time requirements of the network stack, and the ability to perform a carefully designed bit of processing within the handler was sorely missed. The framework for innovation concept led us to better support for building (via atomic sections) the low-level concurrent data structures that cleanly integrate information from the asynchronous external world up into local processing. This is particularly true for low-level real-time operations that cannot be achieved without sophisticated handlers.

TinyOS differs strongly from most event-driven embedded systems in that concurrency is structured into modular components, instead of a monolithic dispatch constructed with global understanding of the application. Not only has this eased the conceptual burden of managing the concurrency, it has led to important software protocols between components, such as split-phase data acquisition, data-pumps found between components in the network stack, and a power-management idiom that allows hardware elements to be powered-down quickly and easily. In a number of cases, attention to these protocols provided the benefits of integrated-layer processing while preserving clean modularity.

TinyOS is by no means a finished system; it continues to evolve and grow. The use of language tools for whole-system optimization is very promising and should be taken further. Currently, components follow implicit software protocols; making these protocols explicit entities would allow the compiler to verify that components are being properly used. Examples of these protocols include the buffer-swapping semantics of the networking stack and the state sequencing in the control protocols. Parallels exist between our needs and work such as Vault [13] and MC [17].

Richer means of expressing composition are desirable. For instance, while developing a routing architecture, we found that layers in the stack required significant self-consistency and redundancy in their specifications. A simple example is the definition of header fields when multiple layers of encapsulation are provided in the network stack. We have explored *template wiring*, which defines a skeleton structure, behaviors of composition, and naming conventions into which stackable components can be inserted. A template wiring produces a set of modules and configurations that meet the specification; it merges component composition and creation into a single step. We expect to incorporate these higher-level models of composition into NesC and TinyOS as they become more clear and well defined.

We continue to actively develop and deploy sensor network applications; many of our design decisions have been based on our and other users' experiences with these systems in the field. Sensor networks are still a new domain, filled with unknowns and uncertainties. TinyOS provides an efficient, flexible platform for developing sensor network algorithms, systems, and full applications. It has enabled innovation and experimentation on a wide range of scale.

References

1. ANSI/IEEE Std 802.11 1999 Edition.
2. TinySec: Link Layer Security for Tiny Devices. http://www.cs.berkeley.edu/nks/tinysec/.
3. G. Asada, M. Dong, T. Lin, F. Newberg, G. Pottie, W. Kaiser, and H. Marcy. Wireless integrated network sensors: Low power systems on a chip. 1998.
4. B. Atwood, B.Warneke, and K. S. Pister. Preliminary circuits for smart dust. In *Proceedings of the 2000 Southwest Symposium on Mixed-Signal Design*, San Diego, California, February 27–29, 2000.
5. L. Barello. Avrx real time kernel. http://www.barello.net/avrx/.
6. D. Beuche, A. Guerrouat, H. Papajewski, W. Schröder-Preikschat, O. Spinczyk, and U. Spinczyk. The PURE family of object-oriented operating systems for deeply embedded systems. In *Proceedings of the 2nd IEEE International Symposium on Object-Oriented Real-Time Distributed Computing*, 1999.
7. Bluetooth SIG, Inc. http://www.bluetooth.org.
8. N. Bulusu, V. Bychkovskiy, D. Estrin, and J. Heidemann. Scalable, ad hoc deployable, rf-based localization. In *Proceedings of the Grace Hopper Conference on Celebration of Women in Computing*, Vancouver, Canada, October 2002.
9. D. W. Carman, P. S. Kruus, and B. J. Matt. Constraints and approaches for distributedsensor network security. *NAI Labs Technical Report #00-010*, September 2000.
10. Center for Information Technology Research in the Interest of Society. Smart buildings admit their faults.
   http://www.citris.berkeley.edu/applications/disaster_response/smartbuildings.html, 2002.
11. A. Cerpa, J. Elson, D. Estrin, L. Girod, M. Hamilton, and J. Zhao. Habitat monitoring: Application driver for wireless communications technology. In *Proceedings of the Workshop on Data Communications in Latin America and the Caribbean*, Apr. 2001.
12. L. P. Clare, G. Pottie, and J. R. Agre. Self-organizing distributed microsensor networks. In *SPIE 13th Annual International Symposium on Aerospace/Defense Sensing, Simulation, and Controls (AeroSense), Unattended Ground Sensor Technologies and Applications Conference*, Apr. 1999.
13. R. Deline and M. Fahndrich. Enforcing High-level Protocols in Low-Level Software. In *Proceedings of the ACM SIGPLAN '01 Conference on Programming Language Design and Implementation*, June 2001.
14. L. Doherty, B. Hohlt, E. Brewer, and K. Pister. SLACKER. http://www-bsac.eecs.berkeley.edu/projects/ivy/.
15. Egnite Software GmbH. Nut/OS. http://www.ethernut.de/en/software.html.
16. J. Elson, L. Girod, and D. Estrin. Fine-grained network time synchronization using reference broadcasts. In *Fifth Symposium on Operating Systems Design and Implementation (OSDI 2002)*, Boston, MA, USA., dec 2002.
17. D. Engler, B. Chelf, A. Chou, and S. Hallem. Checking system rules using system specific, programmer-written compiler extensions. In *Proceedings of the Fourth Symposium on Operating Systems Design and Implementation.*, Oct. 2000.
18. D. Estrin et al. *Embedded, Everywhere: A Research Agenda for Networked Systems of Embedded Computers*. National Acedemy Press, Washington, DC, USA, 2001.

19. M. Flatt and M. Felleisen. Units: Cool modules for HOT languages. In *Proceedings of the ACM SIGPLAN '98 Conference on Programming Language Design and Implementation*, pages 236–248, 1998.

20. B. Ford, G. Back, G. Benson, J. Lepreau, A. Lin, and O. Shivers. The flux OSKit: A substrate for kernel and language research. In *Symposium on Operating Systems Principles*, pages 38–51, 1997.

21. L. F. Friedrich, J. Stankovic, M. Humphrey, M. Marley, and J. J.W. Haskins. A survey of configurable component-based operating systems for embedded applications. *IEEE Micro*, May 2001.

22. D. Ganesan. TinyDiffusion Application Programmer's Interface API 0.1. http://www.isi.edu/scadds/papers/tinydiffusion-v0.1.pdf.

23. D. Ganesan, B. Krishnamuchari, A. Woo, D. Culler, D. Estrin, and S. Wicker. An empirical study of epidemic algorithms in large scale multihop wireless networks. citeseer.nj.nec.com/ganesan02empirical.html, 2002. Submitted for publication, February 2002.

24. D. Gay, P. Levis, R. von Behren, M.Welsh, E. Brewer, and D. Culler. The nesC language: A holistic approach to networked embedded systems. In *Proceedings of Programming Language Design and Implementation (PLDI)*, June 2003.

25. I. Gupta and K. Birman. Holistic operations in large-scale sensor network systems: A probabilistic peer-to-peer approach. In *Proceedings of International Workshop on Future Directions in Distributed Computing (FuDiCo)*, June 2002.

26. T. Ha, B. Blum, J. Stankovic, and T. Abdelzaher. AIDA: Application Independant Data Aggregation in Wireless Sensor Networks. Submitted to *Special Issue of* ACM TECS, January 2003.

27. J. S. Heidemann, F. Silva, C. Intanagonwiwat, R. Govindan, D. Estrin, and D. Ganesan. Building efficient wireless sensor networks with low-level naming. In *Proceedings of the 18th ACM Symposium on Operating Systems Principles*, Banff, Canada, October 2001.

28. D. Hildebrand. An Architectural Overview of QNX. http://www.qnx.com/literature/whitepapers/archoverview.html.

29. J. Hill and D. E. Culler. Mica: a wireless platform for deeply embedded networks. *IEEE Micro*, 22(6):12–24, nov/dec 2002.

30. J. Hill, R. Szewczyk, A. Woo, S. Hollar, D. E. Culler, and K. S. J. Pister. System architecture directions for networked sensors. In *Architectural Support for Programming Languages and Operating Systems*, pages 93–104, Boston, MA, USA, Nov. 2000.

31. J. Hui, Z. Ren, and B. H. Krogh. Sentry-based power management in wireless sensor networks. In *Proceedings of Second International Workshop on Information Processing in Sensor Networks (IPSN '03)*, Palo Alto, CA, USA, Apr. 2003.

32. N. C. Hutchinson and L. L. Peterson. The x-kernel: An architecture for implementing network protocols. *IEEE Transactions on Software Engineering*, 17(1):64–76, 1991.

33. C. Intanagonwiwat, D. Estrin, R. Govindan, and J. Heidemann. Impact of network density on data aggregation in wireless sensor networks. In *Proceedings of the International Conference on Distributed Computing Systems (ICDCS)*, July 2002.

34. C. Intanagonwiwat, R. Govindan, and D. Estrin. Directed diffusion: a scalable and robust communication paradigm for sensor networks. In *Proceedings of the International Conference on Mobile Computing and Networking*, Aug. 2000.

35. Integrated Chipware, Inc. Integrated Chipware icWORKSHOP. `http://www.chipware.com/`.
36. Jason Hill. Integrated μ-wireless communication platform. `http://webs.cs.berkeley.edu/retreat-1-03/slides/Mote_Chip_Jhill_Nest_jan2003.pdf`.
37. C. Karlof, Y. Li, and J. Polastre. ARRIVE: Algorithm for Robust Routing in Volatile Environments. Technical Report UCB//CSD-03-1233, University of California at Berkeley, Berkeley, CA, Mar. 2003.
38. B. Karp and H. T. Kung. GPSR: greedy perimeter stateless routing for wireless networks. In *International Conference on Mobile Computing and Networking (MobiCom 2000)*, pages 243–254, Boston, MA, USA, 2000.
39. O. Kasten and J. Beutel. BTnode rev2.2. `http://www.inf.ethz.ch/vs/res/proj/smart-its/btnode.html`.
40. B. Kauler. CREEM Concurrent Realitme Embedded Executive for Microcontrollers. `http://www.goofee.com/creem.htm`.
41. E. Kohler, R. Morris, B. Chen, J. Jannotti, and M. F. Kaashoek. The Click modular router. *ACM Transactions on Computer Systems*, 18(3):263–297, August 2000.
42. B. Krishanamachari, D. Estrin, and S.Wicker. The impact of data aggregation in wireless sensor networks. In *International Workshop of Distributed Event Based Systems (DEBS)*, Vienna, Austria, Dec. 2002.
43. P. Levis and D. Culler. Mat'e: A tiny virtual machine for sensor networks. In *International Conference on Architectural Support for Programming Languages and Operating Systems, San Jose, CA, USA*, Oct. 2002.
44. P. Levis, N. Lee, A. Woo, S. Madden, and D. Culler. Tossim: Simulating large wireless sensor networks of tinyos motes. Technical Report UCB/CSD-TBD, U.C. Berkeley Computer Science Division, March 2003.
45. D. Liu and P. Ning. Distribution of key chain commitments for broadcast authentication in distributed sensor networks. In *10th Annual Network and Distributed System Security Symposium*, San Diego, CA, USA, Feb 2003.
46. J. Liu, P. Cheung, L. Guibas, and F. Zhao. A dual-space approach to tracking and sensor management in wireless sensor networks. In *Proceedings of First ACM International Workshop on Wireless Sensor Networks and Applications*, September 2002.
47. C. Lu, B. M. Blum, T. F. Abdelzaher, J. A. Stankovic, and T. He. RAP: A real-time communication architecture for large-scale wireless sensor networks. In *Proceedings of IEEE RTAS 2002*, San Jose, CA, September 2002.
48. LynuxWorks. LynxOS 4.0 Real-Time Operating System. `http://www.lynuxworks.com/`.
49. S. Madden, M. J. Franklin, J. M. Hellerstein, and W. Hong. TAG: A Tiny Aggregation Service for Ad-Hoc Sensor Networks. In *OSDI*, 2002.
50. S. Madden, W. Hong, J. Hellerstein, and M. Franklin. TinyDB web page. http://telegraph.cs.berkeley.edu/tinydb.
51. A. Mainwaring, J. Polastre, R. Szewczyk, D. Culler, and J. Anderson. Wireless sensor networks for habitat monitoring. In *ACM International Workshop on Wireless SensorNetworks and Applications (WSNA'02)*, Atlanta, GA, USA, Sept. 2002.
52. Microsoft Corporation. Microsoft Windows CE. `http://www.microsoft.com/windowsce/embedded/`.
53. Microware. Microware Ariel Technical Overview. `http://www.microware.com/ProductsServices/Technologies/ariel_technology_bri%ef.html`.

54. Microware. Microware OS-9. http://www.microware.com/ProductsServices/
Technologies/os-91.html.
55. Millenial Net. http://www.millennial.net/.
56. R. Min, M. Bhardwaj, S.-H. Cho, N. Ickes, E. Shih, A. Sinha, A. Wang, and
A. Chandrakasan. Energy-centric enabling technologies for wireless sensor networks. 9(4), August 2002.
57. D. Mosberger and L. Peterson. Making paths explicit in the Scout operating
system. In *Proceedings of the USENIX Symposium on Operating Systems Design and Implementation 1996*, October 1996.
58. Nilsen Elektronikk AS. Nilsen Elektronikk Finite State Machine Operating System. http://www.ethernut.de/en/software.html.
59. R. Nowak and U. Mitra. Boundary estimation in sensor networks: Theory
and methods. In *Proceedings of 2nd International Workshop on Information
Processing in Sensor Networks*, Palo Alto, CA, April 2003.
60. Palm, Inc. PalmOS Software 3.5 Overview. http://www.palm.com/devzone/
docs/palmos35.html.
61. A. Perrig, R. Szewczyk, V. Wen, D. Culler, and J. D. Tygar. Spins: Security
protocols for sensor networks. *Wireless Networks*, 8(5):521–534, Sep 2002. Previous version of this paper appeared as PSWCT2001.
62. S. Ping. Something about time syncronization. XXX Lets get this written up
as an Intel tech report.
63. J. Polastre. Design and implementation of wireless sensor networks for habitat
monitoring. Master's thesis, University of California at Berkeley, 2003.
64. N. B. Priyantha, A. Miu, H. Balakrishnan, and S. Teller. The Cricket Compass
for contextaware mobile applications. In *Proceedings of the 7th ACM MOBI-
COM*, Rome, Italy, July 2001.
65. QNX Software Systems Ltd. QNX Neutrino Realtime OS. http://www.qnx.
com/products/os/neutrino.html.
66. Red Hat, Inc. eCos v2.0 Embedded Operating System. http://sources.
redhat.com/ecos.
67. A. Reid, M. Flatt, L. Stoller, J. Lepreau, and E. Eide. Knit: Component com-
positionfor systems software. In *Proc. of the 4th Operating Systems Design and
Implementation (OSDI)*, pages 347–360, 2000.
68. C. Sharp. Something about the mag tracking demo. XXX Lets get this written
up as anIntel tech report.
69. C. Sharp et al. NEST Challenge Architecture. http://www.ai.mit.edu/
people/sombrero/nestwiki/index/.
70. J. A. Stankovic, H. Wang, M. Humphrey, R. Zhu, R. Poornalingam, and C. Lu.
VEST: Virginia Embedded Systems Toolkit. In *IEEE/IEE Real-Time Embedded Systems Workshop*, London, December 2001.
71. Symbian. Symbian OS – the mobile operating system. http://www.symbian.
com/.
72. uClinux Development Team. uClinux, The Linux/Microcontroller Project.
http://www.uclinux.org/.
73. University of California at Berkeley. 800-node self-organized wireless sensor
network. http://today.cs.berkeley.edu/800demo/, Aug. 2001.
74. T. von Eicken, D. E. Culler, S. C. Goldstein, and K. E. Schauser. Active messages: a mechanism for integrating communication and computation. In *Proceedings of the 19th Annual International Symposium on Computer Architecture*, pages 256–266, May 1992.

75. B. Warneke, M. Last, B. Leibowitz, and K. S. J. Pister. Smart dust: Communicating with a cubic-millimeter computer. *IEEE Computer*, 32(1):43–51, January 2001.
76. K. Whitehouse. The design of calamari: an ad-hoc localization system for sensor networks. Master's thesis, University of California at Berkeley, 2002.
77. K. Whitehouse and D. Culler. Calibration as parameter estimation in sensor networks. In *ACM International Workshop on Wireless Sensor Networks and Applications (WSNA'02)*, Atlanta, GA, USA, Sept. 2002.
78. Wind River Systems, Inc. OSEKWorks 4.0. `http://www.windriver.com/products/osekworks/osekworks.pdf`.
79. Wind River Systems, Inc. pSOSystem Datasheet. `http://www.windriver.com/products/html/psosystem_ds.html`.
80. A.Woo and D. Culler. Evaluation of Efficient Link Reliability Estimators for Low-Power Wireless Networks. Technical report, UC Berkeley, 2002.
81. A. D. Wood and J. A. Stankovic. Denial of service in sensor networks. *IEEE Computer*, 35(10):54–62, Oct. 2002.
82. M. D. Yarvis, W. S. Conner, L. Krishnamurthy, A. Mainwaring, J. Chhabra, and B. Elliott. Real-World Experiences with an Interactive Ad Hoc Sensor Network. In *International Conference on Parallel Processing Workshops*, 2002.
83. W. Ye, J. Heidemann, and D. Estrin. An energy-efficient mac protocol for wireless sensor networks. In *Proceedings of IEEE Infocom 2002*, New York, NY, USA., June 2002.
84. W. Ye, J. Heidemann, and D. Estrin. A flexible and reliable radio communication stack on motes. Technical Report ISI-TR-565, USC/ISI, Aug. 2002.
85. K. M. Zuberi, P. Pillai, and K. G. Shin. EMERALDS: a small-memory real-time microkernel. In *Symposium on Operating Systems Principles*, pages 277–299, 1999.

A Service-Based Universal Application Interface for Ad Hoc Wireless Sensor and Actuator Networks

M. Sgroi, A. Wolisz, A. Sangiovanni-Vincentelli, and J.M. Rabaey

Abstract. A set of services and interface primitives to be offered to an application programmer of an ad hoc wireless sensor and actuator network (AWSAN) is described. As the definition of sockets has made the use of communication services in the Internet independent of the underlying protocol stack, communication medium and even operating system, the proposed application interface, called the "sensor network services platform" (SNSP), identifies an abstraction that is offered to any sensor network application and supported by any sensor network platform. The SNSP builds on the query/command paradigm already used in several sensor network implementations and further adds time synchronization, location and naming services that support the communication and coordination among application components.

1 Introduction

Ad hoc Wireless Sensor and Actuator Networks (AWSANs) are an essential factor in the implementation of the "ambient intelligence" paradigm. This paradigm envisions smart environments aiding humans to perform their daily tasks in a non-intrusive way [1]. The wide deployment of sensor networks also is set to change dramatically the operational models of traditional businesses in application domains such as home/office automation [2], power delivery [3], and natural environment control [4].

The potential applications of AWSANs can be classified in two broad categories: those that *monitor* and those that *control* the environment in which they are embedded. Monitoring applications gather the values of parameters of the environment, process them, and report the outcome to external users. Control applications, in addition to monitoring, influence the environment so that it follows a reference behavior. Both types of applications require the deployment of networks with a large number of nodes that are capable of capturing different physical phenomena (sensors), making control decisions (controllers), and acting on the environment (actuators). The following characteristics of the AWSAN design problem pose serious challenges to designers:

- the nodes cooperate to satisfy their data gathering and control functionality; hence, *inter-node communication, mostly over RF links, plays an essential role;*

- the requirement that the network operation continue for very long periods of time without human intervention makes *low energy consumption of paramount importance;*
- to meet the stringent power, cost, size and reliability requirements, optimization is needed at all the steps of the design process, including application software development, choice of the layers of the communication protocol stack and hardware platform selection.

Currently, AWSAN designs are tailored to the specific application both in the choice of the network protocols and in the implementation platform. *Today, it is virtually impossible to start developing applications without the previous selection of a specific, integrated hardware/software platform.* Coupling applications with specific hardware solutions slows down the practical deployment of sensor networks: potential users hesitate to invest in developing applications that are intrinsically bound to specific hardware platforms available today. To unleash the power of AWSANs, standards are needed that favor

- the incremental integration of heterogeneous nodes, i.e., nodes that perform a variety of different functions in different ways;
- the development of applications in a way that is largely or even fully independent of the implementation platforms.

To address the first concern, several efforts have been recently launched to standardize communication protocols among sensor network nodes. The best known standards are BACnet and LonWorks, developed for building automation [2]. They are geared towards well-defined application areas, and are built on top of specific network structures. Hence, they are not well suited for many sensor network applications. ZigBee [6] defines an open standard for low-power wireless networking of monitoring and control devices. It works in cooperation with IEEE 802.15.4 [5], which addresses the lower protocol layers (physical and MAC). ZigBee defines the upper layers of the protocol stack, from network to application, including application profiles. Yet, it is our belief that efforts, like Zigbee, created in a bottom-up fashion do not fully address the essential issue: how to allow interoperability between the multitudes of sensor network operational models that are bound to emerge. In fact, different application scenarios lead to different requirements in terms of data throughput and latency, quality-of-service, use of computation and communication resources, and network heterogeneity. These requirements ultimately result in different solutions in network topology, protocols, computational platforms, and air interfaces.

The second concern was partially addressed in the automation and manufacturing community where networks of sensors (mostly wired) are widely deployed. To achieve interoperability between different manufacturers, the IEEE 1451.2 [7] standardizes both the essential sensors (and actuators) parameters and their interface with the units that read their measures (or set their values). In particular, the standard defines the physical interface between the Smart Transducer Interface Module (STIM), which includes one or more transducers, and the Transducer Electronic Data Sheet (TEDS) containing the list of their relevant parameters, and the Network Capable Application Processor (NCAP), which controls the access to the STIM.

Realizing that the present efforts lack generality and, to a certain degree, rigor, we propose in this chapter an approach for the support of true *interoperability between different applications as well as between different implementation platforms.* We advocate a top-down approach similar to the one adopted very successfully by the Internet community and propose a universal application interface, which allows programmers to develop applications without having to know unnecessary details of the underlying communication platform, such as air interface and network topology. Hence, we define *a standard set of services and interface primitives (called* **the Sensor Network Services Platform or SNSP)** *to be made available to an application programmer independently on their implementation on any present and future sensor network platform.* Furthermore, we separate the virtual platform defined by the logical specification of the SNSP services from the physical platform (*called* **the Sensor Network Implementation Platform or SNIP**) that implements it and determines the quality and cost of the services.

As the definition of sockets in the Internet has made the use of communication services independent of the underlying protocol stack, communication medium and even operating system, the application interface we propose identifies an abstraction that is offered to any sensor network application and supported by any sensor network platform. Yet, while conceptually similar, the application interface needed for sensor networks is fundamentally different from the one defined in the Internet space. In the latter, the primary concerns are a seamless set-up, use, and removal of reliable end-to-end communication links between applications at remote locations. In the former, sensor network applications require communication services to support queries and commands [8] among the three essential components of the network (sensor, monitor/controller, and actuator), and need also other services for resource management, time synchronization, locationing and dynamic network management.

TinyDB [9] is the existing approach closest to our effort. It views a sensor network as a distributed database and defines an application-level abstraction based on the Query/Command paradigm to issue declarative queries. However, its main goal is defining the interface and the implementation of a specific service, the query service. Hence, the TinyDB abstraction lacks of several auxiliary services needed in many sensor network applications. In addition, several decisions such as the one made for the service description appear to be driven by implementation considerations.

This chapter is structured as follows. First, we introduce the functional components of an AWSAN and create a framework in which we can define the services offered by the distributed Service Platform. Next, we present the services offered by the SNSP, including the essential Query/Command Service and some auxiliary services such as locationing, timing, and concept repository. A brief analysis of the Sensor Network Implementation Platform follows. Finally, we present a case study, the Demand Response system for Energy Management in residential areas.

2 AWSAN Functional Architecture

The functionality of an AWSAN is captured best as a set of distributed compute functions (typically called *controllers* or *monitors*, given that most of the applications are either control or monitor oriented), cooperating to achieve a set of common goals. AWSANs interact with the *Environment* in the form of spatially distributed measurements and actuations (Fig. 1a). The interactions with the environment are carried out by an array of sensors and actuators, interacting with the controllers via a communication network (Fig. 1b). The ideas presented in this chapter are about formalizing and abstracting the interaction and the communications between the Application and the distributed sensors and actuators through the SNSP and its Application Interface.

2.1 The Application

An AWSAN **Application** *consists of a collection of cooperating algorithms (which we call controllers) designed to achieve a set of common goals, aided by interactions with the Environment through distributed measurements and actuations.*

Controllers are components of AWSANs that read the state of the environment, process the information and report it or apply a control law to decide how to set the state of the environment. A controller is characterized by its desired behavior, its input and output variables, the control algorithm and the model of the environment. In addition, to ensure proper operation a controller places some constraints on parameters expressing the quality of the input data such as timeliness, accuracy and reliability.

2.2 The Sensor Network Services Platform (SNSP)

The **Sensor Network Services Platform** *(SNSP) decomposes and refines the interaction between controllers and the Environment, and among controllers into a set of interactions between control, sensor, and actuation functions.*

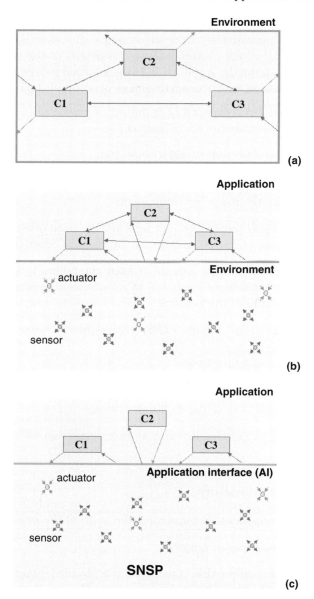

Fig. 1. Functional model of an AWSAN as a set of controllers (**a**) interacting with the environment and among each other; (**b**) interacting with the environment through a set of spatially distributed sensors and actuators; (**c**) interacting with the environment through a unified Application Interface (AI). Observe that interactions between the controllers themselves are now supported through the same paradigm

The services that the SNSP offers the Application are used directly by
the controllers whenever they interact among each other or with the Environment. This approach abstracts away the details of the communication
mechanisms, and allows the Application to be designed independently on how
exactly the interaction with the environment is accomplished (Fig. 1c).

> The **Application Interface (AI)** *is the set of primitives that are used by
> the Application to access the SNSP services.*

The SNSP is a collection of algorithms (e.g., location and synchronization), communication protocols (e.g., routing, MAC), data processing functions (e.g., aggregation), I/O functions (sensor, actuation). The core of the
SNSP is formed by the following components:

- *Query Service (QS)* (controllers get information from other components)
- *Command Service (CS)* (controllers set the state of other components)

In addition, the SNSP should provide at least the following supporting services, essential to the correct operation of most sensor network applications
and the ad-hoc nature of the network:

- *Time Synchronization Service (TSS)* (components agree on a common
  time)
- *Location Service (LS)* (components learn their location)
- *Concept Repository Service (CRS)* (components agree on a common definition of the concepts during the network operation and maintain a repository of the capabilities of the deployed system)

This is not an exhaustive list and more services can be built on top of
these basic ones.

2.3 Sensors and Actuators

> *A **sensor** is a component that measures the state of the environment.*

We identify two types of sensors:

1. *Simple sensors:* Devices that directly map a physical quantity into a data
   value and provide the measure upon request. Devices that measure physical parameters, such as temperature, sound, and light, as well as input
   devices such as keyboards or microphones that allow external users to
   enter data or set parameters are examples of simple sensors.
2. *Virtual sensors:* Objects that perform abstractly the same task as simple
   sensors in the sense that they provide data upon an external request but
   consist in general of a number of different components. Virtual sensors are
   defined by the list of parameters that can be read and by the primitives
   that are used for reading them. Examples of virtual sensors are:

- A sensor that provides an indirect measure of a certain environment condition by combining one or more sensing functions with processing (e.g., transformation, compression, aggregation).
- A controller when it is queried for the value of one of its parameters.
- An external network providing data through gateways. Let us consider a sensor network accessing an external information service provided by a global network (such as, for instance, weather forecasts or energy prices available on an Internet web-server). External services are accessed through a gateway that interfaces the sensor network with the global network. Through the gateway, the global network appears to the rest of the sensor network as a virtual sensor or a virtual actuator; thus, it might be queried or set by a command. In the example of the query of the weather forecast over the Internet, the gateway and the rest of the global network define a virtual sensor, which delivers data to the sensor network application.

> *An **actuator** is a component that sets the state of the environment.*

We identify two types of actuators:

1. *Simple actuators:* devices that map a data value into a physical quantity and may return an acknowledgment when the action is taken. Examples of actuators are devices that modify physical parameters, such as heaters and automatic window/door openers, as well as output devices, such as displays and speakers.
2. *Virtual actuators:* objects that perform abstractly the same operations of actuators in the sense that they receive values to set some parameters but consist in general of a number of different components. Virtual actuators are defined by the list of parameters that can be set and by the primitives that are used for setting. Examples of virtual actuators are:
   - An actuator that provides an indirect way of controlling a certain environment condition by combining one or more physical actuators with processing (e.g., transformation, and decompression).
   - A controller whose parameters are set by other controllers.
   - A network getting commands to take actions through gateways.

A detailed list of the relevant parameters of sensors and actuators is given in [10].

The most important feature, **differentiating sensors and actuators from controllers, is that the former are purely reactive components**; that is, they read or set the state of the environment upon a request or command of a controller. Hence, only controllers can initiate events or sequences of events.

3 The Query/Command Services: The Core of the SNSP

To ensure generality and portability, our proposal formalizes only the primitives that allow the Application to access the services of the SNSP, and does not define the architecture of the SNSP itself. In this section, we outline the functionality and the interface primitives for the two core services of the SNSP: Query and Command. Due to space limitations, we provide a detailed description only of the Query Service primitives. Details of the other services are given in [10].

3.1 Naming

In AWSAN applications, components communicate with each other because of specific features, often without knowing a priori which and how many components share those specified features. For example, a controller may want to send a message to all the sensors measuring temperature in a region, say in the kitchen. In this case, the group of temperature sensors located in the kitchen may be named and addressed for message delivery using the attributes "temperature" and "kitchen" rather than using the list of IDs of all the individual sensors having those attributes [11].

A **name** is an attribute specification and scope pair.

An **attribute specification** is a tuple $((a_1, s_1), (a_2, s_2), \ldots (a_n, s_n), expr_1, expr_2, \ldots expr_l)$, where a_i is an attribute; s_i is a selector that identifies a range of values in the domain of a_i; $expr_k$ is a logical expression defined by attribute-selector pairs and logical operators.

For example, a name can be defined by the pairs "(temperature, $>30°C$)", "(humidity, $>70\%$ R.H.)", "(sound, $<20\,dB$)", and the expression "((temperature, $>30°C$) OR (humidity, $>70\%$ R.H.)) AND (sound, $<20\,dB$)". Examples of attributes commonly used for naming in sensor networks are the physical parameter being measured by sensors or modified by actuators (e.g., temperature, humidity).

Attribute specifications are always understood within a scope.

A **scope** is a tuple $(O_1, O_2, \ldots O_n, R_1, R_2, \ldots R_m)$, where O_i is an *organization* unit and R_j is a *region*.

A **region** is a set of locations. Two types of regions can be identified:

- a *zone* is a set of locations identified by a common name; e.g., the kitchen or San Francisco.
- a *neighborhood* represents a set of locations identified by their closeness to a reference point; e.g., all nodes within a radius of $10\,m$ from a given location.

An **organization** is an entity that owns or operates a group of nodes.

Organizations are essential to differentiate between nodes that operate in the same or overlapping regions, and belong to different organizations (for instance, the police and the fire department).

In general, **names do not have to be unique**. In addition, names may change during the evolution of the network because of the movement of nodes or the modification of some attributes. An essential assumption underlying the SNSP is that all the functions participating in the network (that is, controllers, sensors and actuators) always have a sense of their location within the environment.

3.2 Query Service

> *The Query Service (QS) allows a controller to obtain the state of a group of components.*

A *query* is a sequence of actions initiated by a controller (*query initiator*), which requests specific information from a group of sensors (*query targets*). If the requested information is not available, QS always returns a negative response within a maximum time interval, which has been set previously. Figure 2 visualizes the interactions of QS with the query initiator and the query target. In sensor network applications, query targets are typically sensors that provide controllers with the requested measures but in some applications the target may be a group of controllers (considered as virtual sensors) that are requested their current state.

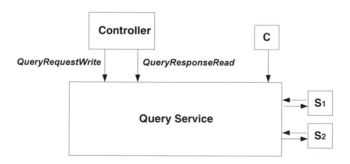

Fig. 2. Query Service interactions

Queried Parameters. In addition to the physical data being measured by a sensor, a controller may query other parameters related to the sensor, such as time (when the measure was taken), location (where the measure was taken), accuracy (how accurate was the measure), and security (if the measure comes from a trusted source). If no parameter is indicated in a query request, by default the response returns the data measured by the target sensors.

Table 1. Query Service primitives

QSRequestWrite (Target, Parameter, QueryClass, ResponseType, Reliability)

Initiates a query of the type indicated in *QueryClass* to obtain a *Parameter* from the components addressed by the name *Target*. It returns a *QueryId* as a descriptor of the query or Error.

QSResponseRead (QueryId)

Returns the value of the parameter, requested by the query identified by *QueryId*, if available. If it is not available, it returns a special value indicating the response has not arrived yet.

QSClassSetup (Accuracy, Resolution, Timeliness, MaxLatency, Priority, {Loc, Time} Tag, Operation, Security, ...)

Creates a *QueryClass*, configuring one up to all the parameters in the list. It returns the descriptor of the query class or error

QSClassUpdate (QueryClass, Accuracy, Resolution, Timeliness, MaxLatency, Priority, {Loc, Time} Tag, Operation, Security, ...)

Updates one up to all the set of parameters of a *QueryClass* previously defined.

QSStopQuery (QueryID)

is used by a query initiator to stop a query and release the *QueryID* for use in future queries. It returns OK or Error.

The primitives of the Query Service are summarized in Table 1.

The QS primitives use the following arguments.

QueryID. Multiple queries, coming from the same or different controllers, can occur concurrently. QS uses a *QueryID* number to relate a query request with the corresponding responses. An arbitrary integer chosen by the QS (for example the *timestamp* indicating the time when the query request is sent) can be used as *QueryID*. The *QueryID* assigned to a query is always released as soon as the corresponding query terminates.

Response Type. In a query, the controller specifies the frequency of the responses it expects from the sensors. Three types of response patterns are especially relevant:

- *one-time* response
- *periodic* responses with interval period p
- *notification* of events whenever an event specified by an event condition occurs

Reliability. A query is *reliable* if the query initiator is guaranteed to receive *at least one* response, which has not been corrupted. In all other cases, the query is said to be unreliable. The default case is the unreliable query. If a reliable query is requested this has to be explicitly specified. The support for reliable queries is provided, within the query service, by means of specific reliability assuring mechanisms.

Query Class. A given query can be subject to a wide range of constraints such as accuracy, timeliness. One option is to repeat all these constraints with every QueryRequest. This would make the query messages traveling through the network quite heavy, which does not fit well with the energy-efficiency and light-weight requirements typically imposed on sensor networks. The *Query Class* allows for a one-time definition of the context of a query by defining and constraining the response scope. The following parameters can be set:

- *Accuracy and resolution* of the sensor measures.
- *Timeliness, MaxLatency and Priority* define response time constraints on the Query
- *{Loc, Tim} Tags* indicate if the query response should include time and location tags.
- *Operations* such as *max, min, average.* They indicate the type of operation to be performed on multiple measures from the same source.
- *Security.* It indicates if the data must be secure or not.

All the query instances belonging to a certain class must follow the parameters of that class.

QS Operation. Figure 3 plots a sequence of primitive function calls associated with a query. The QS execution follows the ***client-server model***. First, the query class parameters are initialized using *QSClassSetup*. Then, the controller calls the *QSRequestWrite* function to initiate individual queries. QS returns the query descriptor (*QueryID*) to the controller which is blocked waiting for an immediate answer on whether the query can be initiated. If the answer is negative, an error message is returned. If the query is successfully initiated, QS begins the procedure of getting the parameter requested by the application.

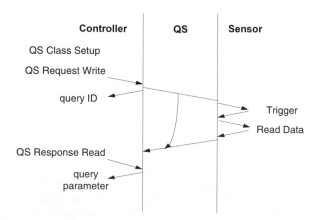

Fig. 3. Query Service execution. Controller/QS interactions are defined by the AI, while the QS/Sensor interface is implementation dependent and might follow for example the IEEE 1451.2 standard

Observe that the service does not specify how the data is obtained. For instance, instead of getting a new sensor measure, the QS may reuse local information previously gathered if the queried parameter is still available from a previous query and satisfies the timeliness requirement. In addition, the QS may perform additional functions such as aggregation to process the data coming from multiple sensors.

To access the values of the queried parameter, the controller calls the *QSResponseRead* primitive specifying the QueryID parameter. A query with a one-time response or an event notification terminates either when the response arrives or when a timeout of duration max_time set by the query initiator expires. The timeout prevents a query from staying active for an unnecessarily long time, especially when it is not known a priori how many responses will be received. It also allows the corresponding QueryID to be released. In the case of a periodic-response query, the query can be terminated at any time by the application simply calling the *QSStopQuery* primitive.

Additional Primitives Used by Virtual Sensors. Virtual Sensors necessitate the introduction of the following primitives, reading the parameter being queried, and writing the corresponding value, respectively acting as counterparts for the corresponding functions in the controllers. As a special case, controllers acting as virtual sensors use them to interact with other controllers that have queried their parameters. These primitives substitute the implementation dependent QS/Sensor interface used for simple sensors.

Table 2. Primitives used for the interaction with Virtual Sensors

QS Request (parameter)
provides a virtual sensor with the name of the parameter being requested
QSResponseWrite (parameter, value)
provides the value of the requested parameter

3.3 Command Service

*The **Command Service (CS)** allows a controller to set the state of a group of components.*

A command is a sequence of actions initiated by a controller (*command initiator*), which demands a group of components (*command targets*) to take an action. The Command Service operates completely symmetrical to the Query Service defined above – its primitives are hence omitted for the sake of brevity. One major difference should be pointed out – while requests are self-confirming (that is, a set of values is returned), commands are not. Often, there is a need to know if a command has reached its targets. Hence, we differentiate between commands of the type *confirmed* (where actuators return an acknowledgment to the controller after the action is taken) or *unconfirmed* (if they do not send any acknowledgment).

4 Auxiliary Services

4.1 Concept Repository Service

> The **Concept Repository Service (CRS)** *maintains a repository containing the lists of the capabilities of the network and of the concepts that are supported.*

The CRS plays a key role in the network operation because it allows distributed components to refer to common notions of concepts such as names, attributes and regions. Moreover, it maintains agreement on these concepts in the presence of changes occurring dynamically during the network operation (e.g., new nodes join the network, or existing nodes move across region boundaries). Below, we do not prescribe how the CRS is implemented. For instance, the CRS may be centralized or distributed over the network. In both cases the correct operation of the system requires that the repository be updated in a timely manner when parameters change. For example, a node that moves from the kitchen to the living-room must update the region in which it is located and therefore must check which region includes its new coordinates. While the Location Service (Sect. 4.3) gives a node its new spatial coordinates, the CRS provides the node with the association of these coordinates with the "living-room" region. In addition to supporting the operation of a single network, CRS supports the interoperation of multiple networks (discussed in Sect. 5) since it provides a complete and unambiguous definition of the capabilities of each network.

The CRS holds the definition of the following concepts:

1. *Attributes*, used to define names. Examples of attributes, common especially in environment monitoring applications, are temperature, light, and sound. Attributes can be added to the repository either by the application or automatically ("plug-and play") by the CRS, when the platform is augmented with sensors (or actuators) that read (or write) a new attribute.
2. *Regions*, used to define the scope of a name. The name of a *zone* is added to the repository together with the zone boundaries, expressed in terms of the spatial coordinates. During the network operation, a component that knows its spatial location can get through the CRS also the name of all the zones that include its location. A single location can belong to multiple zones. Zones may (but do not have to) form hierarchical structures.
3. *Organizations*, used to define the scope of a name. Organizations indicate who owns or operates certain groups of nodes (e.g., the police, the fire department). The capability of distinguishing nodes by organization is necessary when there are nodes performing the same function (e.g., sensor nodes of the same type) in the same region, but are deployed and used to achieve a different task or belong to a different owner.
4. *Selectors, logic operators and quantifiers*, used to define names and targets of queries and commands. The following types are the most commonly used:

- *selectors*: > n, < n, even, odd, =.
- *logic operators*: OR, AND, NOT
- *quantifiers*: all, at least k, any

The CRS primitives allow to *Add* or *Delete* an attribute (or a region or an organization), *Get the list* of all the defined attributes (regions, organizations) and *Check* if a given attribute (region, organization) is currently present in the repository.

4.2 Time Synchronization Service

The **Time Synchronization Service (TSS)** *allows two or more system components to share a common notion of time and agree on the ordering of the events that occur during the operation of the system.*

Typical application scenarios that require time synchronization are "heat room at 6 pm" or "send me the temperature within 5 Seconds".

The TSS is used to measure time and check the relative ordering among the events in the system. If the events to be compared belong to the same component (e.g., "retransmit message if acknowledgment is not received within 10 seconds") only local resources such as clock and timers are used. If they belong to different components, TSS uses a distributed synchronization algorithm to ensure that their clocks are aligned in frequency and time.

A component can be in *synchronized* or *not-synchronized* state. If it is in not-synchronized state, time is measured by the local clock and is called individual time. If it is in synchronized state, the time value is agreed with one or more other components, which share a common reference. Synchronizing multiple components may require a time interval, called *synchronization time*, and can be achieved up to a certain specified accuracy. The *synchronization scope* of a component is defined by the set of components with which it is synchronized.

TSS primitives allow to setup the resolution and accuracy of synchronization (*TSSSetup*), activate or deactivate the synchronization with components specified in a given scope (*TSSActivateSynchronization*), get the time (*TSSGetTime*), and set a timer to expire after a given number of time units (*TSSSetTimer*).

4.3 Location Service (LS)

The **Location Service (LS)** *collects and provides information on the spatial position of the components of the network.*

The operation of sensor networks commonly uses location as a key parameter at several levels of abstraction. At the application level location is used for example to define the scope of names in queries (e.g., "send me the temperature measures from the kitchen"). Depending on the use, location information can be expressed as a point in space or by a region where the node is located.

A *point location, or simply **location**, is defined by a reference system and a tuple of values identifying the position of the point within the reference system.*

LS supports the definition of location in a *Cartesian Reference System*, where location is expressed as a triple (x, y, z), with x, y and z representing respectively the distance from the origin along the axis x, y, and z. Regions can be easily expressed within this framework. A zone can have the form of a block, a sphere, a cylinder, or a more complex shape. A block is represented in terms of the coordinates of four vertices, while a sphere is represented by the center and the length of its radius. Neighborhood is a region defined by the proximity to a reference point. Proximity can be expressed in terms of Euclidean distance (spherical region) or by the number of routing hops.

LS primitives allow to setup the resolution, accuracy and reference system (*LSSetup*), get the location of a component (*LSGetLocation*), and get the list of the regions including a specified location (*LSGetRegions*).

It is often useful for a controller to know what type and how many nodes are located in a given region. Also, it maybe useful to know if these nodes are static (that is, have not moved for a long time), or are mobile. This information can be readily obtained using the primitives defined in the Query Service querying for parameters "location" and "mobility". To find the number of temperature sensors in the kitchen region, it suffices to launch a query for parameter "location" with name "(temperature, kitchen)" and to determine the cardinality of the returned list.

5 Inter-Networking

A large number of applications require interoperation between sensor AWSANs and global networks, such as the Internet. This is motivated for instance by the need of remote observation or management of the AWSAN. This section presents how the services offered by the SNSP can be exported across network boundaries.

5.1 Interconnection of AWSANs

AWSANs may be interconnected together and provide applications with a common set of services if they have the same understanding of the naming conventions. Each of the individual AWSANs that are interconnected may, however, support a different set of concepts and service parameters (e.g., different levels of accuracy, reliability, security ...). A controller can issue a query for any region or organization, also belonging to another network.

When a network is not able to provide a service requested by an application (e.g., queries of a certain parameter, reliable/secure queries . . .), it returns a negative message to the requesting application. As a result, an application gets service support only from some of the networks and receives negative answers from the networks that cannot support it.

AWSANs can be interconnected as shown in Fig. 4(a) directly through a gateway, or (b) through a global network. When the interconnection is through a gateway, the service request messages issued by an application in one network are forwarded to the other network after the gateway has checked that the second network is able to support the service. When the interconnection is through a global network, packets carrying requests and responses are encapsulated and tunneled over the transport layer of the global network (e.g., UDP if the global network is the Internet). In this case, the gateway that interfaces the global network to the target AWSAN (G2 in Fig. 4b) is responsible to check whether the latter is able to support the requested service.

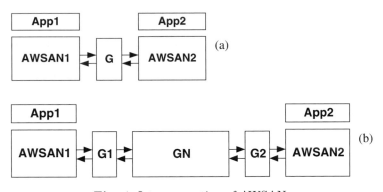

Fig. 4. Interconnection of AWSANs

5.2 Access to Global Services from AWSANs

An AWSAN may want to use services offered by a global network. An example is for instance to display information obtained by a controller on a computer terminal connected to the Internet, or vice-versa to issue control commands to the controller. Another example might be to access a global weather report published on a web-server. Global services are accessed through a gateway that interfaces the AWSAN with the global network. Through the gateway, the global network appears to the rest of the sensor network as a set of virtual sensors and/or a set of virtual actuators: thus, it might be queried or set by a command. In the example of the query of the weather forecast over the Internet, the gateway and the rest of the global network define a virtual sensor, which delivers data to the sensor network application (Fig. 5). How

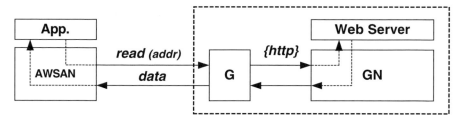

Fig. 5. Access to global services

the virtual sensor is implemented is out of the scope of this chapter. In Fig. 5 standard http is used.

5.3 Usage of AWSANs for Provision of Global Services

A global network may use AWSAN services to support its own services (Fig. 6). A typical application example is the remote monitoring from an Internet host. For example, consider the scenario of a user who wants to keep track of the status of the utilities and check the security in his home from a remote location. The AWSAN includes sensors that are associated with the utilities and monitor their current energy and temperature level, sensors that are associated with doors and windows and detect the arrival of intruders. A home gateway connected and accessible from the Internet collects this information and makes it available to the remote user.

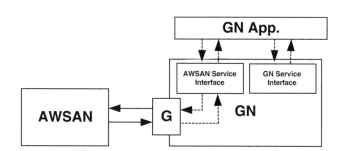

Fig. 6. Accessing an AWSAN from a Global Network

An application located in the global network has direct access to the primitives of the AWSAN service, and knows the address of the gateway that interfaces the global and the sensor network. The service request issued by the global network application component is tunneled to the gateway of the target AWSAN. The gateway forwards the request to the sensor network, gathers the responses (possibly doing some aggregation), and sends them to the requesting application. Note that this scheme requires within the global network only one address for the gateway of a specific AWSAN.

6 A Bridge to the AWSAN Implementation

The SNSP and the AI, as described, are purely functional entities, which are totally disconnected from the eventual implementation. The only nod to performance metrics is that queries and commands can identify constraints on timeliness, accuracy. However, one of the differentiating features of AWSANs is that implementation issues such as the trade-off between latency and energy-efficiency play a crucial role. Conserving independence from implementation, while being sensitive and transparent to implementation costs, is an essential part of this proposal. To accomplish these goals, a second platform is defined.

> *The **Sensor Network Implementation Platform (SNIP)** is a network of interconnected physical nodes that implement the logical functions of the Application and the SNSP described earlier.*

Choosing the architecture of the SNIP and the mapping of the functional specification of the system properly is a critical step in sensor network design. The frequency of the processor, the amount of memory, the bandwidth of the communication link, and other similar parameters of the SNIP ultimately determine *the quality and the cost of the services* that the network offers. Figure 7 visualizes the concept of mapping the Application and the SNSP functions onto the SNIP nodes N_i. The shaded boxes and the dotted arrows associate groups of logical components with physical nodes.

An instantiated node binds a set of logical Application or SNSP functions to a physical node.

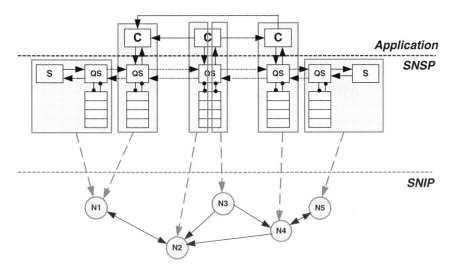

Fig. 7. Mapping Application and SNSP onto a SNIP

Physical parameters such as cost, energy efficiency, and latency can only be defined and validated after mapping.

The operation of the Application or the SNSP may depend upon the value of a physical parameter of the SNIP (for instance, amount of energy level, quality of the channel, etc). To make this information available in a transparent fashion, an additional service called the **Resource Management Service (RMS)** is provided, which allows an **application or other services of the SNSP to get or set the state of the physical elements of the SNIP**.

Typically, the RMS can access physical parameters such as the energy level of a node, the power management state, the quality of the radio channel, the energy cost of certain operations and their execution time, the transmission power of the radio. The RMS can be accessed using the same primitives offered by the QS and the CS. The main difference is that the parameters being set or queried do not belong to the environment but to the physical nodes of the network.

7 A Simple Case Study: Demand Response for Energy Management

The concepts of the SNSP and SNIP are best illustrated with a practical example. In a project funded by the California Energy Commission, a number of faculty at UC Berkeley are exploring how the usage of AWSANs can help to deploy the demand response (DR) concept in residential settings and as a result reduce the average cost of energy to the residential user as well as the overall energy usage [12]. In today's residential arena, the cost of electrical power in $/kWh is a constant to the user and is independent of the time of usage. For the energy provider, however, the cost may vary dramatically. Electrical power at 5 pm in the afternoon may be one or more orders of magnitude more expensive than in the middle of the night (or even the middle of the day). In fact, the cost versus demand curve is almost exponential – once a certain threshold is exceeded, extra capacity comes at a huge penalty. The Demand Response (DR) approach attempts to break this disconnect between actual and invoiced cost by linking the cost of electricity as charged to the user directly to the time of usage. Making the actual cost visible to the consumer may result in a shift of power usage to off-peak hours, hence reducing the average energy cost. Even further, climate control systems and appliance can become energy-cost aware and adapt their behavior accordingly.

The success of this undertaking hinges on the belief that advances in computation, communication, and sensing technology will reduce cost, size and power of the individual components (energy meters, thermostats, temperature meters, etc.) so that ubiquitous deployment becomes possible and eventually a no-brainer. Over the last year, the researchers at UC Berkeley have demonstrated using a number of simple prototypes that the concept is indeed possible and that the technology to accomplish the mission is definitely emerging. However, the approach adopted by the current effort suffers from the typical early-adoption syndrome. AWSANs present a fundamentally new paradigm in computing. As a result, the high-level abstractions needed to shield the application programmer from the intricacies and details of the underlying hardware architecture are not yet in place. This forces the programmer not only to understand the needs of the applications, but also to be architecture and implementation-savvy. In a way, the situation is similar to the time that users of microprocessors had to program in assembler, and were hence exposed to the micro-architecture of the processor. Similarly, before the advent of operating systems, application programmers had to understand how the processor communicates with the peripheral devices. And before the advent of the internet protocol stacks and sockets, communications between distributed processors had to execute explicitly and programmers had to be aware of the network structure. The disadvantage of programming at the "implementation level" is that the produced code is hardware- or architecture-dependent, and cannot be ported to new platforms. This results in legacy stovepipe systems that limit extensibility and stifle innovation. The introduction of high-level languages (such as C) or operating systems (such as VXWorks) helped to orthogonalize between function and implementation, enabling portability and extensibility.

The AWSAN offers a new paradigm in the sense that the functionality lies not in the individual nodes, or the network of the individual nodes, but in the ensemble of the nodes. Together, the sensor, controllers and actuators perform a set of functions such as managing the HVAC or scheduling the energy usage. To make the programming of such an environment seamless, it is essential that programs are developed at this level of abstraction. If not, closed systems are created that depend upon a single vendor, a particular wireless protocol, or single network architecture. This is exactly where AWSANs are today. Application programmers have to be acutely aware of both node and network architecture. The introduction of the TinyOS operating system [13] has brought as a long way in abstracting away the details of the individual node (also called mote), yet communication between the nodes is still pretty much left over to the programmer.

The architecture of the current prototype system is shown in Fig. 8. While fully functional and effective in demonstrating how DR would be installed in a residence, the system itself is "hacked together" (colloquially speaking) from off the shelf components and the system architecture is quite rigid. Functions are hardwired to specific hardware nodes and network links are fixed and bound to specific protocols. Porting this prototype to another set of nodes or adding extra functionality would require substantial re-write efforts and may even require a complete revision of the system. This nulls out one of the real attraction points of DR based on AWSANs: adding new functionality (such as lighting control, or gas metering) or upgraded functionality (such as smarter control) should be relatively easy. This is however not the case in the current implementation. Ideally, as hardware platforms change and expand the fundamental applications already developed by DR users must "transport" from one platform to another, and be "scalable" to the (n + 1 ... etc.) next generation hardware, software and networking.

The "Universal Application Interface for AWSANs", proposed in this chapter, raises the abstraction level for the programmer such that the details of both node and network architecture are hidden. Consider again the DR network of Fig. 8. By removing all implementation details, the abstract diagram of Fig. 9 is obtained. At the core, we have one or more control functions. This controller queries the environment (through sensors) obtaining values such as the cost of power, current power usage, temperature in the different rooms, etc. Based on this information, specific actions can be taken, such as displaying the current cost of power on a display, turning on a red light on an appliance, turning on the fans, etc. In short, all communications are performed as a set of queries and commands.

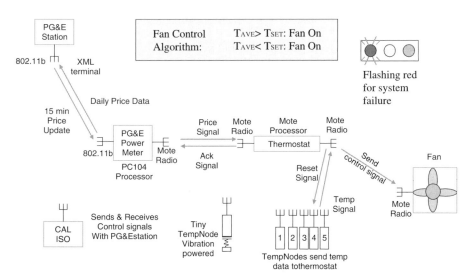

Fig. 8. Communication occurs as follows: Incoming DR pricing is sent from power utility company to the meter on the left and then to the users thermostat in the center of the figure. When prices are low (e.g., 20 cents per kilowatt hour) the user can enjoy unlimited cooling with the thermostat set for 73 degrees Fahrenheit. However, the thermostat also includes a comfort/cost slider which can be set: (**a**) to the left for a consumer/user who does not care how much it costs, (**b**) to the extreme right for a user who wants to curtail use as soon as the price jumps and who by definition does not care how hot the house becomes or, (**c**) for a middle of the road user say at 77 degrees meaning they will tolerate a modicum of discomfort prior to cycling the AC around the 77 degree value

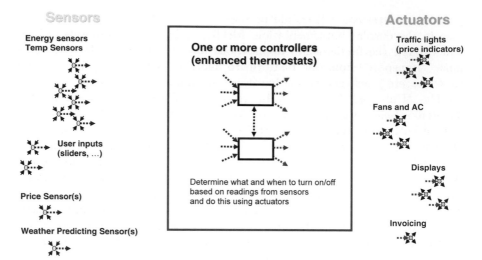

Fig. 9. Abstracted view of Demand Response System, where all implementation details (computation and communication) have been removed

For example, a simple control (application) program at this level of abstraction could be formulated as follows (in pseudo-code):

Every 10 minutes, perform the following {
        Get the price of energy from the energy price sensor;
        If (price > Cost_Thresh) {
             Set red light on washing machine to on;
             Set orange light on all other appliances to on;
    }
}
Every minute, perform the following {
        Get temperature in the living room;
    If (temperature > Temp_Thresh and cost < Cost_Thresh)
        Set fan in the living room to on;
}

Observe how all communications between distributed elements are abstracted away in this description and have been translated into *"get"* queries and *"set"* commands. Also, operations are not bound to specific physical nodes, but are referred to through their physical location (*"living room"*) or attribute (*"price of energy"*). The advantage of this approach is that the application description is orthogonal to the actual implementation architecture and is easily portable. Observe also the use of meta-objects such as living-room and washing machine, which requires the availability of a concept repository service, as described earlier.

These observations lead to a top-down design approach to application design similar to the one adopted very successfully by the Internet community. One can easily see that the application, as described above, can easily be ported to a wide range of implementation platforms (or SNIPs).

8 Summary

A service-oriented platform for the implementation of AWSAN applications was presented. We believe that by defining an application interface at the service layer, it is possible to develop and deploy sensor network applications while being agnostic about the actual network implementation yet still meeting the application requirements such as timeliness, reliability and lifetime.

Some of the concepts introduced in this chapter have broader applicability than AWSANs. For instance, a concept such as the CRS would also be useful in the operation of ad-hoc multimedia networks. In fact, the development of a service-based application interface for this emerging class of applications in a style similar to the one presented here is a logical next step. The potential success of ambient intelligence hinges on the simple and flexible deployment of both ad-hoc multimedia and sensor networks and the interoperability between the two.

References

1. F. Boekhorst, "Ambient intelligence: The next paradigm for consumer electronics", Proceedings IEEE ISSCC 2002, San Francisco, February 2002.
2. D. Snoonian, "Smart Buildings", IEEE Spectrum, pp. 18–23, September 03.
3. J. Rabaey, E. Arens, C. Federspiel, A. Gadgil, D. Messerschmitt, W. Nazaroff, K. Pister, S. Oren, P. Varaiya, "Smart Energy Distribution and Consumption Information Technology as an Enabling Force," White Paper, http://citris. berkeley.edu/SmartEnergy/SmartEnergy.html.
4. G. Huang, "Casting the Wire", Technology Review, pp. 50–56, July/August 2003.
5. IEEE 802.15 WPAN^{TM} Task Group 4 (TG4), http://www.ieee802.org/15/pub/ TG4.html.
6. The Zigbee Alliance, http://www.zigbee.org/.
7. IEEE 1452.2 "Standard for a Smart Transducer Interface for Sensors and Actuators – Transducer to Microprocessor Communication Protocols and Transducer Electronic Data Sheet (TEDS) Formats", IEEE, 1997.
8. C. Srisathapornphat, C. Jaikaeo, C. Shen, "Sensor Information Networking Architecture", in Proceedings of the International Workshops on Pervasive Computing, Toronto, Canada, August 2000.
9. S. Madden, "The Design and Evaluation of a Query processing Architecture for Sensor Networks", Ph.D. Dissertation, UC Berkeley, 2003.

10. M. Sgroi, A. Wolisz, A. Sangiovanni-Vincentelli, J. Rabaey, "A Service-based Universal Application Interface for Ad-hoc Wireless Sensor Networks", White Paper, http://bwrc.eecs.berkeley.edu/.
11. J. Heidemann, F. Silva, C. Intanagonwiwat, R. Govindan, D. Estrin, D. Ganesan, "Building Efficient Wireless Sensor Networks with Low-Level Naming", in Proceedings of the Symposium on Operating Systems Principles (SOSP 2001), Lake Louise, Canada, October 2001.
12. E. Arens et al, "New Thermostat, New Temperature Node and New Meter," Project California Energy Commission, 2003.
13. Jason Hill, Robert Szewczyk, Alec Woo, Seth Hollar, David Culler, Kristofer Pister, "System architecture directions for network sensors," ASPLOS 2000.

Locationing and Timing Synchronization Services in Ambient Intelligence Networks

J. van Greunen and J. Rabaey

Abstract. Awareness of space and time is an essential component of our daily living experience. Hence, any ambient intelligence environment must have a sense of location and time built into it. This information, however, needs to be provided with a minimum of infrastructural overhead. In addition, there services should come with low energy consumption, robustness to changing environmental conditions, and adapt themselves to dynamically varying networks. This chapter presents a set of algorithms for the localization and synchronization of sensor network services, developed with the above design principles in mind. For the first service, localization, several existing localization methods are evaluated. The trade-off between accuracy and overhead (in terms of density, infrastructure and ranging needs) is discussed. For the second service, synchronization, it is demonstrated that accuracy can be sacrificed for gains in energy efficiency. The algorithm starts by constructing a spanning tree across the network. Once the spanning tree structure is constructed, nodes perform pair-wise synchronization along the edges of this tree. The tree-structure reduces the number of pair-wise synchronizations that need to be performed and is thus energy-efficient. The tree can be constructed in a centralized (reference node initiated) or distributed (sensor node initiated) fashion. Simulation results for the different variants of the synchronization algorithm are presented. The chapter is concluded with a number of reflections and perspectives.

1 Introduction

The goal of an ambient intelligent network is one in which a node can automatically configure itself, form connections with other nodes in an ad-hoc fashion, participate in gathering useful information about the environment, and finally relay that information so that intelligent control decisions can be made. In the previous chapter, the role of standardized services in enabling this model of operation was discussed. This chapter explores concrete implementations of such services. Implementations for two services are presented in this chapter: localization and synchronization. These services are crucial to a number of applications. The applications of interest include industrial automation, environmental monitoring, utilities and building automation, etc. Several industrial players have already started forming consortiums like Zigbee [1] to standardize sensor networks. The California building association (CABA) is also particularly interested in developing its own standards. Currently these industries are focusing on deploying custom, proprietary systems.

The goal of academic research is to create an open standard the specification and implementation of services.

2 Localization

In this section, several localization algorithms will be introduced. The goal is to create criteria for the algorithms and then to evaluate these algorithms and compare them to each other. One solution to the localization problem is to equip each node with a Global Positioning System (GPS) receiver. Unfortunately, this is a poor solution in terms of power consumption and per-unit cost, and, even more, does not work within buildings and covered areas. A more practical solution is to have a few nodes that know their position *a priori* (these nodes are referred to as known/beacon/anchor nodes). The unknown nodes in the network can then calculate their positions by "measuring" (in terms of hops or distance) their respective distance from these nodes.

Localization in sensor networks poses several challenges. First, the complexity of the localization algorithm must not grow faster than the network size. Second, the network has constrained communication and computation resources. A good algorithm should try to optimize for both these factors. Third, a solution has to be tolerant to large errors in the distance (or range) measurement.

2.1 Criteria for Algorithm Evaluation

Despite the implementation-specific constraints of a sensor network and the diverse requirements of each application running on the network, a set of global performance criteria can be devised. For specific applications, only a subset of the criteria may be relevant. The following criteria encompass both necessary and desirable properties of any localization algorithm:

- *Scalability*. A scalable algorithm keeps the required per-node computation constant as the network size grows.
- *Communication energy*. In general, it is desirable to keep the total energy spent on communication as low as possible. However, certain applications may require higher accuracy, resulting in increased communication energy.
- *Computation energy*. It is similarly desirable to minimize total computation energy in the network. However, there is often a trade-off between computation energy, speed of convergence, and communication requirements.
- *Accuracy*. Some applications may require a bound on the estimation error.
- *Error tolerance*. The localization algorithm must tolerate large and sporadic range measurement errors. Range errors occur because the hardware used to build sensors typically cannot support the power and speed required by accurate ranging methods [2].

- **Convergence speed**. Applications may need fast or bounded convergence times, for example a time-critical surveillance sensor network.
- **Consistent performance across networks**. The algorithm should perform well on different implementations of sensor networks. That is, the algorithm should not depend on a specific network topology or specialized hardware.
- **Estimate of error**. The algorithm should provide an estimate of the position accuracy at all times. This will facilitate an algorithm that can terminate early when the current error estimate is below a specified threshold.

2.2 Overview of Algorithms

This section provides an introduction to seven existing localization algorithms, along with a discussion of their advantages, disadvantages, and energy costs. In the next section, each of these algorithms will be evaluated and compared.

2.2.1 Centralized LP

The centralized linear algorithm described in [3] uses connectivity constraints to solve the localization problem. This algorithm is based on the assumption that if two nodes can communicate with each other, they must lie within the communication radius R of each other. Figure 1 depicts the nodes within communication range. This ability to communicate is mathematically equivalent to the following 2-norm constraint on the node positions.

$$||a - b||_2 < R \Rightarrow \begin{bmatrix} I_2 R & a - b \\ (a - b)^T & R \end{bmatrix} \tag{1}$$

Fig. 1. LMI communication range, the center node can communicate with both white nodes

Note: the region of communication for a node must be a convex circle of radius R. This is not true for real RF transmissions.

The communication constraints of all the nodes can be combined to form a convex feasibility problem with the following mathematical formulation:

$$
\begin{aligned}
Minimize &: c^T x \\
Subject\ to &: Ax < b\,.
\end{aligned}
\tag{2}
$$

Where x is a vector of (x, y) coordinates for each node in the network. The LP needs to be solved for each unknown node in the network; to solve for the corners of the feasible bounding box around the unknown node, c_{2k} or c_{2k-1} is set to 1 or –1 while all other entries are 0. If the objective function, $c^T x$, is omitted, it has the same effect as selecting a random point within the bounding box. To find the constrain, $Ax < b$, Schur complements can be used to transform the expression in (1) into a linear inequality. A standard LP solver can then be used solve this convex problem.

Advantages. This algorithm is theoretically well defined and easy to analyze because the LP problem has been studied. It is well suited to a heterogeneous sensor network where certain nodes have more computational power than others. The communication time from the central node to the furthest node in the network (plus the computation time) determines the convergence time of this algorithm. This algorithm achieves good accuracy because it incorporates all inter-node information.

Disadvantages. The pooling of information toward a central point will deplete network resources around that point. Moreover, the computation required for solving the linear program scales as the square size of the network. Thus the algorithm is not scalable and cannot easily be implemented on a low-power, low-cost node. In addition, the convex linear program fails when the regions of communication are not convex. In order for the regions to be convex[1] the reachable communication boundary of a node must form a convex hull. Thus, if there is an obstacle on one side of a transmitter that interrupts transmission, the communication region will be distorted and may no longer be convex.

Cost. The following notation is used for all cost calculations in this chapter: D is the physical dimension of the problem (i.e. either 2D or 3D), c is the average degree (number of neighbors) of nodes in the network. N is the total number of nodes in the network and A is the number of anchors in the network.

The average degree of each node and the average distance of nodes from the central node determine the communication cost for the centralized LP algorithm. Computation cost is determined by the number of matrix operations to solve the LP. For a $m \cdot n$ matrix the number of operations is roughly $O(mn^2 + n^3/3)$. In this problem $n = D \cdot N$ and m is approximated by average degree c.

	Energy consumption
Total Computation	$c \cdot (D \cdot N)^2 + (D \cdot N)^3/3$
Total Communication	$pathlen \cdot c \cdot N$

---

[1] A hull is convex if any two points on the hull can be connected by a straight line that lies below the hull boundary.

2.2.2 Rectangular Intersection

This rectangular intersection algorithm [4] is a distributed version of the centralized LP. It is assumed that all nodes are placed within a square region S. The region S is then further subdivided into n^2 *cells* of area $(s/n)^2$. Instead of the "continuous" communication model used in [3], this approach uses a discrete communication model. In the discrete communication model, the communication radius P is given in terms of a number of *cells*. In other words, the communication region is no longer a circle of radius R, it is now a square with side length P cells. For example, if node k and j are one-hop neighbors (nodes j and k can communicate with each other), k lies within a square centered at j. With the discrete model, the combination of connectivity constraints reduces to calculating the intersection of squares (see Fig. 2).

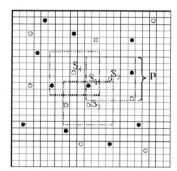

Fig. 2. Intersection of rectangle

The following algorithm is executed at every unknown node S_i in the network:

Step A: Gather position information of one-hop neighbors with known positions

Step B: Compute estimation of position by doing rectangular intersection

The rectangular intersection gives a feasible rectangle for a node's location. Rectangular intersection for a node with K_n known neighbors at locations: $(x_1, y_1), \ldots, (x_n, y_n)$ is:

$$[A, B] \cdot [C, D] = [\max(x_1, \ldots, x_n) - P, \min(x_1, \ldots, x_n) + P] \\ [\max(y_1, \ldots, y_n) - P, \min(y_1, \ldots, y_n) + P] . \tag{3}$$

Advantages. This algorithm requires very little computation and communication energy because nodes only communicate with their neighbors. The algorithm converges rapidly and completes in a single step. It can also be shown that as the number of known neighbors goes to infinity, the expected rectangular intersection for an unknown node is accurate to a single cell. Due to its distributed nature, the algorithm is scalable. In fact, the complexity of the algorithm at each node is completely unrelated to the network size.

Disadvantages. The main drawback of this approach is that each unknown node needs to communicate with several known neighbors. Thus, if there are few known nodes in the network this algorithm performs very poorly. This algorithm also relies on the convexity of the regions of communication, and will fail if they are not convex.

Cost. The computation consists of finding D maxima and minima resulting from the intersections. For each intersection there are c comparisons and $2c$ additions. Thus the total computation cost will be $3 \cdot N \cdot D \cdot c$. On average, each node needs to communicate once with its c neighbors for a total of $N \cdot c$ across the network.

	Energy consumption
Total Computation	$3 \cdot N \cdot D \cdot c$
Total Communication	$N \cdot c$

2.2.3 DV-Hop

This algorithm was developed in [5] to prevent to propagation of ranging error across multiple hops. Thus it computes the location of unknown nodes based on an "average" distance between nodes. The known nodes flood the network with their positions. Each unknown node stores a list of the positions of known nodes and the number of hops they are from these known nodes. Once an unknown node has its 'hop-distance' (number of hops·average distance) from more than three non-collinear known nodes it can compute its position by triangulation (solving $Ax = b$).

Advantages. This algorithm is fairly simple, easy to implement, and tolerant to range measurement errors. It takes a medium amount of time to converge because the hop information must propagate through the network. DV-hop does prevent errors from accumulating in the network. The algorithm is scalable because each node needs only information from at least three anchors.

Disadvantages. DV-hop can achieve only limited accuracy because of the averaging effect. This algorithm also has medium computation cost due to the maximum likelihood triangulation. DV-Hop does not perform very well when the network topology is irregular and the actual distances deviate greatly from the average distances used in the algorithm.

Cost. To compute the triangulation, the hop information first needs to be linearized. This requires $D \cdot (A-1)$ subtractions $(A-1) \cdot (2 \cdot D+1)$ additions and $2 \cdot (A-1) \cdot (D+1)$ multiplications. Once the data has been linearized, the added complexity from the least-squares algorithm is $O((A-1) \cdot D^2 + D^3/3)$. The communication cost consists of flooding the hop counts through the network, thus it is $A \cdot N$.

	Energy consumption
Total Computation	$N \left[A \cdot (3D+1) + O \left(A \cdot D^2 + \frac{D^3}{3} \right) \right]$
Total Communication	$A \cdot N$

2.2.4 Grid of Beacons [6]

The beacons or known nodes are placed on a regular grid. These known nodes periodically transmit their locations. If an unknown node receives above *CMthresh* percentage of the transmitted messages from the beacon, the unknown node is "connected" to the beacon. The unknown node then computes its position as the centroid of all the beacons that it is connected to.

$$(X_{\text{est}}, Y_{\text{est}}) = \left(\frac{\sum_{i=1}^{k} Xi}{k}, \frac{\sum_{i=1}^{k} Yi}{k} \right). \quad (4)$$

Advantages. This algorithm is scalable, distributed, requires very low computation and communication energy, and it converges rapidly.

Disadvantages. In order to achieve reasonable accuracy, there needs to be a regular and dense grid of beacon nodes. This may increase the cost of the network and may be infeasible under certain conditions when the positioning of beacon nodes cannot be controlled.

Cost. The average number of anchors that each node can communicate with (A) is less than c, the average degree of nodes. Thus the total computation in the network is $N \times D \cdot c$. Each node also only communicates with its beacon neighbors. Thus the total communication in the network is $N \cdot c$.

	Energy consumption
Total Computation	$D \cdot c \cdot N$
Total Communication	$N \cdot c$

2.2.5 Kernel-Based Learning Localization

This algorithm, described in [7], uses kernel-based regression and classification algorithms to treat sensor network localization as a pattern recognition problem. The basic tenet of the problem is as follows: Use the known nodes as training data to construct a hyper-plane (classifier) separating points in space. Then this classifier can be used to predict where the unknown points lie. The connectivity information (i.e. node j is in the radio range of node k) serves as input to the kernel function[2]. The algorithm constructs one hyper-plane for a specific region. To get finer localization, the regions need to be small and overlapping. Then the hyper-planes form a cell/grid in which the nodes are located. A bound for the error is derived in [7]; for a network of size $L \times L$ with A known nodes with radio range R the localization error is on the order of $O\left(L^{\frac{1}{3}} R^{\frac{2}{3}} A^{\frac{-1}{6}}\right)$.

Advantages. This is a statistical algorithm so it is more robust to noise errors. In general, learning (SVM) methods use few "support vectors" thus, the computational overhead for this algorithm is low. Each sensor needs to communicate only with the A known nodes.

Disadvantages. This algorithm will perform poorly if there are not many known nodes. The complexity depends in part on the topology of the network.

Cost. The training phase costs $O(A^3 k^2)$ where k is the number of disks covering the network. The computation in the classifying step is proportional to the number of support vectors. Simulations from [7] suggest that this is on the order of $O(R^2 A k^2)$. The communication cost is NA because all nodes need to communicate with the known nodes.

	Energy consumption
Total Computation	$O(A^3 k^2 + R^2 A k^2)$
Total Communication	NA

2.2.6 Start-Up and Refinement

This algorithm [9] consists of two stages: start-up and refinement. The start-up phase provides an initial estimate of each node's position using DV-hop. During the following phase (refinement), the nodes try to improve their initial position estimate. The nodes achieve this improvement by measuring the distances to all of their one-hop neighbors and updating their own position accordingly. All positions in this algorithm are calculated with linearized

---

[2] For more information on statistical learning and kernel functions see "An Introduction to Support Vector Machines: & Other Kernel-based Learning Methods" by N. Christianini and J. Shawe-Taylor.

maximum likelihood triangulation [9] . The refinement method improves the initial position estimate if the errors in the initial position estimates are uncorrelated.

Two additional measures are taken to improve the performance of the refinement stage. First, ill-connected nodes are prevented from participating. An ill-connected node is a node that does not have independent references, i.e. it receives hop-information from less than three neighbors. Second, confidence estimates are used as weights during triangulation. Unknown nodes start with a confidence estimate of 0.1 while known node have a confidence of 1.0. When an unknown node updates its position estimation it also updates its confidence estimate to the average of its neighbors confidence estimates.

Advantages. Start-up and refinement achieves reasonable accuracy. The network can control the number of iterations – thus there is a potential for real-time optimization of energy spent versus accuracy achieved. This algorithm is fairly tolerant to range errors if the network connectivity is high. Start-up and refinement is also a scalable algorithm because nodes mainly communicate only with one-hop neighbors.

Disadvantages. This algorithm is somewhat computation intensive due to the iterative triangulation. The algorithm may not converge to an accurate solution if the initial position estimations are very inaccurate or the errors are correlated.

Cost. The computation cost of this algorithm is very similar to that of DV-hop. The only difference is that in the refinement stage the least-squares triangulation is computed s times with c neighbors (the average degree of nodes in the network).

	Energy consumption
Total Computation	$N \left[(3D + 1)(A + s \cdot c) + O \left((A + s \cdot c)D^2 + \frac{D^3}{3} \right) \right]$
Total Communication	$A \cdot N + c \cdot N \cdot s$

2.2.7 Multilateration (Kalman Filtering)

The multilateration approach [10] is similar to start-up and refinement in that a distributed optimization problem is solved by iterative likelihood estimation. The main difference between the two is that multilateration utilizes Kalman filtering [11]. The multilateration algorithm consists of three stages. During the first phase, all ill-connected nodes are pruned and the well-connected nodes are placed in collaborative sub-trees. Well-connected nodes have the following three properties. First, they must have more than three neighbors. Second, if unknown, they must have non-co-linear reference points. Third, if pairs of unknown nodes use each other as constraints in position estimation, their other constraints must contain diverse reference groups.

In the second stage, the participating nodes obtain initial position estimates. The smallest bounding box around a node is found by forwarding distances on a minimum hop path from anchor nodes. Figure 3 shows the bounding box for an unknown node R in the x-dimension. The last stage in the multilateration algorithm is also called refinement. During this stage each node measures distances to its neighbors and executes a recursive Kalman filter. This is an approximation of a fully decentralized Kalman filter because the nodes do not exchange covariance information.

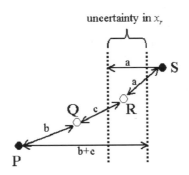

Fig. 3. Node R is using distance measurements along a minimum hop path to obtain an initial position estimate

Advantages. Multilateration achieves good accuracy provided reasonable initial position estimates. The number of filter iterations can be controlled so there is also a potential for real-time energy-accuracy optimization. The computation is scalable because nodes mainly communicate with one-hop neighbors.

Disadvantages. Similar to start-up and refinement, this algorithm may not converge to an accurate solution given inaccurate initial position estimations or correlated errors. There are also potential scalability problems in the way the collaborative sub-trees are constructed and maintained. The Kalman filter update is also computation intensive.

Cost. The computation cost of this algorithm is higher than that of start-up and refinement because more calculations are required in the Kalman filter update. The Kalman filter gain, covariance matrix, measurement noise matrix, and Jacobian of the "blending" matrix must be calculated.

Energy consumption	
Total Computation	$4N(3D + 1)(A + s \cdot c) + N \cdot O\left((A + s \cdot c)D^2 + \frac{D^3}{3}\right)$
Total Communication	$A \cdot N + c \cdot s \cdot N$

2.3 Comparison

Based or the analysis of the previous section, we can compare the presented algorithms according to the outlined criteria. The most important criterion for localization is scalability. If an algorithm is not scalable it cannot be implemented on a real network. With the exception of the centralized LP algorithm, all algorithms presented in the previous section are scalable in the number of nodes. The kernel learning algorithm does not scale well as the physical network size (number of covering disks) increases. Consistency across networks is another criterion. The rectangular intersection, kernel learning, and grid algorithms require many nodes that know their own location, and will increase the cost of the network. The other algorithms do not place such constraints on the network.

Another important consideration is the accuracy of the algorithm. The rectangular intersection and grid algorithms will only have good accuracy if there are many known nodes in the network. The kernel learning algorithm also performs well when there are many know nodes, or when the network size is small. In contrast, DV-hop and multilateration achieve moderate accuracy in a variety of network conditions. Start-up and refinement is more accurate than DV-hop. Multilateration will be the most accurate when the initial estimate is good because it uses information from all the previous position estimates in the current estimate. However, both multilateration and start-up and refinement is sensitive to errors in initial position estimate.

Finally, choosing an algorithm that minimizes the energy spent during the localization is desirable. The rectangular intersection and grid algorithms are the most energy efficient because they communicate only with local neighbors and have very simple computation requirements. Network coordinate, multilateration, and start-up & refinement are moderately efficient because both algorithms mainly use information from immediate neighbors and have complex but limited size calculations. In contrast, the centralized LP is very inefficient because the linear program grows with N^2. A summary of the above comparisons is given in Table 1.

3 Lightweight Time Synchronization

Applications require that network nodes synchronize their *clocks* (i.e. local times) to a global time. Traditional synchronization algorithms [12] have focused on minimizing the synchronization error and achieving maximum accuracy, without regard to the computation- and communication energy expended by the algorithm. In AWSANs, however, energy is a highly constrained resource. Thus, traditional synchronization algorithms are not adequate as they place a heavy burden on the network resources. In the following section, synchronization schemes are introduced that sacrifice accuracy

Table 1. Comparison of localization algorithms

Criterion	Centralized LP	Rectangular Intersection	Kernel Learning	DV-hop	Grid	Start-up & Refinement	Multilateration (Kalman Filter)
Scalability	No	Yes	Moderate	Yes	Yes	Yes	Yes
Communication energy	$(D \cdot N^2)(D \cdot N /3 + c)$	$3 \cdot N \cdot D \cdot c$	$O(A^3k^2 + R^2 Ak^2)$	$N \cdot O(A \cdot D^2 + D^3/3)$	$D \cdot c \cdot N$	$N \cdot O[(A + s \cdot c) D^2 + D^3/3]$	$N \cdot O[(A + s \cdot c)D^2 + D^3/3]$
Computation energy	$pathlen \cdot c \cdot N$	$N \cdot c$	$A \cdot N$	$A \cdot N$	$N \cdot c$	$A \cdot N + c \cdot N \cdot s$	$A \cdot N + c \cdot N \cdot s$
Accuracy	Good	Good with many known nodes, else poor	Medium	Medium	Good with many known nodes, else poor	Good	Good
Error tolerance	Medium	Good	Good	Good	Medium	Medium	Good
Convergence speed	Proportional to size squared (matrix calculation)	Fast	Proportional to size	Proportional to size	Fast	Proportional size & # of iterations	Proportional to size & filter iterations
Consistency across networks	No	No	No	Yes	No	Yes	Yes
Estimate of error	No	No	Yes	No	No	No	No
Network cost	High	High	Depends on physical size	Low	High	Low	Low

by performing synchronization less frequently and between fewer nodes. The efficiency of these schemes can be adjusted to perform to the desired accuracy.

The *lightweight tree-based synchronization* (LTS) algorithms presented in this section are designed to work with generic low-cost sensor nodes. The algorithms focus on minimizing overhead (energy) while being robust and self-configuring. In particular, the algorithms operate correctly in the presence of node failures, dynamically varying channels, and node mobility. The synchronization algorithms were previously published in [13].

3.1 Related Work in Sensor Network Time Synchronization

A number of synchronization algorithms for sensor networks have been published, among which the Reference Broadcast (RBS) [14], TINY/MINI-SYNC [15], and Level synchronization [16] schemes deserve special attention. In the first scheme, RBS, an intermediate node is used to synchronize the local time of two nodes. The intermediate node transmits a "reference packet" to the two nodes. The two nodes record the time that they received the packet and then exchange this recorded time to find the difference. The accuracy of RBS is mostly determined by the amount of time it takes either node to receive and process the reference packet. RBS has a complexity of 4 received and 3 transmitted messages for two nodes. For n nodes and m reference broadcast packets, RBS has a complexity of $O(mn)$ – for each of the m received reference packets, a node exchanges information with all other $n-1$ receivers. A major goal of the algorithm presented in this chapter is to lower multi-hop complexity.

In [14] "multi-hop clock conversion" is added to the RBS algorithm. Multi-hop clock conversion is designed to synchronize groups of nodes that are already synchronized to different broadcast nodes. Let A and B be two broadcast nodes. It is assumed that at least one node belongs to both group A and group B. Synchronization is then performed by finding the statistical best fit of the receiver time differences of these dual-group nodes. The multi-hop RBS algorithm relies on effective clustering of the nodes around the broadcast nodes. This clustering service adds considerable overhead to the RBS algorithm. Further, Elson et al. [17] proposes post-facto synchronization. Post-facto synchronization utilizes RBS synchronization, but nodes are synchronized only after a time-sensitive packet has been transmitted.

The second type of synchronization, TINY/MINI-SYNC, is based on the assumption that the nodes' clock drifts are of the following linear form: $t_i = a_i t + b_i$ where t_i is the local clock of node i, a_i and b_i are drift parameters, and t is "real" time. Under this assumption, the offset between two nodes is also linear. TINY/MINI-SYNC nodes exchange time-stamped packets (as described in Sect. 3.2). These exchanged packets are used to estimate the best-fit offset line between the two nodes. As more packets are exchanged the computation complexity required for calculating the best-fit line increases.

Each node performs this pair-wise synchronization scheme with each of its neighbors.

The third scheme, called level synchronization [16] introduces the pair-wise synchronization that forms the base of the algorithms introduced in this chapter. The scheme was chosen because it is extremely simple and computationally efficient. For the multi-hop component, the level-based scheme assigns a logical level to each node. This level indicates a node's distance from the chosen leader node. The assignment is fixed for the lifetime of the leader. The static nature of the level hierarchy reduces the robustness of this solution.

Before discussing the multi-hop schemes in detail we first present the technique for pair-wise time synchronization.

3.2 Pair-Wise Synchronization

The following section describes a basic scheme to synchronize pairs of nodes. Nodes j and k can synchronize their local time by exchanging two packets with the following procedure:

- Node j transmits the first packet at time t_1 with respect to its local time.
- Node k records the time t_2 when it receives the first packet. Time t_2 is equal to t_1 plus the transmission time D from node k to j plus the offset d between node j and k's clocks. Generally the transmission time D is unknown and is a function of the distance between the nodes and signal propagation characteristics.
- Next, node k transmits a second packet to j that contains t_1 and t_2. This packet is also timestamped by k at time t_3.
- Node j receives the second packet at time $t_4 = t_3 + D - d$. See Fig. 4 for a graphical depiction of the exchange.
- The offset d can be calculated at node j by subtracting t_4 from t_2

$$t_2 - t_4 = t_1 - t_3 - D + D + 2d$$
$$d = {}^1/_2(t_2 - t_4 - t_1 + t_3) \tag{5}$$

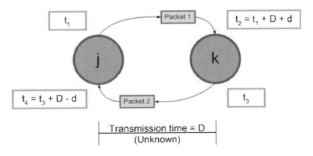

Fig. 4. Packet exchange for pair-wise synchronization

- The two nodes are synchronized once node j has calculated the offset d. However, a third message is required if the offset d must also be communicated to node k.

The underlying assumption is that the transmission time is the same from j to k and k to j, that is $D_1 = D_2$. Of course D_1 and D_2 are not exactly equal and this introduces some error in the synchronization. In order to analyze this difference, a brief overview of the components that make up the transmission time D and the error contribution of each component is presented. Kopetz and Schwabl [18] have divided the transmission time into four parts:

- **Send time**. The time spent assembling the message at the sender, which includes processing and buffering time. The message can be timestamped after the send time has completed so send time does not contributed to the difference in transmission times.
- **Propagation time**. The time for the signal to propagate across the physical medium between the two nodes. The propagation time is a function of the distance between the nodes. The propagation time is the same in both directions and thus does not contribute to the difference between transmission times.
- **Receive time**. The processing time required for the receiver to receive a message from the channel and notify the host of its arrival. Elson et al. [14] characterized the receive time using a testbed of "COTS MOTES," a narrowband radio and sensor platform developed by Warneke, Atwood, and Pister [19]. Results from receive delay differences showed that the distribution of inter-receiver variability was Gaussian with zero mean and a variance of $11.1\,\mu s$. The experiment validates the use of a Gaussian distribution to model the variability of receive time.
- **Access time**. The delay associated with accessing the channel, including carrier sensing. The differences in access times arise in much the same way that the difference in receive time do because the packets go through the same physical and MAC layers of the radio. Thus, as with receive time, it is assumed that differences in access times among nodes are also Gaussian with zero mean.

When the error from the receive- and access time are combined, their variance increases at most four times (depending on the degree of correlation between them). Thus, when two nodes synchronize with each other, there is 99% confidence that the accuracy will be within $(2.3)*(4\cdot)$ or $9.2*$. For the COTS MOTES [19] this is an accuracy of 0.1 ms.

In the next section the pair-wise (single-hop) synchronization algorithm is extended to *multi-hop* synchronization. In addition to accuracy of the synchronization, the stability (drift of clocks over time) and resynchronization requirements are discussed.

3.3 Multi-Hop Synchronization

Multi-hop synchronization is an extension of the pair-wise synchronization algorithm. In a straightforward extension of pair-wise synchronization, a group of n nodes requires n^2 pair-wise synchronizations. Due to the relatively low accuracy requirements of our sensor network, this n^2 factor will be avoided by linearizing the synchronization by performing pair-wise synchronization only along network edges that form a spanning tree structure, described later.

There are several important considerations for multi-hop synchronization that influence the design of an efficient algorithm:

- **Global reference**. It is assumed that at least one node in the network has access to a global time reference. Also, the global time kept by any reference node is orders of magnitude more accurate than the accuracy achievable by the single-hop synchronization. All nodes with reference to global time are stationary.
- **Selective synchronization**. Multi-hop synchronization can keep all nodes synchronized, or synchronize a select number of nodes. In other words, the algorithm can synchronize only the nodes that are transmitting time-sensitive data.
- **Resynchronization rate**. Due to clock drift, the nodes will periodically need to be resynchronized. A bounded clock drift is assumed. In [20] a node's clock denoted by $H(t)$ is defined to be "ρ – bounded" provided that for all real time t,

$$\frac{1}{(1+\rho)} \leq \frac{dH(t)}{dt} \leq 1 + \rho \,. \tag{6}$$

A clock $H(t)$ that drifts at a constant rate will have $\rho = 1.0$.
- **Error estimation & limitation**. The synchronization algorithm should keep track of nodes' accuracy and errors from clock drift. When the node clocks have (or might have) drifted, a resynchronization scheme should be invoked.
- **Robustness**. There should not be a single point of failure in the system (except maybe the reference node) and the algorithm should be robust to node failures.
- **Mobility**. Synchronization should work for both stationary or mobile nodes

In the next sections two algorithms for multi-hop synchronization are set forth. The first algorithm is a centralized approach in which the synchronization and periodic updates are generated from a reference node. The second scheme is a distributed multi-hop synchronization method where the nodes (and not the reference nodes) are responsible for initiating and performing resynchronization.

3.3.1 Centralized Multi-Hop LTS

Centralized multi-hop synchronization is a simple linear extension of the single-hop synchronization. The basis of the algorithm is the construction (either offline or dynamic) of a low-depth spanning tree T comprising the nodes in the network. In general, a new spanning tree is constructed each time the algorithm is performed. In order to synchronize nodes in the tree, pair-wise synchronizations are performed along the edges of T. In *centralized* multi-hop synchronization, the reference node initiates the synchronization by synchronizing with all immediate (single-hop) children in T. Next, each child of the reference node synchronizes with their subsequent children. This process continues until the leaf nodes of T are reached. The algorithm terminates when all the leaf nodes have been synchronized. The running time of the algorithm is proportional to the depth of the tree.

Error Analysis. The variance of the synchronization error increases along each branch of the tree as a linear function of the number of hops. This is because the errors resulting from the respective pair-wise synchronizations are independent and thus additive. As assumed in the pair-wise synchronization discussion, the synchronization error between two adjacent nodes is a Gaussian random variable with a variance of four times the receiver variance. Thus, for a node at depth d in the spanning tree, the expected error is zero but the variance of the error is $4 \times d$.

The following example illustrates the effect of error accumulation on accuracy. Consider a network with 1000 randomly distributed nodes in a rectangular region of size 140 m * 140 m with a radio range of 10 m and a reference node in the center of the region. If the spanning tree is created by breadth-first search, the longest possible path in the network is along a diagonal of length $70*2^{1/2}$ which consists of at most 10 hops. Thus the maximum resulting variance at the leaf nodes will be 40. If the nodes are COTS MOTES then with 99% confidence the leaf nodes will be accurate to within 2.3*40, which is an accuracy of 1 ms.

The error accumulation is highly associated with the spanning tree used for synchronization, specifically its depth. Thus, algorithms for constructing low-depth spanning trees in a distributed manner are crucial. An optimal tree is one with minimum depth. This minimizes the running time of the algorithm and also the accumulation of error along the tree. Breadth-first-search (BFS) fulfills this criterion, however, it is difficult to distribute and has a higher communication overhead compared to other tree-construction algorithms. The communication complexity of BFS can be reduced to $10 \times n \times m^{1/2}$ where n is the number of nodes and m is the number of edges between them.

There are other tree-construction algorithms with other desirable properties. Distributed depth first search (DDFS) developed by Awerbuch [21] is a computationally efficient algorithm. Figure 5 describes the algorithm in pseudocode. Savings in communication arise because each node informs its

```
Start the algorithm at node u the initiator:
visited_u := true ;
    for all x  Neigh_u do send <visit> to x;
    for all x  Neigh_u do receive <ack> from x;
    for some w  Neigh_u do send <dfs> to w; statusu[w] :=
cal end
Upon receipt of <visit> from v:
    statusu[v] := done ; send <ack> to v
Upon receipt of <dfs> from v:
    if not visited_u then
        begin visited_u := true; status_u[v] := father;
            begin forall x  Neigh_u \ {v} do send <visit> to
x;
            forall x  Neigh_u \ {v} do receive <ack> from x;
        end;
        if there is a w  Neigh_u with status_u[w] = unused
        begin send <dfs> to w; status_u[w] := cal
        else if there is a w  Neigh_u with status_u[w] = father
        begin send <dfs> to w end
    else (* initiator *) stop
```

Fig. 5. Awerbuch's distributed depth-first search (algorithm taken from [21])

neighbors when it is visited the first time, before it continues the recursive search among its children. Thus, DDFS eliminates the return calls along non-tree edges. The communication complexity of the algorithm is 4^*m (where m is the number of edges) and the time complexity is bounded by $4n-2$. A second tree-exploration algorithm is called the "Echo" algorithm. This algorithm is efficient in practice; in most cases it has complexity $O(d)$, where d is the depth of the tree.

Efficiency. The communication cost of the multi-hop synchronization algorithm arises from the spanning tree construction and the pair-wise synchronization along the tree's $n-1$ edges. Pair-wise synchronization has a fixed overhead of 3 messages per edge for a total of $3n-3$ messages. The overhead for constructing the spanning tree depends on the complexity of the algorithm used to construct the tree. If DDFS is employed, the total overhead for centralized multi-hop synchronization is $3n-3+4^*m$ per network synchronization.

Clock Drift and Resynchronization. It is desirable to keep all clocks accurate (with high probability) to within τ units of global time. In the centralized multi-hop algorithm, the reference node must periodically resynchronize the network. Two parameters are required by the reference node in order to calculate a good resynchronization interval: the instantaneous accuracy obtained by synchronizing the entire network, and the maximum rate of clock drift.[3] Thus, in centralized multi-hop synchronization, the maximum depth

---

[3] The assumption is made that the running time of the algorithm is negligible compared to the clock-drift and pair-wise synchronization accuracy.

of the spanning tree must be communicated to the reference node when nodes synchronize. This introduces the overhead of forwarding depth information back along the spanning tree when synchronization has completed. Given this maximum depth, a single synchronization session is accurate to within $9.2*d*$(where is the variance per hop in units of time) with 99% probability.

By assuming a ρ-bounded clock, the expected clock drift rate will not exceed ρ. Thus, in order to maintain time accuracy to within τ units of global time, the reference node must resynchronize at a rate of at least $(\tau - 9.2*d\times)/\rho$. (The numerator represents the amount of time that the clock can drift and the denominator represents the drift rate). In this work, it is assumed that all nodes know their clock-drift ρ. This is a reasonable assumption because the clock drift of oscillators can be found in standard specification sheets and can easily be programmed on the nodes during assembly or during a network initialization phase. If the instantaneous synchronization accuracy $11.1\,\mu s$ as for the COTS nodes, the drift ρ is the drift of a typical quartz crystal, which is 20–50 parts per million, a depth of 5 and an accuracy of 0.5 seconds, the resynchronization rate would be approximately 0.1 MHz or once every 9900 seconds. The reference node can calculate this rate and periodically generate a resynchronization.

Robustness. The centralized multi-hop synchronization algorithm is robust in the following ways. First, although the algorithm is sensitive to failures in the reference node, backup or multiple reference nodes can be used. Second, given that a new spanning tree is created every time the network is synchronized, the algorithm is robust to dynamic channel variations, changes in topology, changes in size, and node mobility. In particular, channel characteristics in sensor networks with mobile nodes are assumed to be constant relative to the time required to synchronize the network. The multi-hop algorithm can also keep the network synchronized to the required accuracy τ in the presence of network changes. This is because the reference node can calculate the maximum possible tree depth based on the radio range and network size so that updates occur frequently enough to maintain accuracy.

3.3.2 Distributed Multi-Hop LTS

This algorithm performs node synchronization in a distributed fashion and does not make use of an overlay spanning tree to direct the pair-wise synchronizations. This algorithm also moves the resynchronization responsibility from the reference node to the nodes themselves. An individual node's resynchronization rate can be determined using the same parameters as the reference node uses in the centralized case. Therefore, to determine their resynchronization rates, nodes will need to obtain the following information: the desired accuracy τ, their distance d (in number of hops) from a reference node, their clock drift ρ, and a record of the time that has passed since they were synchronized. A particular node j needs to resynchronize at a rate of at least $(\tau - 9.2^* d_j)/\rho_j$. When a node j determines that it needs to be re-synchronized, j will send a resynchronization request to the closest reference node. In order for j to resynchronize, all nodes along the routing path from the reference node to j will be synchronized in a pair-wise fashion.

Assuming that the clock drift ρ is the same for all nodes in the network, the nodes furthest from the reference node will have the greatest synchronization error and correspondingly the greatest synchronization rate. Therefore, the synchronization will be driven by these edge-nodes along paths that almost look like a reverse tree. An advantage of this algorithm is that certain nodes may not require frequent synchronization. If a node's rate of event observation is significantly lower than its required synchronization rate, it may not always need to be synchronized to the required accuracy. In other words, it is better to synchronize only when the node has a data packet to transmit. Thus, the nodes can opportunistically synchronize.

Avoiding Cycles. When a synchronization request is forwarded from a leaf node to the reference node it is possible for a cycle to occur. A cycle occurs when the node at the head of the synchronization chain requests synchronization from a node that is lower down in the same request chain. Cycles occur because the routing is dynamic and a node requesting synchronization may not know the entire routing path at the time of the request. When cycles occur they cause deadlock, because the nodes mutually depend on each other for synchronization. It is impossible to avoid cycles due to the asynchronous and distributed nature of the synchronization requests. Once a cycle has occurred a distributed graph-searching algorithm can be used to detect it. These algorithms present a considerable overhead.

An alternative approach, which does not rely on detecting cycles but does avoid potential cycles, is proposed. The approach works as follows: when a node sends a synchronization request to one of its neighbors it sets a timer that is proportional to its distance from the reference node. If the timer expires before a synchronization response from the neighbor arrives, the node simply initiates another synchronization request with a different neighbor. This scheme does not prevent cycles from occurring but reduces their impact at an overhead cost of additional synchronizations.

Algorithm Enhancements. Although some nodes have relaxed synchronization rates, in the distributed approach there are some potential inefficiencies of synchronization requests. For example, two adjacent nodes may attempt to resynchronize and send two separate synchronization requests. However, because the nodes are adjacent, it is more efficient to aggregate the requests. In general, duplicate requests can be eliminated by having each node keep track of pending requests from itself and other nodes. If a node k wishes to resynchronize or is forwarding a request from another node, it is beneficial for k to query each of its adjacent nodes to discover if any have pending requests. If so, k can forward the request to a node with pending requests, which aggregates the request.

Another way to increase efficiency is through path diversification. This is best described with an example. Let k and j be two nodes that are relatively close to a reference node and must each be resynchronized at the same rate. Assume that when other nodes wish to resynchronize with the reference node, they tend to favor a forwarding path that includes k, but not j. Thus, k itself never needs to send a resynchronization request because it is frequently resynchronized by other requests. On the other hand, j rarely "sees" new synchronization information by virtue of being in the path of other requests and must occasionally generate its own resynchronization request. Path diversification allows requests to be sent through both k and j in proportion to how frequently they require resynchronization. Path diversification can be implemented as follows. Each node knows when it next requires resynchronization. When a node x is forwarding or generating a synchronization request, it forwards the synchronization request to a participating neighbor with the earliest resynchronization deadline.

3.4 Simulation and Results

Simulation Setup. Results are based on Omnet++ [22] simulations of connected ad-hoc networks consisting of 500 nodes. The nodes are placed uniformly at random within a 2-dimensional 120 m*120 m rectangular area. The radio range is set to 10 m. Additionally, the network contains a single reference node that keeps accurate time. This reference node is placed in the center of the rectangular area. All nodes in the network are aware of their own locations, the location of the reference node and the locations of their single-hop neighbors. Location information is used only to construct the spanning tree for multi-hop synchronization and to route synchronization requests toward the controller. In the simulation, the depth-first search algorithm is employed to construct the spanning tree.

The simulation was executed for 36,000s or 10 hours. A very simple channel model is employed. At each attempted packet transmission, independent of all other communication attempts, the success probability for a packet transmission to a neighboring node that is within radio range is Bernoulli with parameter p. In the simulation p is either 0.95 or 0.65.

The sender and receiver delay is modeled as a Gaussian variable with a mean of 0.0001 seconds of and a standard deviation of 11 microseconds. The required accuracy is 0.5 seconds. The drift of the clocks is 50 ppm, which is typical for quartz crystals. The required accuracy combined with the drift and synchronization accuracy results in an average inter-synchronization time of about 1000 seconds.

Results. First the efficiency of the synchronization algorithms in terms of the number of pair-wise synchronizations required to keep the network synchronized is investigated. Figure 6 shows the total number of pair-wise synchronizations for all nodes in the network using different algorithms. When all nodes are participating the centralized solution has the fewest number of

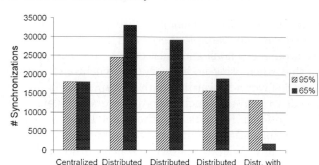

Fig. 6. Number of synchronizations for different algorithms and channel qualities

synchronizations. The 18000 synchronizations that occur in the centralized algorithm translate, on average, to 36 synchronizations for each node over the course of 10 hours.

Depending on the channel quality the distributed algorithm adds about 40%–100% overhead compared to the centralized algorithm. Adding options (algorithm enhancements described in Sect. 3.4) reduces the overhead of the distributed algorithm to 15%–60%. The channel quality (packet transmission success rate) affects the number of synchronizations performed by distributed algorithms greatly because the synchronization requests take longer paths. This leads to an increase the depth of the synchronization trees. As the depths of the synchronization trees increase more frequent synchronization is required. The increase the average depth of the synchronization tree can be seen in Fig. 7.

When only 60% of the nodes are participating the distributed algorithm with enhancements significantly outperforms the other solutions, especially if the channel is good. The use of the distributed algorithm is justified if only a portion of nodes is participating in synchronization.

Next, the accuracy of synchronization was investigated. Figure 8 shows the average accuracy, (offset from the reference time) directly after synchronization, as a function of the depth at which the synchronization occurred. The expected linear increase is evident in the figure, confirming that the accuracy of synchronization deteriorates linearly as the hop-distance from the reference node increases.

In addition to post-synchronization accuracy, the offset or error before synchronization is an important parameter. The offset before synchronization gives a bound on the maximum inaccuracy of the nodes. The average offset before synchronization is shown in Fig. 9. The average offset peaks at about 0.4 seconds, which is within the accuracy bound of 0.5 seconds. In the distributed case (especially when the enhancement options are used) the average offset for nodes close to the controller is smaller. In the distributed

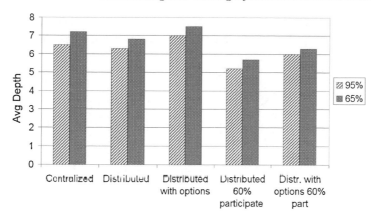

Fig. 7. Average depth of synchronization tree for different algorithms

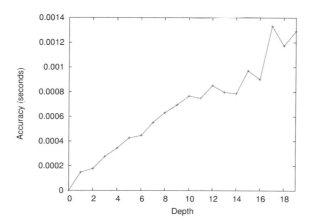

Fig. 8. Accuracy of synchronization as a function of node depth in the tree

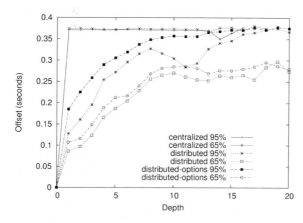

Fig. 9. Average time offset before synchronization as a function of node depth in tree

case the nodes near the controller synchronize more frequently because they can have many separate synchronization requests from more distant nodes.

4 Summary

In this chapter, a number of techniques to add a space and time dimension to the Ambient Intelligence experience were discussed. In general, it is fair to state that this can be done at a reasonable expense in communication and computational resources. Yet, issues such as accuracy, robustness and reliability remain. There is definitely room for further exploration in this arena, which is a sense, is a true example of collaborative distributed computing.

References

1. Zigbee: www.zigbee.org.
2. S. Klemmer, S. Waterson, and K. Whitehouse "Towards a Location-Based Context-Aware Sensor Infrastructure".
3. L. Doherty, K. Pister, and L. El Ghaoui, "Convex Position Estimation in Wireless Sensor Networks," in IEEE Infocom 2001, Anchorage, AK, April 2001.
4. S. N. Simic, S. Sastry "Distributed Localization In Wireless Ad Hoc Networks".
5. D. Niculescu and B. Nath, "Ad-hoc Positioning system," IEEE GlobeCom, Nov 2001.
6. N. Bulusu, J. Heidemann, V. Bychkovskiy, and D. Estrin, "Density-Adaptive Beacon Placement Algorithms for Localization in Ad-hoc Wireless Networks," in *IEEE Infocom 2002*, New York, NY, June 2002.
7. X. Nguyen, M. Jordan, and B. Sinopoli, "A Kernel-based Learning Approach to Ad-hoc Sensor Network Localization" class project, UC Berkeley.
8. N. Christianini, J. Shawe-Taylor *An Introduction to Support Vector Machines: & Other Kernel-based Learning Methods* Cambridge University Press March, 2000.
9. C. Savarese, J. Rabaey, and J. Beutel, "Locationing in Distributed Ad-hoc Wireless Sensor Networks," in IEEE International Conference on Acoustics, Speech, and Signal Processing (ICASSP), pages 2037–2040, Salt Lake City, UT, May 2001.
10. Savvides, H Park, and M. Srivastava, "The Bits and Flops of the N-hop Multilateration Primitive For Node Localization Problems" Proceedings of the First ACM International Workshop on Sensor Networks and Applications, Atlanta, Septmber 28, 2002.
11. Kalman, A New Approach to Linear Filtering and Prediction Problems, Transactions of the ASME–Journal of Basic Engineering, 82, 35–45, 1960.
12. D. Mills, Network Time Protocol (Version 3) Specification, Implementation and Analysis, from *http://www.faqs.org/ftp/rfc/rfc1305.pdf*.
13. J. van Greunen and J. Rabaey, Lightweight time Synchronization for sensor networks 2nd ACM international conference on Wireless sensor networks and applications, San Diego, CA, September 2003.

14. J. Elson, L. Girod, and D. Estrin, Fine-Grained Network Time Synchronization using Reference Broadcasts, *Proceedings of the Fifth Symposium on Operating systems Design and Implementation,* Boston, MA. December 2002.
15. M.L. Sichitiu, C. Veerarittiphan, Simple, Accurate Time Synchronization for Wireless Sensor Networks. *IEEE Wireless Communications and Networking Conference,* 2003.
16. Saurabh Ganeriwal, Ram Kumar, Sachin Adlakha and Mani Srivastava, "Network-wide Time Synchronization in Sensor Networks," *Technical Report* UCLA, 2002.
17. J. Elson and K. Römer, Wireless Sensor Networks: A New Regime for Time Synchronization, *Proceedings of the First Workshop on Hot Topics In Networks (HotNets-1),* Princeton, New Jersey. October 28–29, 2002.
18. H. Kopetz, W. Schwabl. Global time in distributed real-time systems. *Technical Report 15/89,* Technishe Univesität Wien, 1989.
19. Warneke, B. Atwood, K.S.J. Pister, Smart Dust Mote Forerunners, *Proceedings of the Fourteenth Annual International Conference on Microelectromechanical Systems (MEMS 2001),* Interlaken, Switzerland, January 21–25, 2001, pp. 357–360.
20. E. Anceaume and I. Puaut, A Taxonomy of Clock Synchronization Algorithms, *Research report IRISA, NoPI1103,* July 1997.
21. B. Awerbuch, A new distributed depth first search algorithm, *Inf. Proc.* 1985.
22. A. Varga, "The OMNeT++ Discrete Event Simulation System," in European Simulation Multiconference (ESM'2001), Prague, Czech Republic, June 2001.

Security for Ambient Intelligent Systems

I. Verbauwhede, A. Hodjat, D. Hwang, and B.-C. Lai

1 Introduction

Providing security in a distributed, ambient intelligent system is a huge challenge. The main reason is that the traditional security model is not valid anymore. Traditional security assumes that there is a vulnerable channel between communicating parties, where eavesdropping, modification of messages or denial of service attacks can occur. But it also assumes that the sender and receiver operate in some form of secure environment. Hence all models of attacks focus on the channel. Due to the distributed nature of ambient intelligent systems, the attack can be anywhere on the communication channels and on the devices. The attacker has the advantage that he can choose the easiest entry.

For instance, a traditional concept like a firewall assumes that there is a trusted "inside" and a distrusted "outside" with a clear boundary between the two. In an ambient intelligent system, there is no inside and outside. Both the user and the attacker are inside the system.

Similar arguments can be made for Virtual Private Networks (VPNs). VPNs are a sophisticated layer of software on top of existing infrastructure, like the Internet, to allow sender and receiver to communicate in a secure way. It typically provides end-to-end privacy and authentication for e-commerce, financial transactions, confidential information exchange in business environments, etc. It does not address the fundamental property of ambient intelligent systems, namely that the information is distributed in the system and not centralized in the endpoints that communicate, like a customer on a home PC talking to a bank. Even if every node could be provided with VPN software, it will not work, because its computation requirements are too high, and it would deplete the limited energy supply of typical nodes in an ambient intelligent system. Secondly, it does not protect against denial-of-service attacks and sleep-deprivation attacks, two possible attacks in ambient intelligent systems. It is also too expensive for many typical set-ups of ambient intelligent systems. For instance, individual monitoring nodes, such as for temperature or seismic activity, do not contain a lot of useful information and the individual sensor readings do not need privacy protection as provided by VPNs. It is the combined knowledge of a large set of nodes or a whole area that provides useful information, such as to track activity in the environment. The privacy of that information is not even protected by a traditional VPN.

Hence it is our opinion that security in an ambient intelligent system and in sensor networks is a *system design* problem. Security is difficult to provide, because a system is as secure as its weakest link. The attacker has the advantage that he can choose his entry point. Hence in an embedded context, all levels of abstraction need to be investigated regarding security. This includes system and protocol level, algorithm level, architectural level and physical level. We will give examples of each of these levels and illustrate this with examples.

2 Overview of the Chapter

Representing a domain by an abstraction pyramid is a useful tool to detect the layers of abstraction and the interaction between the different components [14,26]. In [14], a hierarchy of Y-charts is used to reason about modeling effort, evaluation speed and accuracy of the evaluation. Here we use it to show the different layers of abstraction involved for embedded security. It is illustrated in Fig. 1. It is well known that design decisions at the top of the pyramid have large consequences for the realization of the system in general. We will

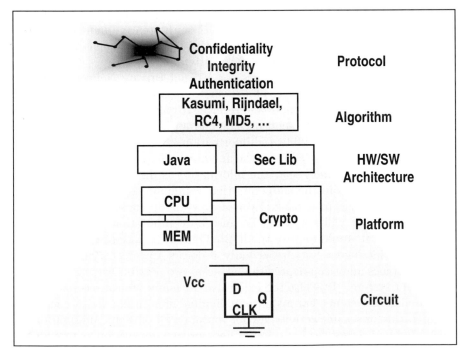

Fig. 1. Embedded security pyramid

use this pyramid in this chapter, to discuss different topics associated with the security design problem. It will also be used to illustrate the interaction and interdependencies between the different layers of abstraction.

In Sect. 3, we start with an overview of some known security attacks and basic security requirements. In the next section, the protocol level is addressed. Energy evaluations of existing protocols are made and new scalable key distribution protocols are proposed. In the Sect. 4, the algorithm level is discussed. This includes an energy comparison between different public key and secret key algorithms. The architecture level is discussed in Sect. 5. Secure physical implementation is addressed in Sect. 6. The conclusions are in Sect. 7.

3 Security Attacks and Security Requirements

The distributed nature of ambient intelligent environments has both advantages and disadvantages when considering its security problem. The advantage is that information is distributed and hence attacking one node does not disclose the complete system. The disadvantage is that every node and the system as a whole need protection. Many types of attacks are possible. To get some insight and classify the attacks, it is instructive to map the attacks and weaknesses onto the security pyramid of Fig. 1.

One example is a denial of service attack by flooding the network with requests. This is not unique to sensor networks. Similar attacks exist in other environments such as the Internet [3, 5, 27]. The main difference is that for sensor networks, energy will be drained and the nodes can die irrevocable while in general networks the performance will drop or stop, but is usually restored after the source of attack is removed. Secondly, the very nature of distributed sensor networks makes that the devices are physically accessible at every node. A distributed denial of service attack needs to be addressed at the protocol level with the assumption that a certain percentage of the nodes might be compromised. Measures like event counters, introduction of trust levels, are means to address this problem. Any protocol needs to be investigated for its computation resources, i.e. energy and memory requirements, and for its communication requirements, i.e. radio transmission and reception.

At the algorithm level, one should choose strong cryptographic algorithms that at the same time can be implemented on the limited resources available in the network. For instance, RC4 requires few computation resources, yet weaknesses have been found [7]. For resource constrained devices, there is going to be a computation, security, communication, trade-off.

There can be attacks on the communication channels. In a wireless environment it is reasonably easy to place a jamming device in the proximity of the sensor network. If no security is provided, this jamming device can block any communication through this wireless network (= denial-of-service attack). Addressing this problem might require solutions at the protocol level, such as re-routing information, or introducing sufficient redundancy in the network down to solutions at the radio level, such as using a frequency hopping system. Another attack on the communication channel is eavesdropping, which is fairly easy in wireless networks. This means that without security an adversary could easily extract useful information from conversations between nodes. Even if the conversation is encrypted, the fact that there is activity might disclose some information. Eavesdropping again needs to be addressed at all levels: the protocol level might decide to introduce some fake activity. The individual links can include encryption of the links. Wood and Stankovic [36] also enumerate a series of attacks on the communication channels. The attacks are classified based on the traditional network layers: physical layer attacks (e.g., jamming), link layer attacks (e.g., interrogation attack), network and routing layer attacks (e.g., misdirection) and transport layer attacks (e.g., flooding).

There can also be attacks on the individual nodes themselves. One example is battery power exhaustion. Battery life is the critical parameter for the nodes and many techniques are used to maximize it; in one technique, for example, nodes try to spend most of the time in a sleep mode in which they only turn on the radio receiver, or even the processor, once in a while. In this environment, energy exhaustion attacks are a real threat: without sufficient security, a malicious node could prohibit another node to go back to sleep causing the battery to be drained. This "sleep deprivation torture attack" is also a type of denial-of-service attack [28, 29].

The so-called side-channel attacks are another attack on the nodes [15,16]. The nodes are observed while in operation and the timing, power or electromagnetic variations are measured. This leakage of information through side-channels is a consequence of the energy dependency of the calculations on the data. Thus all techniques to address side channel attacks should address the issue of masking the operations or making the energy consumption and execution times of sensitive calculations independent of the actual data.

The distributed nature of the ambient intelligent system also requires a new approach in addressing the privacy issue. Individual sensor nodes might not carry much data, such as simple temperature measurements. Therefore, both the trust level (i.e. what is the error margin on this temperature?) and the individual security risk (i.e. has the sensor node been tampered with) are considered low. Useful information is obtained by combining the data of a subset of sensor nodes. A friendly monitoring node can collect the data from neighboring sensor and actuator nodes and extract useful information. This will increase the trust level and reduce the tampering risk but it does not address the privacy issue. It actually aggravates the privacy issues. Indeed, the intruder can also collect the data and process it on its own processors to extract useful information. It is even possible that the intruder has a larger energy supply to perform the calculations than the friendly monitoring nodes.

The above classification of attacks is only a first attempt in classifying the problem. An important warning is that the attacker is very creative and will find yet another way to attack the system. As engineers, we have to make it difficult and cost inefficient to attack the system, as perfect security is in general not achievable or not economical.

This relates to the problem of risk management: increasing the security provisions has a price in terms of power consumption, processing speed, ease of use, flexibility and the lifetime of the system. Medical applications will need the highest level of security, while climate control in buildings might settle for a lower level of security (it might only result in discomfort of the occupants). Yet, monitoring for hazardous materials (e.g., biological or chemical weapons) will again require the highest level of security for authentication purposes. As with any security application, risk management includes weighing the extra costs against the potential risks and damages. On top of this, risk management should not be statically decided but should be dynamic during operation: it will need to include some negotiating protocol to decide on the key length, encryption algorithm, etc. This decision can be based on the required level of security, available energy, CPU power and available bandwidth (~bit rate).

3.1 Fundamental Security Concepts

As is clear from the above examples, the basic security requirements are still present, albeit in a different form. The basic security requirements are [19]: confidentiality, data integrity, authentication and non-repudiation. For distributed systems, availability is added to this list.

Availability means ensuring that the service offered by the node will be available to its users when expected. As mentioned above, the sleep deprivation torture is a real threat and has to be prevented.

Authenticity of origin (access control) is ensuring that the principals with whom one interacts are the expected ones. In most security sensitive applications, authenticity is essential. Granting resources to, obeying an order from, or sending confidential information to a principal whose identity is unsure is not the best strategy for protecting the other security properties. Assuring correct authentication is the most challenging task in an ad-hoc environment, due to the absence of an online server. When a new node comes within range, it cannot connect to an authentication server (like in the Kerberos system) to check the validity of a ticket or certificate: the traditional solutions no longer apply. Other constraints imposed by power budgets or computation restrictions may also prohibit the use of public-key techniques and certificates.

Besides authentication of the origin, it is also necessary to guarantee the authenticity (or integrity) of the data. We need a means to assure that the data we receive is valid (say temperature, location, humidity or toxin level) and fresh (no replay of old data). An adversary could, for example, maliciously alter a sensor node to send out incorrect values. This can be avoided in theory by making the nodes tamper-resistant, something easier said than done, especially within the cost constraints of these small sensor nodes. The redundancy of the system should help in detecting "false nodes" because they are inconsistent with the surrounding nodes. Non-repudiation insures that once an entity sends a message, it cannot later claim to have not sent the message. This issue becomes important when a dispute arises and one entity denies that certain actions were taken. In this case some means of resolving the situation is necessary.

Confidentiality or privacy is a matter of encrypting the messages with a key that is usually made available by the authentication process. So the real issue is authentication: it is pointless to attempt to protect the secrecy of a message without first ensuring that one is talking to the expected principal.

Other objectives, such as anonymity, certification, receipts, etc. can be derived based on the previous security goals.

3.2 Protocol Level

At the top level in the security pyramid is the choice of a security protocol. A protocol describes a series of steps to be executed by two or more participants in the network to obtain a particular set of security goals.

The unique features of an ambient intelligent system need to be taken into account during the development of these protocols. First, both the computation and the communication energy need to be taken into account. This is different from traditional environments where only computation cost is taken into account and exchange of data items is considered "free". Secondly, nodes need to establish secure connections with neighbors in an ad-hoc fashion without the availability of trusted secure servers. One cannot assume that every node has a direct connection to a base station or a more secure node.

In this section, first a series of energy measurements are shown for existing protocols. This is interesting for two reasons. First, the energy of the underlying cryptographic algorithms is examined. And secondly, the radio transmission energy is taken into account. In the second section, new proposals for key distribution protocols are made that target the distributed ad-hoc nature of ambient intelligent systems.

3.3 Energy Cost of Existing Key Set-Up Protocols

Traditional key-set up protocols are either based on public-key algorithms or secret key algorithms [19]. It is important to look at these from a computation-communication trade-off. Public-key based protocols tend to have less communication traffic, but more computation cost. The opposite holds for secret-key based protocols.

To investigate this, we implemented two distinct existing key-exchange protocols on the WINS sensor node [35], which includes a StrongArm processor. The protocols are compared based on energy consumption for both the computation part and the communication (i.e., radio transmission) cost. The first protocol is the key agreement part of the Kerberos protocol [30] listed as protocol 12.24 in [19]. It assumes the availability of a central trusted server and it uses a secret key algorithm, in this experiment the AES algorithm. The second is the public key Diffie-Hellman key agreement protocol. The basic version of the Diffie-Hellman key agreement protocol is adapted from protocol 12.47 of [19]. We base it on the Elliptic Curve public-key encryption technique (ECC) [IEEE99] because it is the most promising public key algorithm in terms of low energy.

The power consumption of the radio transmission on the WINS node is calculated using the expressions presented in [24]. The radio's power consumption varies between 396 to 711 mW depending on the transmission power level. This corresponds to a consumption of 771 to 1080 mW for the whole sensor node. The power consumption of the receive mode is 376 mW for the radio and 751 mW for the whole node. All these numbers are at a transmission rate of 100 kbits/s.

3.3.1 Diffie-Hellman Key Agreement Protocol (Public Key Based)

This protocol consists of the following basic steps:

Set-up phase: a common elliptic curve and a specific point P on the curve are chosen and publicly know. Alice and Bob each generate a random initial key a and b, which they keep secret. Alice calculates $a.P$ and Bob calculates $b.P$. These are point multiplications on the elliptic curve.

Protocol messages:

$$A- > B : a.P$$
$$B- > A : b.P$$

Protocol calculations: A receives $b.P$ and calculates $a.b.P$, B receives $a.P$ and calculates $b.a.P$. Thus both can generate a shared key $a.b.P$, but the eavesdropper E, only knows $a.P$ or $b.P$. This information is not sufficient to calculate $a.b.P$. The strength of the elliptic curve cryptography is based on the difficulty of the discrete logarithm problem: knowing the curve and the Point P, it is practically infeasible to obtain the secret value a or b. In this protocol, each participant makes two point multiplications, one transmission and one reception.

3.3.2 Kerberos Key Agreement Protocol (Secret Key Based)

Alice and Bob want to set-up a secure communication channel. Initially they do not share a secret, but both share a secret key with Trent, the trusted server. E is a symmetric encryption algorithm, NA is a nonce chosen by A; TA is a timestamp from A, K is the session key chosen by Trent, to be shared by Alice and Bob. L indicates the lifetime. The protocol runs as follows:

Set-up phase: A and T share a key KAT; similarly, B and T share a key KBT.

Protocol messages:

$$A->T : A, B, NA$$
$$T->A : E_{KAT}(K, L, T, B, NA), E_{KBT}(K, L, T, A)$$
$$A->B : E_{KBT}(K, L, T, A), E_K(A, T+1)$$
$$B->A : E_K(T+1)$$

Protocol calculations: Alice and Bob want to agree on a common session key. At set-up, both receive a secret key to communicate securely with Trent. Alice initiates the protocol by sending a message to Trent asking for a secret session key. This message includes Alice's and Bob's identities and a nonce NA. Trent generates the random session key K and creates two messages. The first is encrypted with KAT, the second with KBT. The first message for Alice includes the secret K, along with timestamp T, lifetime L and Bob identity and repeats the nonce NA. The second message for Alice contains the secret key K, lifetime L and timestamp T and the identity of Alice. Alice will receive this message but she cannot decrypt it: she will pass it to Bob in the next step.

Both messages are sent to Alice. Alice can decrypt her received message to recover K. She checks if the identity of Bob and the nonce NA corresponds to what she had sent before. Then she makes a message including her identity and a new timestamp $T+1$ and encrypts it with the shared key K. This results in $E_K(A, T+1)$. She sends both E_{KTB} (K, L, T, A) and $E_K(A, T+1)$ to Bob. Now Bob can decrypt the received messages and recover K. He also verifies Alice's identity, because it sits in both messages. He can also verify the timestamps and the lifetime, and then forms a message with $T+1$ and encrypts it with K and sends it back to Alice. Alice decrypts this message and verifies the timestamp $T+1$. Now Alice and Bob share a secret common session key K.

In this protocol there are four data encryptions on the transmitted data and four decryptions on the received data. The total number of transmissions and receptions is six each.

The results of these protocols on the WINS sensor node are summarized in Table 1.

Although the communication energy is higher in the Kerberos protocol, overall the total energy is still an order of magnitude lower power than the Diffie-Hellman protocol. The main reason is that the elliptic curve public key algorithm requires 2 to 3 orders of magnitude more energy to perform one point multiplication compared to one secret key AES encryption. The 300 mJ for one point multiplication reported in Table 1 shows the average for a 128 bit session key. It increases with one order of magnitude for a 256 bit session key. (More details on energy numbers for specific algorithms are in the next section.) One could argue that a dedicated ASIC could save another

Table 1. Energy comparison between key exchange protocols

	Diffie-Hellman with ECC, 128 bit, nodes A, B		Kerberos with AES, 128 bit, nodes A, B, T	
Computation	4 point multiplications	4×300 mJ	4 encryptions 4 decryptions	4×3.4 mJ 4×4.04 mJ
Communication	2 transmits 2 receptions	8.1 mJ 5.6 mJ	6 transmits 6 receptions	6×10.8 mJ 6×7.5 mJ
Total		1214 mJ		140 mJ

order of magnitude for the computation power. Yet, this applies to both algorithms. It is only when the transmission energy becomes in the same order as the energy of the public-key algorithms, that the number of transmissions becomes important. However, distributed systems, like ambient intelligent networks, don't use large transmission ranges instead they use multi-hop systems to relay the information over longer distances [22].

Thus this experiment clearly indicates that there is a request for low energy public key algorithms. This is however an idle demand because the security of public key algorithms is based on the inherent computational complexity of the algorithms and the computational complexity is directly related to energy dissipation. One approach would be to develop domain specific co-processors for certain public key algorithms, because it is proven that dedicated cores can save orders of magnitude on power [13,25]. Another approach would be to develop asymmetric public-key algorithms, where one side, the resource constraint node, needs to perform only limited amount of computations. Unfortunately, this might not work in an ambient intelligent ad-hoc environment, where it is assumed that the nodes are peers and operate under similar conditions.

3.4 Energy-Scalable Key Distribution Protocols

As is clear from the previous section, public-key algorithms are extremely energy hungry. This is a fundamental property of the algorithms. Yet, they are scalable and are sometimes proposed for ad-hoc networks [10].

Secret-key algorithms can be made energy efficient. SPINS is one example of a secret-key based security protocol [21]. It includes two symmetric-key cryptographic protocols, μTESLA and SNEP. μTESLA uses a timed hash-based mechanism for authenticated broadcast from base stations (super nodes) to nodes. SNEP relies upon a shared secret key between each node and a base station, and performs a Kerberos like protocol for key establishment between two nodes, providing for data confidentiality, authentication, integrity and freshness [21]. The disadvantage of using a Kerberos-like system is that it cannot be scaled. Another symmetric key scheme is presented in [2]. In this scheme, all nodes share an initial universal key. This key is used as a root to generate other keys, such as the so-called universal traffic encryption, which is the key that nodes use to communicate with each other in the network. This scheme assumes that the nodes are tamperproof, and if one node is compromised, the whole system is.

In this section, we introduce a new energy-scalable key establishment protocol, called cluster key grouping [11]. The goal is to take the resource limitations into account and to set up a framework, called the security-memory-energy curves, to evaluate and quantify the multi-metric tradeoffs involved in security design.

In cluster key grouping, a key pool of P keys is generated off-line as shown in Fig. 2. Prior to deployment, each node is programmed with C clusters of keys, each cluster having a width of W keys per cluster; the total number of keys stored in each node is $K = W.C$. All nodes are programmed with the same number of clusters C, with each cluster having the same width W. The key ring model presented in [6] is the specific case where $W = 1$ and thus $K = C$, as shown in Fig. 2b. In the scenario in Fig. 2c, each node is programmed with $C = 2$ clusters, each of $W = 4$ keys; in the scenario in Fig. 2d, each node is programmed with only one large cluster of width $W = 10$.

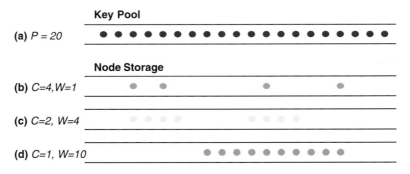

Fig. 2. Cluster key grouping pre-deployment scenarios

Upon deployment, each node broadcasts the starting address of each of its C clusters. The remaining $W - 1$ addresses of each cluster are not broadcast since they are implicitly known from the starting address. If two nodes share at least one key between them, then a connection can be established based on the shared key and a secure link is said to be formed between them.

The probability that the entire network is securely connected is related to Pc, the graph connectivity. The graph connectivity is a function of P, the probability two nodes share at least one key between them, also called the overlap probability. Given N nodes in a network, the desired graph connectivity can be calculated using the equations [6]:

$$P_c = \lim_{N \to \infty} \Pr[G(N,p) \ is \ connected] = e^{e^{-c}}$$

where c is a real constant, and $p = \ln(N)/N + c/N$. For an N-node network, a P can be specified to meet the required graph connectivity.

3.4.1 Effects of C and W on p

For a specified overlap probability p, we now investigate how C and W are used to generate p. p is derived as a function of the number of clusters C as [11]:

$$p = 1 - \left(\frac{P - 2\,CW + 1}{P - CW + 1}\right)^{2C} \cdot \frac{(P - CW + 1)^2}{P \cdot (P - 2\,CW + 1)} \, .$$

This equation can be used as follows: a desired probability of overlap p is given as a specification. For a certain cluster size C, this equation determines the cluster width W required to obtain the desired p.

Figure 3 shows the total number of keys required ($K = C.W$) for different C values, given $P = 10,000$ and the specified $p = 0.4$. At one extreme is the case presented in [6], where $W = 1$ and hence $K = W.C = C$, as in Fig. 2b. By spreading the keys evenly throughout the entire key space, the lowest $K = 51$ total number of keys is required to obtain the specified p. At the other extreme is the case where only one wide cluster is stored, hence $C = 1$ and $K = C.W = W$, as in Fig. 5d. Since all the keys are forced into one cluster and are not spread over the key pool, this case requires the highest $K = 2001$ total number of keys to obtain the specified p.

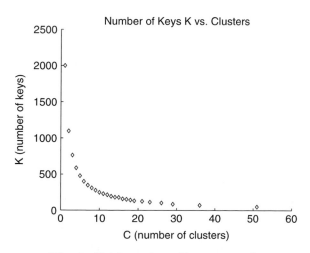

Fig. 3. Total number of keys per node

3.4.2 Energy-Memory Tradeoffs

The cluster-key grouping protocol is energy-scalable. By this we mean that the transmission energy expended to complete a key agreement varies with the value of C. Upon deployment each node must broadcast its C starting addresses to its neighbors at a cost of Eb Joules per bit, requiring a total transmission energy of:

$$energy = C \cdot E_b \cdot \lceil \log_2 P \rceil \ [Joules] \ .$$

It is clear that the required transmission energy increases linearly as C increases. However, by choosing different values of C, the memory requirements of the security architecture are also affected, but with an opposite trend. In this sense, the protocol can also be considered memory-scalable. Each node requires memory to store $K = C.W$ total keys, each of keysize bits (i.e. 64, 128, etc.). Each node also must store the starting address of each of C clusters, with each address requiring $\lceil \log_2 P \rceil$ bits of memory. Hence the total memory requirement of each node is:

$$memory = K \cdot keysize + C \cdot \lceil \log_2 P \rceil \ [bits] \ .$$

The memory requirements fall quickly as C increases (as can also be seen from Fig. 3). Thus for a specified overlap probability p, though energy requirements increase as C increases, memory requirements decrease as C increases. This leads to a tradeoff between the two physical metrics called the weighted memory-energy curve, which is the memory multiplied by the energy, as shown in Fig. 4.

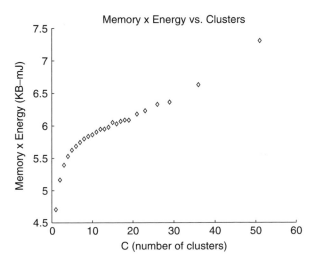

Fig. 4. Memory – energy tradeoffs in cluster key grouping (to achieve $p = 0.4$)

3.4.3 Security Tradeoffs

The above tradeoffs involve only the physical metrics of energy and memory. The question that remains is how security robustness is affected by such tradeoffs. Since security is an abstract concept, we must first formulate a "metric" for security that can be then traded off with the physical metrics. This metric will be called the security leakage factor, which quantifies the significance of a security breech to a system.

We begin by introducing the notion of the compromise factor, which is the number of keys compromised if a single node is compromised divided by the key pool, hence *compromise factor* $= K/P$. Clearly, the greater the key spread over the key pool (i.e. the larger the number of clusters C), the fewer total keys K are required, creating a lower total compromise factor. In order to quantify the effects of such a compromise on the security architecture, the security leakage factor is defined. The security leakage factor (SLF) is a function of the compromise factor:

$$security\ leakage\ factor = 1 + s_w \cdot compromise\ factor$$

where s_w is the security weight, which is any natural number. Security weight can be any positive real number, but for the sake of illustration natural numbers are used. As stated earlier, the security leakage factor – and the security weight in particular – attempts to roughly quantify the importance a security compromise has to a network. For example, if a network for whatever reason is not affected by a compromise or assumes that one cannot be made, then $s_w = 0$ would be assigned. If a network is extremely sensitive to key compromise, then a higher security weight (e.g., $s_w = 5$) would be assigned.

With a security metric in place, a comparison framework can be defined which quantifies the tradeoffs between security, memory, and energy. We call this framework the SME curve, which is defined as security leakage factor · memory · energy. Figure 5 shows SME curves for different security weights. As can be seen, depending on the security weight, there is a minimum of the curve along a particular number of clusters C.

The case of $s_w = 0$ (no security effects) is the memory-energy curve mentioned earlier, whose minimum is at $C = 1$. When security has a low priority ($s_w = 1$), the SME follows a curve similar to the weighted memory-energy function. However, when security leakage is more important ($s_w = 3$), then the curve alters into a reverse bell with a minimum at $C = 13$ clusters. This indicates that a network can have a minimal loss of security coupled with minimal energy consumption and memory requirements by choosing this cluster size. The sudden increase in the metric at smaller values of C is due to the increased importance of the compromise factor, which is large for small cluster numbers. (This characteristic increase begins at the security weight of approximately 1.8). At a security weight of $s_w = 5$ the minimum of the curve shifts to the right (towards a lower compromise factor) and hence a minimum is achieved at $C = 19$ clusters. As s_w increases to even larger

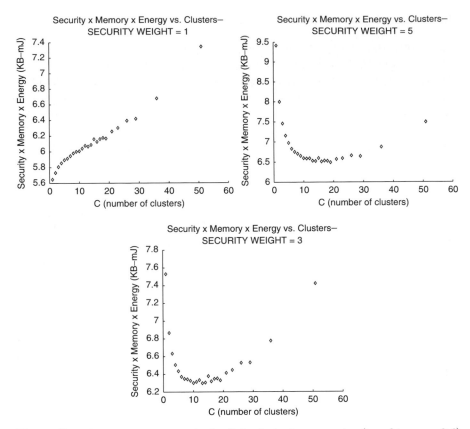

Fig. 5. Security-energy-memory tradeoffs in cluster key grouping (to achieve $p = 0.4$)

values, the "tail" of the curve occurring at larger values of C continues to decrease until eventually the curve mimics the compromise factor itself; for $s_w = 65$ and greater, the minimum SME is always at $C = 51$ clusters.

Therefore, by designing for the minimum of the SME curve instead of the minimum of energy curve, tradeoffs between energy, memory, and security can be taken into account. Though the security leakage factor is specific to the cluster key grouping protocol, it demonstrates the notion that security can be quantified as a metric, which can then be used to perform tradeoffs with traditional metrics such as energy, memory, processing latency, etc.

4 Algorithm Level

Cryptographic algorithms are the building blocks of protocols. Together with low energy protocols, energy efficient implementations of cryptographic algorithms are needed. This requires a comparison between algorithms: public-key and secret key. This relates to two problems: how to compare the security strength of algorithms and secondly, how does the energy consumption vary with the word lengths (for data as well as for keys).

One interesting starting point to compare the computational security of crypto algorithms is the study of Lenstra and Verheul [17]. They derive lower bounds to obtain computationally equivalent key sizes for different algorithms. Security, computationally equivalent to the security offered by DES in 1982, is obtained by using, in the year 2004: symmetric keys of at least 73 bits, RSA moduli of at least 1108 bits and elliptic curve systems over prime fields of at least 138 bits, assuming no cryptanalytic progress will take place. If some data needs to be kept secure for a couple of years (e.g., till the year 2020) key lengths need to be increased accordingly.

It is clear from this study that within the category of public key algorithms, the Elliptic curve public key algorithm requires less energy than the RSA algorithm. Within one class of algorithm there is again a security energy trade-off. I.e. larger word lengths increase the security but will need extra energy to perform the calculations. It is useful to quantify these statements.

In a first experiment, we will quantify the increase of the energy cost with the word lengths (both in input data and key lengths) for a public key algorithm, elliptic curve, and a secret key algorithm, AES. The elliptic curve is chosen because it is most promising for low energy consumption. The AES is chosen because it will be the most widely used secret key algorithm for the foreseeable future.

The AES algorithm is the latest NIST standard for secret key algorithms [4]. It is based on the Rijndael algorithm and it operates on 128 bit data and a choice of 128, 192 or 256 bits of key. The original Rijndael also allows a choice of input widths between 128, 192 or 256 bits. The energy requirements for an implementation on the StrongArm of the WINS sensor node (discussed in a previous section) are shown in Fig. 6. The key schedule varies between 0.11 and 0.32 mJ, the encryption between 0.20 and 0.54 mJ and the decryption between 0.25 and 0.70 mJ. The variation is due to the difference both in word lengths and in the number of encryption rounds. The number of encryption rounds varies between 10 and 14 and grows with the key and data size. The difference between encryption and decryption energy is a result of a larger numbers of shifts in the shift row operation and the larger $GF(2^8)$ elements used in the mix column transformation.

As is clear from Fig. 6, the variation in energy requirements in limited to no more than a factor 3 between the lowest energy and the highest energy combination of key length and data length.

To calculate the energy consumption of the Elliptic curve public-key algorithm, we use the double-add-subtract point multiplication algorithm defined by the IEEE p1363/D1 standard [12]. It refers to calculating $k . P$ where k is the secret integer and P is a point on the Elliptic curve. These calculations are executed in a Galois Field, in this case GF(2^n). In this experiment,

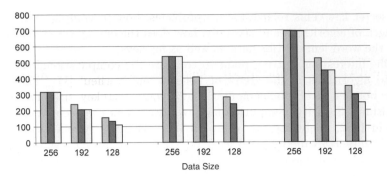

Fig. 6. The energy cost of the Rijndael encryption algorithm on the WINS sensor node, (**a**) key scheduling, (**b**) encryption, (**c**) decryption energy

n is chosen to be 128, 192 or 256 bits. The number of operations and the energy consumption depend on the size and the number of non-zero bits in the secret integer k. On average, the energy consumption is 0.3, 1.07 and 2.34 Joules for the 128, 192 and 256 bits Galois field. This is almost a factor 8 different between a 128 bits field and a 256 bits field. Within one Galois field size, the energy also varies. For a 256 bits field, the energy varies from 0.6 to 5.12 Joules, or almost a factor 9 depending the size of the secret key k.

Please note that this is an energy efficient implementation, but it is prone to side-channel attacks. An attacker could monitor the execution time and guess from that the size of the secret key. To make the algorithm resistant to side-channel attacks, its execution time and its energy profile should be made independent from the secret key k. Algorithmic masking techniques to make the execution time independent of the data values, will mostly result in a higher power consumption.

Comparing a secret key algorithm, AES with 128 bits key and 128 bits data (which is above the recommend 74 bits), to an Elliptic curve algorithm in $GF(2^{\wedge 128})$ (which is less than the recommend 138 bits) shows that secret keys algorithms are at least 3 orders of magnitude more energy efficient than public key algorithms.

5 Architecture Level

Energy efficiency is the main driver for the selection of a hardware architecture for the nodes in an ambient intelligent system. Instead of one central processing unit, embedded systems use a heterogeneous collection of dedicated processing units, connected together by a flexible interconnect system. One example is the Maia platform [22], another example is the RINGS architecture [26]. Dedicated units are also beneficial for cryptographic algorithms. This is illustrated in Table 2, which shows the energy efficiency of several implementations of the same AES algorithm.

Table 2. AES implementation on different platforms

AES 128 bits key 128 bits data	Throughput	Power	Figure of Merit (bits/Joule)	Relative Merit
018 μm CMOS	2 Gbits/s	56 mW	35.7 Gbits/J	1/1
FPGA	1.32 Gbits/s	490 mW	2.7 Gbits/J	1/10
Asm on Pentium III	648 Mbits/s	41.4 W	15 Mbits/J	1/1900
C on Emb Sparc	133 Kbits/s	120 mW	1.1 Mbits/J	$1/33.10^3$
Java on Emb Sparc	450 bits/s	120 mW	3.7 Kbits/J	$1/10^7$

The metric used is the amount of encrypted bits that can be computed with one Joule. As can be seen from the table, a dedicated AES co-processor is one order of magnitude more energy efficient than the same implementation on an FPGA. The effect of software optimizations can be seen from the other experiments. Hand optimized assembly code [18] is one order of magnitude more energy efficient than compiled C code. Two layers of software, i.e. a Java program executed on a Java Virtual machine KVM, that runs on an embedded Sparc results in another 3 orders of magnitude loss in energy efficiency.

This experiment clearly indicates that for energy efficient cryptographical implementations, one must resort to optimized domain-specific co-processors [26]. These processors are programmable within their application domain. We envision that the SoC will consist of a collection of heterogeneous programmable domain-specific units, connected together by a reconfigurable interconnect paradigm, called the RINGS platform [26]. One generic example is shown in Fig. 7. It shows how an embedded SoC can contain multiple application domains, in this case, a signal processing domain, a communications domain to provide a wireless connection and a cryptographic domain. The systems is connected together by means of a reconfigurable interconnect [33].

One main advantage of this partitioning in physically separate processor units is that it improves security. Indeed one can divide the SoC in secure and non-secure parts or even in parts with different security levels. This cannot be done with software only solutions that run on one embedded processor. In the later case, the system is quite vulnerable to software attacks since the data and program memory are shared between the secure and non-secure processing parts.

However, the design space exploration and programming of a RINGS type SoC is a challenge. For this, specialized design and programming environments, like Gezel [34] are required.

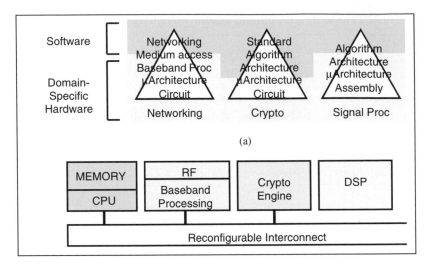

Fig. 7. General RINGS architecture and programming view

6 Physical Implementation Issues

Ambient intelligent systems are very different from regular networked systems and this is certainly the case for their physical protection and integrity. In a regular networked system, one assumes that e.g., the communicating PC's are in a secure place such as an office that can be locked. Distributed systems, like ambient intelligent systems, should assume that the nodes themselves are vulnerable and unattended. Hence, one should look carefully at their physical integrity.

Making the individual nodes tamper resistant [1] is not feasible and not economical for cheap consumer electronic type of applications. But for sensitive monitoring applications, such as the monitoring of airports, bridges and other infrastructure for biological, chemical or other attacks, tamper resistant casings should be considered.

Even when a tamper resistant case is provided, the embedded device might still leak information through the so-called side-channels. The operation of the device is monitored without intrusion and from this, sensitive information can be derived. There are the so-called timing attacks, power attacks, differential power attacks, electro-magnetic attacks. It has been shown that smart cards are very vulnerable for these side-channel attacks [16]. For instance in a simple timing or simple power attack the current drawn by the embedded device is monitored and from that the operation of the device can be guessed. A more powerful attack is the differential power analysis [16].

Many countermeasures have been proposed at the algorithm level, e.g., by the introduction of randomness in the instructions or the use of some masking techniques [8, 20]. However, our idea is to address the problem at the logic level. Indeed, the fundamental reason for the power difference is at the logic level. Static complementary standard CMOS (scCMOS) is the most used circuit style [23]. However it clearly shows the differences between different data inputs and data transitions. The attack is based on the fact that the power consumption of CMOS circuits is proportional to the Hamming distance between consecutive data applied to the system.

In standard complementary CMOS logic (scCMOS), the only transition which causes dynamic power dissipation from the power supply is a 0 to 1 transition. A 1–0 transition causes a stored output capacitance to discharge to ground. During a 0–0 transition or 1–1 transition, no dynamic power is used. To combat this at the circuit level, we have developed a circuit style which has the same dynamic power dissipation regardless of the transition (0–1, 1–0, 0–0, or 1–1), preventing the differential power analysis attack. The circuit style, called SABL (sense amplifier based logic) [31], uses elements of both differential and dynamic circuit styles to form a secure circuit. SABL makes the four output events equal by charging the same capacitance at every event. A sample SABL logic gate that can be used as a NAND or AND is shown in Fig. 8.

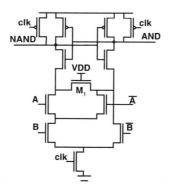

Fig. 8. Dynamic, differential sense-amplifier based logic (SABL) styles

The resulting energy consumption per cycle is shown in Fig. 9 for a typical encryption operation (here, a sample Kasumi S9 box). As seen in the figure, a standard CMOS module widely varies in energy dissipated from 0 pJ to 10.42 pJ per cycle, making it relatively easy to perform a DPA attack. SABL logic dissipates a narrow range of only 11.14 through 11.51 pJ per cycle. Thus, dynamic power *variation* is decreased by 116 times, and the cell essentially dissipates the same energy each cycle, foiling a DPA attack.

However, there are power and area penalties to be paid for this secure circuit technology. Though dynamic power variation is reduced, this comes at a cost of almost doubling average power consumption (11.32 pJ/cycle versus 5.92 pJ/cycle). In addition, cell area is increased by 1.8 times. If the entire

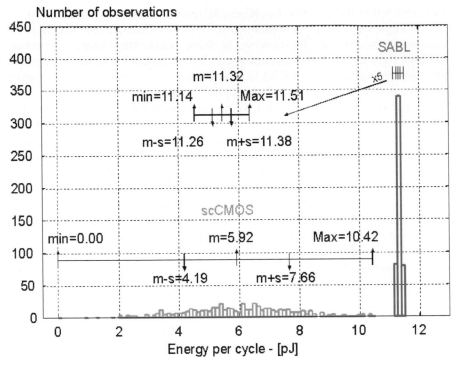

Fig. 9. Energy distribution for scCMOS and SABL logic style

embedded device uses this technology, huge power and area penalties would be paid. However, all the secure functions and sensitive data (that could leak in a DPA attack) reside in a secure module. Hence, security partitioning helps because only the secure portion of the chip needs to be protected by SABL. Table 3 gives an estimated area of a secure embedded device in a 0.18-μm TSMC CMOS technology. It contains an embedded processor, the Leon Sparc and a secure co-processor module. If the entire device is implemented in SABL logic, then the device would be $5.26 \, \text{mm}^2$ (core area). By judiciously partitioning at the architecture and micro-architecture levels, only the co-processor module needs to be secure, resulting in a total area of $3.77 \, \text{mm}^2$, an area reduction of 28%. Hence, by combining the partitioning techniques

Table 3. Area Comparison of scCMOS and SABL

LEON Processor	Secure Coprocessor	Total
$1.86 \, \text{mm}^2$ (scCMOS)	$1.06 \, \text{mm}^2$ (scCMOS)	$2.92 \, \text{mm}^2$
$1.86 \, \text{mm}^2$ (scCMOS)	$1.91 \, \text{mm}^2$ (SABL)	$3.77 \, \text{mm}^2$
$3.35 \, \text{mm}^2$ (SABL)	$1.91 \, \text{mm}^2$ (SABL)	$5.26 \, \text{mm}^2$

at the architecture and micro-architecture levels with security techniques at the circuit level, the entire device can be made robust without wasting area and power dissipation. This again shows that embedded design necessitates consideration of the inter-relationships between security levels.

7 Conclusions

In this chapter, we have given an overview of the security challenges when designing ambient intelligent systems. Security breaches and attacks can occur at all levels of abstraction. Thus the solution will be multi-facetted also. At the protocol level, we have shown that it is possible to develop key distribution protocols that are scalable, energy and memory efficient at the same time. We have proposed a framework that allows to trade-off these conflicting requirements. At the algorithm level, we have shown that public-key algorithms require two to three orders of magnitude more energy than secret-key algorithms. At the architecture level, we have shown that dedicated domain specific co-processors are three orders of magnitude more energy efficient than hand optimized assembly code and five to seven orders of magnitude compared to compiled C or Java code. The proposed RINGS architecture combines flexibility with energy efficiency. At the same time it increases the security by providing a clear hardware partitioning between secure and non-secure parts of the SoC. Distributed systems don't operate in a physically protected environment. Hence they are vulnerable to physical probing attacks, such as side channel attacks. To address some of these issues, we have proposed a circuit style of which the power consumption profile is independent of the data being processed.

Cryptographic techniques and security protocols can and will be developed to address the social, ethical and economical impacts of ambient intelligent systems, as raised in chapter Social, Economic, and Ethical Implications of Ambient Intelligence and Ubiquitous Computing. Security techniques cannot be ignored during the design of ambient intelligent systems and security, energy, performance trade-offs need to be made. However, techniques alone will not address the issues: it will need to be combined with a responsible usage of the new technology. But then this is an issue raised with every new technology.

Acknowledgements

The authors would like to thank the members of Embedded Security group at UCLA (www.ivgroup.ee.ucla.edu). The funding support of NSF, grant CCR-0098361, UC Micro grant 02-079, Panasonic, Atmel and Xilinx is gratefully acknowledged.

References

1. R. Anderson, "Security engineering: a guide to building dependable distributed systems," Wiley 2001.
2. S. Basagni, K. Herrin, E. Rosti, D. Bruschi, "Secure pebblenets," *Proc. 2nd ACM International Symposium on Mobile Ad Hoc Networking & Computing (MobiHOC 2001)*, pp. 156–163, Oct. 2001.
3. R. Comerford, "No longer in denial," IEEE Spectrum, pg. 59–61, January 2001.
4. J. Daemen, V. Rijmen, *The design of Rijndael*, Springer-Verlag, 2002.
5. D. Dittrich, "Distributed Denial of Service (DDoS) attacks/tools resource page," http://staff.washington.edu/dittrich/misc/ddos/, 2004.
6. L. Eschenauer, V. D. Gligor, "A key-management scheme for distributed sensor networks," *Proc. 9th ACM Conference on Computer and Communications security (CCS '02)*, pp. 41–47, Nov. 2002.
7. S. Fluhrer, I. Mantin, A. Shamir, "Weaknesses in the Key Scheduling algorithm of RC4", 8th Annual workshop on selected areas in cryptography, Aug. 2001, LNCS 2259.
8. L. Goubin, J. Patarin, "DES and Differential Power Analysis: The "Duplication" Method," Proc. CHES 1999, LNCS 1717, pg. 158, Jan. 1999.
9. A. Hodjat, I. Verbauwhede, "The energy cost of secrets in Ad-Hoc networks" IEEE CAS workshop on Wireless Communication and Networking, Pasadena, CA, Sept. 2002.
10. J.-P. Hubaux, L. Buttyan, S. Capkun, "The quest for security in mobile ad hoc networks," *Proc. 2nd ACM International Symposium on Mobile Ad Hoc Networking & Computing (MobiHOC 2001)*, pp. 156–163, Oct. 2001.
11. D. Hwang, B.-C. Lai, I. Verbauwhede, "Energy-Memory-Security Trade-offs in distributed sensor networks," Proc. 3rd International Conference on Ad-Hoc Networks and Wireless (ADHOC-NOW 2004), Springer-Verlag LNCS 3158, pp. 70–81, July 2004.
12. IEEE P1363/D1, Standard specification for Public-Key cryptography, November 1999.
13. S. Janssens, J. Thomas, W. Borremans, P. Gijsels, I. Verbauwhede, F. Vercauteren, B. Preneel, J. Vandewalle, "Hardware-software co-design of an elliptic curve public-key cryptosystem," IEEE Workshop on Signal Processing Systems (SIPS2001), Antwerp, Belgium, Sept. 2001.
14. B. Kienhuis et. al, "A Methodology to design programmable embedded systems," LNCS, Vol. 2268, Nov. 2001.
15. P. Kocher, "Timing Attacks on Implementations of Diffie-Hellman, RSA, DSS, and Other Systems," Advances in Cryptology, Proceedings Crypto'96, LNCS 1109, N. Koblitz, Ed., Springer-Verlag, 1996, pp. 146–158.
16. P. Kocher, J. Jaffe, B. Jun, "Differential power analysis," M.Wiener (Ed.), CRYPTO 1999 Proceed-ings, LNCS 1666, Springer-Verlag, 1999.
17. A. Lenstra, E. Verheul, "Selecting cryptographic key sizes," International workshop on Practice and Theory in public key cryptography, PKC2000, Jan. 2000.
18. H. Lipmaa, "AES/Rijndael: speed" http://www.tcs.hut.fi/~helger/aes/rijndael.html.
19. A. Menezes, P. van Oorschot, S. Vanstone, "Handbook of Applied Cryptography," CRC Press, 1997.

20. T. Messerges, E. Dabbish and R. Sloan, "Examining Smart-Card Security under the Threat of Power Analysis Attacks," IEEE Transactions on Computers, Vol. 51, pg. 541–552, April 2002.

21. A. Perrig, R. Szewczyk, V. Wen, D. Culler, J. D. Tygar, "SPINS: Security protocols for sensor networks," *Proc. 7th ACM Mobile Computing and Networks (MobiCom 2001)*, pp. 189–199, July 2001.

22. J. Rabaey, "Wireless beyond the Third generation – Facing the energy challenge", Proc. of the 2001 International symposium on Low Power Electronics and Design," pp. 1–3, August 2001.

23. J. Rabaey, A. Chandrakasan, B. Nikolic, "Digital Integrated Circuits: A design perspective, 2nd edition" Prentice Hall, 2003.

24. V. Rughunathan, C. Schurgers, S. Park, M. Srivastava, "Energy-Aware Wireless micro-sensor networks," IEEE Signal Processing magazine, pg. 40–50, March 2002.

25. P. Schaumont, I. Verbauwhede, "A Reconfiguration Hierarchy for Elliptic Curve Cryptography," Proc. 35th Asilomar Conference on Signals, Systems and Computers, Asilomar, Nov. 2001.

26. P. Schaumont, I. Verbauwhede, "Domain-specific codesign for embedded security," IEEE Computer, pg. 68-74, April 2003.

27. C. Schuba, I. Krsul, M. Kuhn, G. Spafford, A. Sundaram, and D. Zamboni, "Analysis of a denial of service attack on TCP," In Proc. of the 1997 IEEE Symposium on Security and Privacy, pg. 208–223, IEEE Computer Society Press, May 1997.

28. F. Stajano, R. Anderson, "The Resurrecting Duckling: Security Issues for Ad-Hoc Wireless Net-works," B. Christianson, B. Crispo and M. Roe (Eds.) Security protocols, 7th International Workshop proceedings, LNCS, 1999.

29. F. Stajano, "The Resurrecting Duckling – what next?," B. Christianson, B. Crispo and M. Roe (Eds.) Security protocols, 8th International Workshop proceedings, LNCS, 2000.

30. D. R. Stinson, "Cryptography Theory and Practice," First Edition, CRC Press, 1995.

31. K. Tiri, I. Verbauwhede. Securing Encryption Algorithms against DPA at the Logic Level: Next Generation Smart Card Technology. Proc. Of Workshop on Cryptographic Hardware and Embedded Systems, LNCS 2779, pg. 125–136, Sept. 2003.

32. K. Tiri, I. Verbauwhede, "A Logic Level Design Methodology for a Secure DPA Resistant ASIC or FPGA Implementation," Proc. DATE 2004, Paris, Feb. 2004.

33. I. Verbauwhede, M.F.-C. Chang, "Reconfigurable Interconnect for next generation systems," Proc. ACM/Sigda International workshop System Level Interconnect Prediction (SLIP02), ACM Press, pp. 71–74, 2002.

34. I. Verbauwhede, P. Schaumont, Christian Piguet, Bart Kienhuis, "Architectures and design techniques for energy efficient embedded DSP and multimedia processing," Proceedings DATE, Feb. 2004.

35. WINS – Wireless Integrated Network Sensor http://wins.rockwellscientific.com

36. A. Wood, J. Stankovic, "Denial of Service in Sensor Networks," IEEE Computer Magazine, Oct. 2002, pg. 54–62.

30. ...
31. ...
32. ...
33. ...

Low-Cost Wireless Control-Networks in Smart Environments

G. Stromberg, T.F. Sturm, Y. Gsottberger, and X. Shi

1 Introduction

In the recent past, home and office automation systems have been in the main focus of most application scenarios for ubiquitous computing. These systems are particularly expected to increase the comfort of living by providing increased functionality (e.g., autonomously controlling jalousies and lighting depending on insulation, or temperature control) and more ease-of-use by intelligent devices and device interoperation. Further, lower installation costs, increased flexibility and portability by using wireless communication technologies, ecological aspects (e.g., pollution reduction by intelligent climate control in office buildings), and security (e.g., fire alarm, escape route signalling, access control) play an important role. Most of these aspects have been discussed in preceding chapters of this book.

Ubiquitous computing has also been suggested for other domains of our everyday life: In production environments, many devices need to co-operate and react spontaneously to the condition of other units in the plant. To this end, e.g., self-configuring conveyor belts and autonomously co-operating machinery have been proposed. In telemedicine, tracking of vital functions, localization and automatic alarming has been suggested. Many localization and remote metering applications can be found in the logistics sector. Computer and entertainment electronics offer a wide field of application for device interactions and remote control. Overall, the progress in ubiquitous computing will leverage new technologies to satisfy the demands and future needs of our society on the global scale, conserve energy resources and save cost.

The predominant goal of ubiquitous computing is to create a networked environment which allows us to access even complex systems in a simple and intuitive way, relieving us of the details of the underlying technology [19, 34, 39]. A ubiquitous system does not only cover the classical application fields of data processing, but interweaves information technology with our real physical environment. Thereby, it offers access not only to information anywhere and at any time, but also an unobtrusive and transparent access to technical appliances in our direct or remote environment. In order to realize this vision, three fundamental components are essential:

1. Intelligent Sensors and Actuators

    Sensors and actuators are the technical components which interface with the real world. They are needed to control and monitor our physical environment. Control networks in a ubiquitous computing environment must be able to organize themselves dynamically and autonomously, which is in contrast to state-of-the-art sensors and actuators. These are usually connected in a static, inflexible, manually-configured network. Self-configuration is especially important for control networks with a large number of nodes, since a manual configuration is too complicated for inexperienced users which are the primary target group of ubiquitous computing environments.

    To this end, the sensor and actuator nodes must exhibit the intelligence to set up the network connection, advertise and provide their functionalities to the other nodes of the network and access remote functionalities without user interaction.

    In many cases, ubiquitous computing systems need to provide a certain level of fault tolerance. At least a basic set of functionality must be preserved even if some nodes fail. Therefore, this functionality must be implemented in the sensor and actuator nodes themselves, whereas more advanced functionality may be centralized and remotely accessed from specific nodes in the network.

2. Middleware Platform

    The objective of the middleware platform of a ubiquitous computing system is to link the network components on a logical level. The middleware platform provides mechanisms which allow the nodes to discover each other, to offer their functionality to other nodes in a formalized way as so-called services and to access remote services offered by other nodes. Further, the middleware platforms support the so-called life-cycle management of services, which means monitoring, starting and stopping the services provided by the individual network nodes.

3. Semantic Data Processing

    In the future, the components of a ubiquitous computing environment will spontaneously establish networks and interact autonomously. Thereby, they create new services which consist of the joint access of several remote services on several remote components. To this end, more than a simple remote control operation is required. The underlying component-oriented services must be understood, interpreted and accumulated to higher-value services by the ubiquitous computing system. New services arise from the combination of several sub-services. For example, the desire of a user for "more light" will automatically be translated into appropriate commands to the installed lights and to the jalousies, dependent on the time, external weather conditions, the direction of the insolation, or the user preferences.

    This approach requires the semantic analysis of the condition of the environment and of the users' demands, and the translation into appropriate actions.

Recently, several middleware platforms have been proposed which provide a basis for semantic data processing by defining mechanisms for network establishment, and discovery, description and access of services distributed in the network. However, due to excessive demands on data processing, these systems address components which exhibit high computing power and large memory. Thus, small sensors and actuators are typically excluded from the integration into these networks for cost and power reasons.

Concurrently, research on autonomous sensor networks has propelled mainly in academia and has brought forth a number of approaches which are dedicated to the specific features of small sensors and actuators. Many of these approaches are related to static networks which must be configured manually. Few other approaches, e.g., the ones discussed previously in this book, aim at dynamic sensor networks which allow for spontaneous ad-hoc network establishment. However, these approaches more or less ignore the need for semantic data processing and do not provide a service-oriented, transparent middleware which offers mechanisms for service discovery, description and remote access as needed in autonomous device control scenarios introduced above.

In this article, we will focus on the requirements for embedding small sensors and actuators into a middleware-based, "IT-centric" ubiquitous computing environment. At the end, we will present an example on how autonomous interaction of small, low-cost and low-power appliances can be realized utilizing one of the best promoted middleware standards for semantic device interoperation, namely Universal Plug and Play (UPnP). Therefore, the subject considered in this article is complementary to both the standardization pursued in middleware design as well as to the research conducted on autonomous sensor networks.

2 Application Scenarios for Smart Homes

The aim of this chapter is to describe and discuss possible application scenarios for low-cost wireless control-networks. Useful application scenarios for ambient computing environments can be found in the Scenario Paper of the IST program of the European Union [6] and the iAppliances 2001 White Paper [28].

2.1 Surveillance (at Home)

2.1.1 Dial P For Payment

Peter has long awaited AG & E, the Atlantis Gas and Electricity provider, to send him the yearly invoice for his power, water and gas consumption. He has expected them to write a couple of letters and emails in which they ask him to make an appointment with their service assistant so that he can read his power, gas and water meter, and he has also expected him refusing all these appointments because he is usually not at home during their business hours (Peter likes starting work early and golfing or hiking in the afternoon). Even worse, once they finally made an appointment in the past the service man has never been on time, so they missed each other often.

But this time, it all works differently. Yesterday Peter has received an email from AG & E on his palmtop in which they ask him to walk into every room of his apartment. His palmtop would gather all necessary information about his consumption, and later assist him during the payment using his favourite banking software.

"Well", Peter thinks, "most likely I will be paying my neighbours' bills as well!" But reading the email more carefully, he learns that AG & E has updated the gas, power, water and heating meters last year, and each of these sensors can be uniquely identified by state-of-the-art authentication mechanisms. Thus, AG & E know exactly which data is his. Vice versa, his neighbours cannot read the data on his consumption because they did not receive the corresponding key attached to the AG & E email. "Your data is as save as your online banking account" (which Peter, like most of his friends, enjoys for years), the email reads.

But still it is not clear to Peter what exactly he has to do. Just walking through the rooms? There is no program to install (or, to be more precise, three of these for the different types of consumption meters for gas, power and water). No manuals to read? All he has to do is to save the key attached to the email into a certain directory of his palmtop. That's easy.

Well, he acts as requested and first walks into the kitchen. Suddenly, a window pops up on his palmtop, and asks him to confirm the readings of the water meter. The numbers match. He is also requested to confirm the readings of the power and gas meter as well as the heat meters attached to each heater in his apartment.

In the living room, an additional window pops up which says that, in contrast to all other heaters in his apartment, this heater is not entirely cooled down during nights and that this could indicate a malfunction of the valves. As a result, heating power is wasted. Peter checks this request the next day and finally finds that the system correctly detected a malfunctioning valve. The community of new sensors performed an integrated, intelligent system check.

After Peter has walked through all his rooms (and all necessary data have obviously been gathered), his favourite banking software pops up. The bank transfer form has already been prepared, and Peter simply enters his security numbers and proceeds as usual, finally typing in "P" for payment.

2.1.2 Ginger's Green Fingers

One week later, Peter visits his friend Ginger who has moved into a new apartment some days earlier. Still being enthusiastic, he reports about his experience with AG & E, and that a technician has already repaired the damaged valve of his heater in the living room.

Ginger, being amazed by Peter's verve, smiles and explains that her new apartment has all kinds of intelligently connected equipment. The *air condition* and the *window openers* are jointly controlled by *temperature sensors* inside and outside the house to generate a pleasant room climate. *Insulation detectors* gather information about the current solar radiation and control the *jalousies* appropriately (in co-operation with the *air condition*, as she demonstrates proudly). Even her flowerpots are intelligent: some plants cannot withstand direct sun radiation, and each plant feels most comfortable in a different ambience and with its specific amount of water and fertilizer. *Electronic gardeners* take care of that by monitoring sun radiation and earth humidity. They control the jalousies and inform Ginger when to water or fertilize the plants. The *electronic gardeners* already came with her plants when she bought them.

Peter is fascinated. He also wants such features at home. Thus, Ginger and Peter turn to Ginger's PC and link to the Internet to get more information (The jalousies are automatically lowered as the *insolation sensor* in Ginger's screen notices that the sun radiation causes bothersome reflections). Peter learns that in a smart home, many different devices are connected together and exchange information, so that certain additional services become available. Services may be configured by the consumer via an easy-to-handle user interface, and some devices (such as the air condition and the individual heaters) need no user configuration at all. He also learns that significant amounts of energy are saved in smart homes, and that he could easily upgrade many of his home appliances to participate in a smart home environment. Peter is enthusiastic, and decides to order several *electronic gardeners* for his favourite plants.

2.1.3 System Description

The emphasis of the selected scenarios is on intelligent home appliances. There exist numerous concepts which address the home automation domain. We concentrate on small, cheap, wireless sensors, which may be easily integrated into an existing environment. Due to the fact that only moderate data rates are required, the proposed sensors consume little energy. Therefore, the described scenarios differ from those considered by for example the Home Audio Video Initiative [14] or established home installation system providers which focus on high-data rate applications.

Today, numerous devices exist in the home environment. Most of those require interaction of a third person, which comes along with organizing and co-ordinating different appointments. The *gas, water and power meters* addressed in our first application scenario are just rudimentary applications of common peripheral devices which provide user interaction upon user request. We can also foresee many other applications of this kind such as failure reporting of home appliances like *coffee makers, washing machines, refrigerators, television sets* etc. Besides plain failure reporting, the home appliance can also provide an expert system for precise fault diagnostics, and even multi-medial repair instructions for dedicated non-critical error situations.

So far, these peripheral devices usually work without any (e.g., the heat meter) or only rudimentary (e.g., the washing machine) user interface, although in some specific situations the user interface is desired to be more intelligent or multi-medial. Embedding a sophisticated user interface (graphics display, touch screen, or keyboard) into conventional everyday devices would significantly increase the cost of the home appliance while the benefit becomes obvious only in rare situations. Until today, cost reasons have prohibited the widespread use of sophisticated user interfaces in the named examples, but a lot of money must be spent in these rare situations when user maintenance is required.

An intelligent smart environment system, in contrast, offers a dedicated, application specific user interface at minimal additional hardware cost (see Fig. 1). It uses a software and hardware infrastructure that almost entirely already exists, and no additional software drivers are required to run the software application. Further, it is an open system, thus allowing more and different cheap devices to make use of the smart environment idea.

Our second example pictures a maintenance scenario in which user interaction is stimulated by a certain condition of one or a couple of sensors

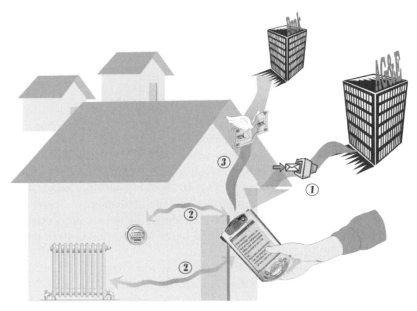

Fig. 1. Automated metering for gas, power, water and heating

attached to a smart transceiver[1]. Again, we can imagine many similar applications. For instance, consider a washing machine which autonomously predicts certain errors like motor or bearing defects, and informs the user in time to call a technician to replace the suspect part (error forecasting).

In this scenario however, the smart transceivers depend on the availability of a terminal which provides computing power and a user interface. There are again many cheap peripheral devices which typically do not need a sophisticated user interface except for specific, rare situations.

Moreover, the communication flow in these examples may differ from the application scenario described above. Since the user interaction is stimulated by the peripheral device, it may be advantageous to initiate the discovery of a terminal by the smart transceiver. This is especially useful if the community of intelligent devices (peripherals and terminals) becomes large or if multi-hop protocols are supported to extend the range of the systems.

The scenarios are fundamentally characterized by the fact that there is a sensor which monitors the condition of an attached device. The sensor may be a heat meter, an electrical power, gas or water meter, or for instance a microphone in the washing machine to monitor the motor condition. Either upon user request or in response to a certain change of the condition of the attached sensor, the smart transceiver enables user interaction by providing an appropriate user interface to the (mobile) terminal. The user interaction itself, however, differs significantly from application to application.

All appliances have in common that the terminal application containing the user interface is capable to connect to the peripheral device. It may use both a unidirectional control flow from the terminal to the peripheral device as well as a bidirectional communication such as required for authentication (both the application as well as the peripheral device authenticate themselves to the respective counterpart). Therefore, the control flow needs to be rather flexible. It particularly depends on the specific application.

In our first example, the user interface makes use of other software components, such as the banking software. Further, certain software components (such as the e-mail client) are capable to invoke and communicate with the terminal application. This imposes another constraint on this application: it must be capable to be embedded into an existing software infrastructure.

2.2 Health and Navigation

2.2.1 Grandpa Jim Goes Shopping

Every second day, Grandpa Jim makes his usual shopping trip. Since Grandpa Jim has a fragile health and sometimes slight mental dropouts, he usually would have to stay in a nursing home. But now he can live in his own house under the remote attendance of a medical service organization. When

---

[1] The term "smart transceiver" refers to an RF transceiver with integrated microcontroller (see Sect. 5.1).

Grandpa Jim starts to leave the house, the *home terminal* asks his *medical shirt* for a special status report and informs the medical service via Internet. Since every medical parameter is within its normal limits, no action is required. The *home terminal* locks the entrance door behind the old man and Grandpa Jim walks down the lane to the street crossing. There, the *smart road sign* confers with Grandpa Jim's *mobile terminal* which tells him the street names over a speaker in his jacket collar. Without any problems, the visually handicapped man reaches the supermarket and buys his usual goods.

2.2.2 Jogging All Along

On every business trip, Jane uses to jog for a while after business hours. Therefore, upon arrival at the hotel, her *mobile terminal* negotiates with the hotel computer system to arrange a route according to her usual preferences. In the evening, Jane starts to jog from her hotel room. Her *mobile terminal* activates the *remote lights* in the hotel corridor in the direction of her route to show her the way. The *smart elevator* opens its doors as she approaches and takes her to the first floor. As she jogs through the nearby park, the *smart signs* there continue to show her the way through the green. After a while, her *health belt* sends a warning report. Jane is warned to slow down by the *mobile terminal* which used her *enhanced wristwatch* as buzzer and display. As she has preconfigured, the *mobile terminal* also sends a status report via Internet to her family doctor who will regard it on her next health check. At the end of her jogging tour, Jane stops by at a coffee shop to take a refreshment which has been pre-ordered by her *mobile terminal*.

2.2.3 System Description

The scenarios described above come from the Personal Area Networking domain. Most of all health care applications receive a boost by the latest innovations in the sensor area. Medical shirts, bio-sensors, or complete wirelessly linked patient data management systems in hospitals are under development. Especially, the transmission of critical data to adequately trained personnel and the security of personal data are important features for these scenarios.

A major functionality of a smart personal environment is to aid, guide, and, if necessary, warn the persons using it. The discussed approach contributes to this goal by providing access to peripheral devices which are used as extended tools or, metaphorically speaking, as eyes, ears and hands of the user. In the presented scenarios, we describe situations in which a digital agent program residing on the *mobile terminal* makes use of different peripheral devices without direct user interaction to generate a higher-level service.

In the scenario *Grandpa Jim goes shopping*, the higher-level service is to enable an aged man to live an ordinary life in his own house. A sophisticated wearable *medical shirt* observes his health, and an unobtrusive digital agent[2] on his *mobile terminal* guides him to the nearby supermarket. Automatically, this agent uses several peripheral devices with smart transceivers to aid and secure Grandpa Jim. In our example, we saw that

- a status report was requested from the *medical shirt*,
- a medical service was informed,
- an entrance door was locked,
- a *smart road sign* was asked for information, and
- a wearable speaker was used.

The second example *Jogging all along* describes a similar situation but with more interaction by the user. Jane had configured a digital agent to determine and guide a jogging tour according to her preferences. Without further interactions by Jane, this agent arranges such a route in every business hotel. It collects local information from the hotel computer system (or the Internet) and plans the route. During jogging, it unobtrusively shows Jane the right direction by using the peripheral devices in the surrounding like *remote lights*, *smart elevators* and *smart signs*, see Fig. 2. It even arranged a stop-over and ordered a refreshment. Additionally, Jane's health is monitored by a *health belt* and feedback is given to her by an *enhanced wristwatch*.

Since the benefit of all these described applications is the intelligent inter-action of different devices and services, we do not foresee a single application to be the one and only "killer application". Furthermore, most applications or smart peripheral devices are not restricted to a single application domain. Remotely controlled lights, for example, can be used in the convenience field as well as in the safety or the navigation field. Thus, the true "killer appli-cation" is the inter-working of different peripheral devices or of the services they provide. Every peripheral device and its provided service is a reusable component to create different high-level user services as described by our scenarios.

3 Enablers for Semantic Control Networks

The following chapter notes some basic technologies which are fundamental for the development of wireless sensor networks. In doing so, we focus on technologies for low-power applications.

3.1 Networking Technologies for Wireless Networks

Table 1 provides an overview of various radio link technologies that are stan-dardized and in use today. These technologies are aimed at different target applications. Consequently, they differ in supported data rate and range.

---

[2] We assume that such a digital agent may be distributed on different computing devices resp. float between devices.

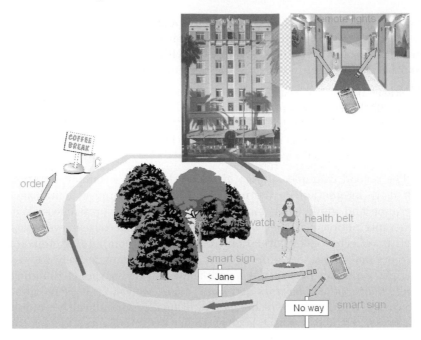

Fig. 2. Route guidance and health surveillance

Short range wireless (SRW) systems typically comply with certain low-level telecommunication standards, e.g., ETSI EN 300 220 [11]. The ETSI EN 300 220 specifies particular constraints on the physical data transmission such as the maximum emitted power, channel utilization (duty cycle, frequency occupation), or carrier frequencies. However, also other ETSI SRW standards

Table 1. Technical Parameters of some In-use Interfaces for Wireless Communication

	Data rate	Range	Memory		Power Dissipation		MIPS
			ROM	RAM	Receive	Idle	
SRW	10–15 kbits/s	<15 m	–	–	27 mW	30 μW[1]/ 15 nW[2]	–
Bluetooth	720 kbit/s	10 m	0.5/1 MB	32 kB	100 mW	50 mW[1]/ 150 μW[2]	∼ 5
GPRS EGDE	100–384 kbit/s	100–3000 m	>1 MB	>500 kB	>1.2 W	40 mW	300–1500
3G	>384 kbit/s	100–3000 m	>1 MB	>500 kB	>1.5 W	>50 mW	>5000
WLAN	1–54 Mbit/s	10–100 m	>500 kB	250 kB	>1 W	150 mW	>2000

[1]Sleep mode, [2]power-down mode.

exist which define higher layers for specific applications like cordless phones, audio data transmission etc.

Bluetooth also supports specific applications by defining so-called profiles [2]. A number of profiles has been specified, each of which covers a certain vertical section of the network stack from the physical layer up to the application. This guarantees compatibility for the applications which are addressed by the profiles supported by Bluetooth devices. Typical applications are wireless connections between office appliances (modems, printers, mice), or networked personal accessories (e.g., wireless headsets). Inter-operability for other applications can only be guaranteed by specifying and implementing additional profiles.

For Wireless Personal Area Networks (WPANs), the IEEE computer society has additionally founded the IEEE 802.15 working group. It is comprised of several task groups, each of which is dealing with specific target applications of WPANs. The fourth task group is working on the IEEE 802.15.4 standard for low data-rate systems [22]. It defines the PHY layer and the MAC layer of the network stack. Further, network-level security is addressed. The supported data rates are either 20 kbit/s (40 kbit/s) in the 868 MHz (915 MHz) band or 250 kbit/s at 2.4 GHz.

Note that the IEEE 802.15.4 standard is also supported by the ZigBee Alliance [44]. The ZigBee Alliance is a consortium of semiconductor manufacturers, technology providers, and OEMs dedicated to providing an application and networking layer to reside on top of the 802.15.4 MAC. However, currently no commercial products are available which comply with either ZigBee or IEEE 802.15.4.

From Table 1, we see that the other listed transmission technologies are inappropriate to provide the basis for smart transceivers in sensor network applications. The main reasons are power consumption and hardware complexity. The requirements regarding computing power are provided in units of million operations per second (MIPS). Compared to small networks like WPANs, this number is substantially higher for wide-area systems like GPRS or 3G due to excessive protocol processing. A significant amount of computing power is also required for data links with high bandwidth such as Wireless Local Area Networks (WLANs). Concurrently to computing power, hardware complexity increases for these systems due to requirements for larger memory.

3.2 Networking Protocols for Wireless Networks

Power in a sensor network can be saved by optimizing the layers of the networking protocol stack. There are two basic approaches: One is minimizing the amount of transmitted data and transmission power, since the RF front end is one of the major power consumers in a sensor node. The other approach is switching the whole or parts of the system into sleep mode, which could save a factor of ten to a hundred in power. So far, much effort has been devoted to exploiting the power saving potential for wireless sensor networks based on these two approaches.

To reduce the payload, the highly correlated data collected by sensors can be aggregated into smaller data units [15]. To decrease the overhead, short MAC addresses can be dynamically assigned and used instead of the long global unique MAC addresses, which dramatically reduces the relative MAC overhead since the payload on a sensor network is usually very small [35]. Dynamically adjusting the transmission power for successful communication can save large amounts of power in the RF front end [25].

Besides, MAC layer duty cycle scheduling is probably the most effective way to save power, which is e.g., adopted by the IEEE802.15.4 WPAN [22] and the Bluetooth park mode [2]. The basic idea of MAC layer duty cycle scheduling is that a node sleeps for most of the time and periodically wakes up for a short time to listen to the channel for communication.

So far, the implementation of a totally passive RF wake-up mechanism for transceivers is a big design challenge. Therefore, a wireless transceiver has to stay in receive mode to sense a possible signal on the air, which is extremely power inefficient. To this end, MAC layer duty cycle scheduling has been introduced. This reduces the time a transceiver is in receive mode. The IEEE802.15.4 standard is a typical example of a MAC layer duty cycle scheduling protocol.

In this standard, the master of a star topology periodically broadcasts a short-length beacon. The slaves spend most of their time in sleep mode and periodically wake up to listen to the beacon. The beacon not only sends information on the beacon period, but also on the slaves the master will communicate with. If a slave has data to transmit or the beacon indicates that the master will send data to this particular slave, the slave stays awake for the rest of the cycle. Otherwise, the slave enters sleep mode until the next beacon. The duty cycle of this mechanism is the ratio between the beacon length and the cycle period: given a specific beacon length, the longer the cycle period, the smaller is the duty cycle, but the higher is the communication delay. Therefore, finding a good trade-off between power saving and delay is a challenge. For networks with time-critical applications for which responsiveness is important, duty cycle reduction is therefore severely limited.

3.3 Middleware Platforms

A prime direction of research on ubiquitous networks focuses on the seamless interaction of end devices by defining semantic mechanisms for device discovery and control. As a consequence, in such an environment the user interacts with very complex systems in a simple and intuitive manner without noticing technical details of the background technologies. An enabling middleware platform has to support spontaneous ad-hoc networking of components without manual installation and has to be flexible enough to integrate low-cost smart sensors and actuators.

CORBA (Common Object Request Broker Architecture) [36,37] is a universal architecture which allows a client application to call functions or methods at a remote server. The kernel of a CORBA system is a so-called Object Request Broker (ORB) which mediates the communication between client and server. The ORB also enables the communication between different computing architectures. Furthermore, the CORBA architecture integrates the functionality of service discovery but an automatic integration of a new component into the network is not supported. Also, since each network component has to facilitate an ORB, CORBA is a too heavy-weighted platform, because small sensor modules are not able to even provide a "slim" ORB implementation [27].

Based on Java, RMI is a middleware which allows for remote calls to functions on servers. Due to the strict focus on Java it does not support any communication between applications which are not Java-based. On the other hand this focus enables the exchange of data (objects) and code between various systems. The central disadvantage of RMI is that the Java-Interpreter (JVM) and the software modules used for RMI are too large. There exist some lean JVMs [12,29] but they do not provide the necessary functionality to attend a RMI system. RMI also lacks possibilities for offering and searching services throughout the net.

One approach to complement the missing functionality of RMI is Jini [13,38]. Jini provides mechanisms for distributing and discovering services and supports code downloading. The core of a Jini system is a lookup-service which facilitates the search for services throughout the Jini-net. Each service has to announce its existence to the lookup-service and to deposit a set of attributes as description of its features. The communication to the lookup-service is based on RMI where the communication between server and client is defined by the drivers and may therefore use any proprietary protocol and format. Due to the RMI-based communication Jini is not dedicated for small systems. Furthermore the lookup-service represents a central unit.

JXTA [31] is based on a SUN initiative and aims for simplifying the configuration and operation of peer-to-peer networks. JXTA provides mechanisms for searching communication partners and for the composition and decomposition of communication channels. Due to the slim protocol design, JXTA is well-suited for small systems. In principle, JXTA offers the possibility to transfer data as well as program code between systems. But it lacks for standardized features for the execution.

OSGi [30] targets on the connection of various components in home networks and the connection to the Internet. Therefore, the home network devices may be controlled via the internet. Also the direct communication of networked devices is possible via an API. Unfortunately, the description of the functionality of small devices is not considered in the features scope. Furthermore, a central device (the gateway) is necessary for the communication.

The most complicated part of the installation of an OSGi network is the configuration of the system which has to be done by a specialized operator.

Salutation [33] and HP's JetSend [16] are protocol architectures for ad hoc networks of typical PC peripheral devices as printers, scanners or fax modems. Salutation defines a set of features and data types which may be used to describe and use services. Therefore, in contrast to Jini, it is not necessary to send driver components via the network. But also the possibilities for using the services are limited due to the rigid description of existing features.

The main attention of the Home Audio Video Initiative (HAVi) [14] is turned to the transmission of digital audio and video data between consumer electronics devices. Typical target devices of HAVi are digital TV's and audio/video recorders. The connection between the devices is established via an IEEE 1394 bus (iLink /Firewire).

PiNet [3] is an architecture which permits small handheld devices wireless access to information. Unfortunately, it is focussed on devices for information presentation only. Small sensors or actuators are not considered. Furthermore, the architecture lacks principles for search and discovery and is based on central PiNet servers.

The Universal Plug and Play (UPnP) Architecture [23] is one of the best established middleware platforms for Ubiquitous Computing. UPnP uses open and standardized protocols based on XML which allow to describe and control various devices. Therefore, UPnP itself integrates into and participates in the steadily evolving community of semantic Web service languages derived from XML. All communication is transferred over TCP/IP or UDP/IP and is thus hardware-independent. The most important advantages of this open concept are semantic interoperability of devices from different vendors and the simple extensibility to future devices. The price which has to be paid is the intense requirement for memory and computing power for UPnP Devices since variable XML-based protocols have to be processed.

For a better understanding, we take a closer look at the steps of the UPnP Device Architecture, see Fig. 3:

- **Addressing**: A UPnP Device requires a network address using a DHCP server or, if none is a available, using Auto-IP.
- **Discovery**: UPnP Devices advertise themselves by multicast messages and react according to search messages.
- **Description**: UPnP Devices and their embedded services are defined by XML description files which have to be provided for HTTP download by the device. Note that these files are static and do not change except for few parameters.
- **Control**: UPnP Devices are controlled by the XML-based SOAP protocol [34] and therefore process XML messages and answer with dynamically created XML messages.

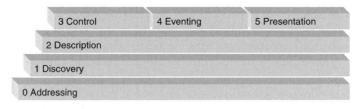

0	Control point and device obtain an address to participate in the network
1	Control point finds all devices and device advertises its availability
2	Control point learns about device capabilities
3	Control point invokes actions on device
4	Control point listens to state changes of device
5	Control point controls device and/or views device status using an HTML UI

Fig. 3. UPnP protocol steps

- **Eventing**: UPnP Devices maintain event subscriber lists, accept subscriptions and keep track of subscription durations. When an event occurs, XML-based event messages with dynamic content are sent to each subscriber.
- **Presentation**: The optional presentation page exposes an HTML-based user interface which may be used for controlling the device by embedded elements like Java applets.

The complexity of UPnP control and eventing prevents a cheap and low-power device to implement the complete UPnP Device Architecture. Nevertheless, such a device is able to do UPnP addressing, discovery, description, and presentation. Thus, it can act as a so-called UPnP Basic Device.

4 Sensor Networks

This chapter describes some existing concepts for distributing sensors and how to connect them in so-called sensor networks. Further, the main application fields and advantages and disadvantages are explained.

Sensor networks, in general, are divided into static and dynamic networks. Static sensor networks are characterized by a fixed communication structure that must be configured manually. Dynamic sensor networks are able to automatically establish connections between the communicating members.

4.1 Static Sensor Networks

The European Installation Bus (EIB) [8–10] as cable-based system is used for many years in in-house networking for building systems. Many different device types, such as sensors (switches, motion detectors, fire alarms, temperature sensors) or switching actuators for lights, jalousies, or complete control units as a heating control are logically linked together by the EIB system. The protocols used are designed in a way that also components with only a few computing resources may execute them. Unfortunately, the final system configuration has to be designed by an operator and cannot be changed dynamically.

A system which is related to the EIB system is the Local Operating Network (LON) [26]. LON has been developed by the company Echelon to provide a universal technical basis for automation in all application domains. It distributes control functions de-centrally over all network nodes. Each LON-enabled device owns a control unit, the so-called Neuron, which consists of three 8-bit microcontrollers. Two of them are handling the LON protocol (LONTalk) and one may be used by the application executed at the node. Similar to EIB, the system configuration has to be entered manually via a central tool. Furthermore, there is no mechanism for transmitting executable code over the network.

Konnex [21, 42] is a system which is based on EIB. This technology provides some extensions to EIB regarding improvements in the configuration management. But it still lacks possibilities for a dynamic and automatic distribution and use of driver modules. Another technology which is related to the LON protocol is BACnet. But also BACnet lacks for mechanisms for dynamic self-configuration.

Besides sensor networks for building automation there have been developed micro sensor systems. These systems are locally distributed and monitor environmental data and forward them to a central execution unit. Within the scope of the Smart Dust project [7, 20, 41] 1 mm^3 sized, cheap prototypes have been developed. Each prototype contains a sensor, an execution unit, a battery, and a communication module which operates on the basis of passive optical devices in the ultra low-power range. But, for data exchange a central active unit is necessary, the so-called Base Station Controller. Furthermore, all sensing units of the network have to be exactly localized, which is a remaining challenge. The smart dust approach addresses the distributed monitoring of measuring data and the transmission of these data to a central unit for execution. The typical features of an ad-hoc network are not realized in the smart dust project.

4.2 Dynamic Sensor Networks

Dynamic distributed micro sensor systems have to provide all functionalities of static sensor networks with the additional feature to autonomously integrate components into the network at any time. Especially, dynamic energy-efficient multi-hop sensor networks are investigated at various research institutes [4].

With PicoRadio [5, 32] a radio-based ad-hoc sensor network is created which monitors data using hundreds of distributed sensor modules. The collected data are handed over in small packets from sensor to sensor until they reach the base station. The hardware architecture of a PicoNode consists of a microprocessor, programmable logic, a dedicated DSP block, an RF module and the sensor(s), altogether integrated in a single chip. An important research aspect is the optimization on system level for the lowest possible power consumption. In the same way as Smart Dust PicoRadio concentrates on a distributed collective acquisition of measured data, but it uses a self-configurable mobile ad-hoc network for data transfer. However, the system does not support standardized discovery and look-up mechanisms or a suitable software infrastructure to obtain and integrate the driver software dynamically.

Bradio [43] is an infrastructure for ubiquitous networks which is designed for the integration of sensor components. The system is based on special modules which allow for a radio-based communication between the components. It has particularly been developed for the use in small battery-powered devices. This system also does not support standardized discovery and lookup features or mechanisms for driver download.

In the Wireless Integrated Network Sensors (WINS) project, radio-based single-chip sensor components are described [1]. The focus of WINS is on the development and integration of low-power MEMS (Micro Electro-Mechanical Systems) technologies. Another architectural proposal for distributed sensor components is given by [24]. This approach is based on a StrongArm 1100 processor core, which exceeds the typical execution capabilities and the power budget of a sensor network node.

5 A Specific Example: The Sindrion Control Network

We have developed a distributed system architecture called Sindrion which features an effective solution for small low-cost appliances. The basic idea is that complex operations are sourced out from the peripherals to dedicated network components, which we call terminals. Sindrion supports the wireless communication and the autonomous inter-operation of these nodes. The system is aimed at small wireless sensors and actuators, information beacons as well as maintenance and control applications for white goods, for which the power consumption and a low price of the nodes are as important as the capability to embed the devices into a well-established IT infrastructure.

5.1 Basic Principle

The basis of the Sindrion system is to set up a wireless link between peripheral devices and dedicated computing terminals. The objective of this connection is to source out complex data processing from the peripherals to the terminals. To this end, the peripheral devices contain small smart transceivers, the so-called Sindrion Transceivers, which are attached to embedded sensors or actuators (see Fig. 4). Typical peripheral devices are environmental sensors, small actuators like switches, or home appliances. They bear very limited or no computing power, and the embedded sensors and actuators can be controlled by simple proprietary analog or digital control lines. These are connected to the input- and output ports (I/O ports) of the Sindrion Transceivers.

The terminal is equipped with an RF transceiver which is compatible with that included in the Sindrion Transceiver. Data and protocol processing are done in the terminal, which features a virtually unlimited amount of processing power and memory compared to the Sindrion Transceiver.

Figure 4 shows the fundamental structure establishing the communication between the terminal and a previously unknown Sindrion Transceiver. The procedure is as follows:

- **Discovery**: The two end devices find each other in the discovery phase. For compatibility reasons with existing IT infrastructure, the UPnP discovery protocol is used [23].
- **Code Download**: If the terminal does not yet contain the control application for the Sindrion Transceiver, the transceiver's application code is downloaded by the terminal. Preferably, this code is written for a middleware platform such as the Java Virtual Machine. This guarantees platform independence and allows for the seamless integration into various terminals. Further, high-level programming languages facilitate application development.
- **Application-Specific Communication**: The following communication between the downloaded service application on the terminal and its counterpart – the transceiver control – on the Sindrion Transceiver may be completely application-specific and does not have to be defined by any standard.

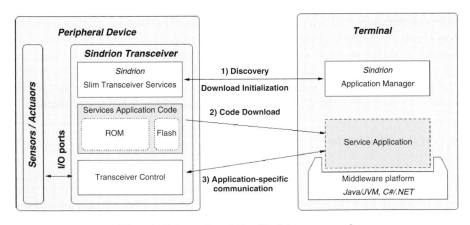

Fig. 4. Schematic of the Sindrion approach

Due to the low-power and low-cost approach of the Sindrion sensor nodes, UPnP control and eventing capabilities will not be implemented in the Sindrion Transceivers. Sindrion Transceivers neither implement the SOAP protocol nor administrate event subscriptions.

On first sight, this excludes cheap and low-power sensors and actuator nodes from participating in a UPnP environment.

On second sight, UPnP bridges or gateways may be used to connect a non-UPnP network to a UPnP environment. In this approach, sensor nodes have to be configured as nodes of a specific network and additionally as UPnP Devices inside the gateway. Therefore, seamless integration without user interaction and complex configuration as envisaged for ambient intelligence applications cannot be accomplished. Furthermore, if just one or few sensor nodes are added to an existing UPnP environment, an additional gateway device has to be installed as expensive overhead.

With Sindrion, we have found a third way to integrate devices into a UPnP environment which is compatible with UPnP and neither requires any gateway nor gateway configuration.

Nevertheless, every Sindrion Transceiver acts as UPnP Basic Device, which is a special device type without embedded UPnP services. The Sindrion Transceiver supports UPnP discovery, description and presentation, which need much less computational effort than UPnP control and eventing. Without prerequisites, a Sindrion Transceiver can therefore be connected to a UPnP network as displayed in Fig. 5. Following the Sindrion approach, the downloaded presentation page contains a Java applet, by which the Sindrion Transceiver and its attached peripheral device can be controlled.

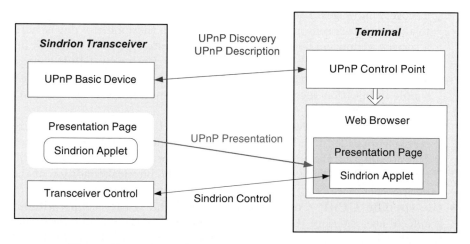

Fig. 5. Basic device functionality of the Sindrion Transceiver

5.2 The Sensor Node in the UPnP Environment

The UPnP Basic Device approach is useful to control a Sindrion Transceiver without prerequisites on the terminal side in a point-to-point fashion. For more complex control or service scenarios, an arbitrary UPnP Device with full control and eventing capabilities is needed with a longer lifetime than a displayed web presentation page inside a browser, e.g., think of a lighting device which is controlled by alarm equipment.

In this kind of situation, a Sindrion Application Manager demon runs on the terminal which is used to start UPnP proxies exported from the Sindrion Transceiver. It also administrates a file repository for the service application Java byte code which was downloaded from Sindrion Transceivers and which can be reused. The Sindrion Application Manager discovers a Sindrion Transceiver by standard UPnP mechanisms and downloads the service application code from the transceiver. Next, the service application is started on a middleware platform like the Java Virtual Machine on the terminal and acts as a UPnP proxy for the connected Sindrion Transceiver. Thereby, it embeds the sensor node into a UPnP environment.

The detailed concept of the embedding is depicted in Fig. 6. The service application shown can be decomposed into two separate applications:

- **the UPnP Proxy**: This application contains the exported functionality of the Sindrion Transceiver. Depending on the mirrored sensor node, the proxy may consist of two UPnP components:
  - **a UPnP Device**: This proxy provides all the needed UPnP services and full control and eventing functionality. It exclusively represents the Sindrion Transceiver and announces the represented device by UPnP discovery messages. For example, if the Sindrion Transceiver is attached to a temperature sensor, the proxy announces a thermometer device with a service action *getTemperature*. Every interested UPnP Control Point can communicate with the announced UPnP Device by UPnP control and eventing. The proxy translates this SOAP based communication into simple application-specific messages and thereby controls the Sindrion Transceiver.
  - **a UPnP Control Point**: The members of the network which utilize services are called UPnP Control Points. In order to implement actuator functionality in the Sindrion Transceiver, the service application can contain a UPnP Control Point which maps the physical condition of the peripheral device (e.g., a switch) on UPnP control messages.
- **the specific UPnP Control Point**: The UPnP Proxy comes without graphical user interface (GUI) and communicates to the outside world only through UPnP mechanisms. The purpose of the optional *specific* UPnP Control Point is to provide a convenient GUI for direct user interaction with the Sindrion Transceiver. Typically, the GUI for a presentation page applet is identical to the GUI of this application. As depicted in Fig. 6, the

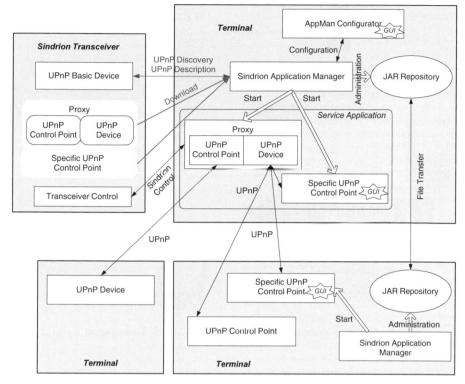

Fig. 6. The Sindrion Transceiver mapped as an arbitrary UPnP Basic Device

code for this UPnP Control Point may be exchanged from one terminal to another. Since the specific UPnP Control Point is fully UPnP compatible, it may run on any networked terminal and may be replaced by any other UPnP Control Point which is able to control the given UPnP Device.

With the given software architecture, a Sindrion Transceiver is seamlessly embedded into a UPnP environment. The underlying scheme of Fig. 6 is opaque for the other members of the UPnP network – the sensor node just appears as any fully enabled UPnP Device as shown in Fig. 7. Further, different nodes can interact by the UPnP interface to form complex services. For example, the already mentioned thermometer device (with proxy) may provide temperature data to a small Sindrion-enabled LCD display (with proxy) in another room, at the same time to a PC to store a temperature timeline, and to an HVAC system for ventilation control.

5.3 Components Overview

Since the Sindrion system is aimed at granting low-cost and low-power access to ambient intelligence systems, power consumption and hardware cost

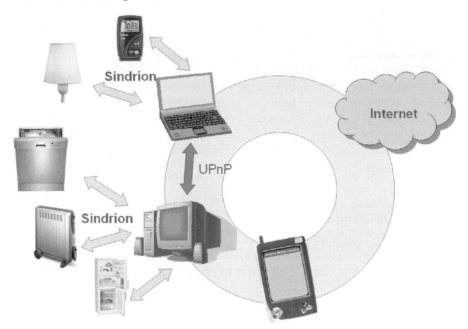

Fig. 7. Sindrion Transceivers as a part of a UPnP Distributed System

of the transceiver are the most rigid constraints of the Sindrion system. Typical application scenarios [6, 28] show that there is a substantial amount of peripherals in the price range of 5\$ to 20\$, to which the transceiver may contribute only a certain fraction. Depending on the particular application, the transceiver will be integrated with a battery or power supply, and some other external components such as an oscillator and an antenna. Additionally, integration costs must be considered. Given the price range mentioned before, we assume that the transceiver cost should not exceed 2\$–3\$. This single-chip transceiver will contain the following subcomponents:

- **RF transceiver**: To reduce the chip count, the RF transceiver should be largely self-contained. Note that low-power wireless systems with short range and low data rate are particularly in our scope. In this field of application, a number of highly integrated wireless solutions are already available which require only few external components.
- **Microcontroller**: The microcontroller should be small in terms of chip area and consume only little power. A simple 8-bit or 16-bit microcontroller with appropriate I/O capabilities to attach the sensor or actuator peripherals suffices. As shown later, the most important characteristic is that the controller effectively supports different power states for sleep conditions.
- **Firmware memory**: The firmware memory is a small memory that contains the firmware for the microcontroller. The firmware handles the net-

working protocols, the control of the transceiver's resources, primarily of the I/O interfaces, and housekeeping tasks. The content of this memory is fixed and common to all Sindrion Transceivers. Therefore, it can be stored in a power and area efficient mask-programmed ROM.

- **Sindrion library memory**: The Sindrion libraries implement common, typical tasks of the downloaded service applications. By providing a high-level application programming interface (API), they facilitate the development of the service application by hiding the underlying hardware realization to the application-level program. Typical functions provided in the Sindrion programming library e.g., support the control of the transceiver's hardware I/O interfaces or UPnP networking. Since the Sindrion programming libraries are common to all transceivers, the Sindrion library memory can also be realized by a mask-programmed ROM.

The service application itself however needs to be implemented by the integrator of the transceiver. Depending on the particular application of the Sindrion Transceiver, the size of this application can vary significantly. Therefore, we presume that the service application will be stored in an external non-volatile memory attached to the Sindrion Transceiver.

5.4 Prototype System

In order to show feasibility and applicability of the Sindrion system, we realize a prototype transceiver module using standard off-the-shelf components. This module is designed in close analogy to a single-chip solution, so that chip area and power consumption of a single-chip solution can be precisely estimated.

In control networks, the main focus of the transceiver system design is reducing the standby power to a minimum since the communication between terminals and transceivers is typically low. Many application scenarios suggest that the transceivers are battery-powered, so that a reasonable design target is that the Sindrion Transceiver's standby power consumption should be less than $1\,\mathrm{mW}$[3]. Note that in this state, the transceiver must be able to respond to incoming messages over the communication channel or react on changes at the input pins. Thus, we will call this state the *sentry state* of the Sindrion Transceiver.

Obviously, most power can be saved by turning off transceiver hardware when not used and turning it on again as needed. Although this is the most favourable behaviour of the transceiver, it is also the most complicated one since it requires intelligent power control schemes. In the Sindrion network, there are, however, only two triggers to wake up the hardware:

- The attached peripheral device, i.e., a sensor, indicates that data is to be processed and transmitted to the terminal.

---

[3] This corresponds to a life-time of a standard AA battery of almost one year.

- The terminal sends a control command or a UPnP packet on which the transceiver must react. In this so-called *com wake-up* scenario, an incoming data packet triggers the wake-up of the transceiver.

It is rather obvious that checking for the first condition is much simpler than checking for the second. In the first case, an electrical signal is already available which may trigger an external interrupt, which wakes up the transceiver. This is a common feature already available in off-the-shelf microcontrollers with advanced power control features.

In the second case, however, there is no electrical wake-up-signal available. The detection of an incoming data packet requires at least that the packet be received, checked against errors and address decoded. This means that the RF signal traverses the entire physical (PHY) layer and parts of the medium access control (MAC) layers. Thus, in order to wake up the transceiver by initiating a communication, some data processing is involved. Further, the RF transceiver must be turned on in order to receive data packets. Correspondingly, the com wake-up also requires the RF transceiver to be turned on at least at these time instances when data is sent from a terminal.

To this end, we utilize a so-called pseudo wake-up by signal scheme: If the Sindrion Transceiver is in sentry mode, the RF transceiver autonomously alternates between a very low-power sleep mode and the so-called data-rate detection (DRD) mode, which allows to detect within a few bit intervals whether a certain data rate is transmitted on the RF channel. When the DRD circuitry detects that data is being transmitted, it generates an output impulse which is used to wake up the block for PHY and MAC layer processing in the Sindrion Transceiver. This function is implemented in an FPGA (see Fig. 8).

The RF transceiver used in the prototype module is an off-the-shelf 433 MHz ISM band transceiver which is typically used as a "remote keyless entry" (RKE) system [18]. It fulfills the applicable ETSI regulations for this type of application [11]. The RF transceiver is optimized for low standby power and natively supports the pseudo wake-up by signal scheme. The maximum supported data rate is 50 kbit/s.

An FPGA is connected to the RF transceiver in which the physical (PHY) and the lower medium access control (MAC) networking layers are implemented. To this end, the FPGA is used for clock (re-)generation, forward error correction (FEC) and error checking by a cyclic redundancy check (CRC) code. During data retrieval, an entire data packet is stored in the FPGA. If the packet is error-free, the FPGA further decodes the address. If the destination address indicated in the received data packet matches the MAC address of the Sindrion Transceiver or if it is a broadcast address, the FPGA wakes up the attached microcontroller for further data processing by a com wake-up-signal.

Although each individual transceiver is addressed only rarely by the terminals, the overall traffic in the respective RF signal band may be high because

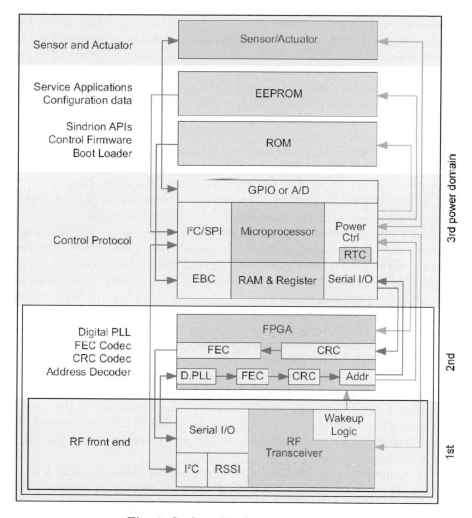

Fig. 8. Sindrion Hardware Architecture

also other participants are allowed to send in the ISM band which is used by the Sindrion RF transceiver. Further, noise can lead the DRD circuit to wrongly indicate that there is an incoming data packet, and noise artifacts can result in non-correctable packet errors in highly disturbed environments. Therefore, we have chosen strict conditions when to generate a com wake-up signal, i.e., a valid and address-decoded packet must have been received for further data processing.

Finally, upper layer protocol processing[4] is implemented in a 16-bit microcontroller [17]. This controller supports individual subcomponents like the digital interfaces, the A/D converter, etc. to be individually turned off in order to save power. The microcontroller further supports a power-down mode with very low power consumption in which the internal RAM is preserved.

As we see from the above description, the design of the transceiver module relies on the proper partitioning of hardware components into different power domains, see Fig. 8. The first power domain is simply comprised of the RF transceiver alternating between sleep and DRD mode when the Sindrion Transceiver is operated in sentry mode. All other components are turned off. Thus, the Sindrion Transceiver module only consumes roughly 440 μW in the first power domain. This power consumption corresponds to that in sentry mode which is to be optimized for this kind of control networks.

When data is detected on the channel, the second power domain is activated which is comprised of the RF transceiver and the FPGA. The activation of the FPGA via a power switch is almost instantaneous since the FPGAs used in this project do not require a boot-up procedure. The active power consumption of the FPGA is however quite high in the range of 10–20 mW. The power consumption of the second power domain is therefore approximately 37 mW to 47 mW. When powered off, the FPGA will consume negligible power. Note, that an on-chip realization of the dedicated hardware will be much more power efficient.

The third power domain finally covers the entire Sindrion Transceiver module. Thus, the microcontroller, the FPGA and the transceiver are active. Assuming a clock frequency of 9 MHz, the power consumed by the microcontroller with all peripherals turned on is less than 33 mW, resulting in an overall maximum power consumption of the entire transceiver in the third power domain of 80 mW.

These numbers show that the concept of several power domains is very effective. However, it needs support from both hardware and software. The most important factor in this regard is the proper design of the MAC networking protocol which must handle the com wake-up. Its discussion and evaluation is the subject of the next section.

5.5 Wake-Up-By-Signal Scheme

In order to utilize the com wake-up mechanism, a certain data rate must be generated on the channel to which the DRD logic of the Sindrion Transceiver is sensitive. To this end, each data packet sent from the terminal to a transceiver must be preceded by a so-called wake-up-signal which exhibits

---

[4] The upper layer protocol processing tasks involve the processing of the upper MAC layer parts, the IP network and the TCP transport layer, and the application layer. The latter includes the UPnP discovery, description and download, as well as the processing of the control messages.

the appropriate data rate. Note that the data rate of the wake-up-signal must differ from that of the data frames so that a regular data transmission does not wake up all transceivers in the vicinity of the sender.

Since the RF transceiver alternates between DRD and sleep mode in programmable time intervals, the length of the wake-up-signal must be adjusted so that the transceiver is in DRD mode at least once during this signal. Further, it must be able to complete the detection of the wake-up-signal. In total, this takes about 2.66 ms. Therefore, to achieve a duty cycle of 1%, a cycle period of 266 ms is required, which guarantees a response time of below 300 ms for the Sindrion Transceiver. Usually, a 500 ms response time can fulfil the demands of most Sindrion applications, so that the duty cycle of the RF transceiver can be around 0.6%. As shown in Fig. 9, this dramatically decreases the power consumption of the RF transceiver, which is one of the biggest power consumers in the system.

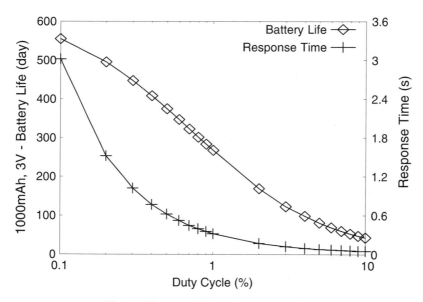

Fig. 9. Battery Life vs. Response Time

6 Conclusion

In this article, we have provided a brief overview on the wide field of academic research and industrial standardization efforts on deploying low-cost wireless control networks in smart environments. The convergence of embedded systems technologies – wireless data transmission, low power digital design, small embedded non-volatile memories – has leveraged the field of intelligent wireless sensors and actuators. These are the essential participants of control networks.

The successful deployment of control networks necessitates an agreement not only on the networking layers, but also on layers related to the semantics of the device operation. This means that sensor and actuator nodes need standardized mechanisms for discovery and description of the functionality they provide as well as methods for accessing remote services. These issues are covered by so-called middleware platforms for semantic device interaction, of which some are currently in the standardization process.

Typically, these middleware platforms have been designed with computer-like systems in mind. Processing the associated protocols is therefore too complex for resource-limited smart embedded sensors and actuators. As a compromise, we have proposed a system which balances the load between small sensors and computing devices typically available in the environment. By establishing proxies on the remote computing devices, small and energy-efficient smart sensors and actuators can be integrated into Universal Plug and Play, which is one of the best established middleware platforms in this application domain.

References

1. Asada G, Dong T, Lin F, Pottie G, Kaiser W, Marcy H (1998) Wireless integrated network sensors: Low power systems on a chip. In: European Solid State Circuits Conference, The Hague, Netherlands http://citeseer.nj.nec.com/278712.html.
2. Bluetooth SIG, Inc. (2001) The Bluetooth V1.1 Core Specification http://www.bluetooth.org/specifications.htm.
3. Carmeli B, Cohen B (2001) PiNet: Wireless Connectivity for Organizational Information Access Using Lightweight Handheld Devices. In: IEEE Personal Communications.
4. Chandrakasan A, Amirtharajah R, Cho S, Goodman J, Konduri G, Kulik J, Rabiner W, Wang A (1999) Design considerations for distributed microsensor systems. In: Proceedings of the IEEE 1999 Custom Integrated Circuits Conference, pp. 279–286.
5. Da Silva Jr. JL, Shamberger J, Ammer MJ, Guo C, Li S, Shah R, Tuan T, Sheets M, Rabaey JM, Nikolic B, Sangiovanni-Vincentelli A, Wright P (2001) Design Methodology for PicoRadio Networks. In: Design, Automation and Test in Europe (DATE).
6. Ducatel K, Bogdanowicz M, Scapolo F, Leijten J, Burgelman JC (2001) Scenarios for Ambient Intelligence in 2010. Draft Final Report, Information Society Technologies Advisory Group (ISTAG).
7. Ecreasing Computing Device Smart Dust: Communicating with a CubicMillimeter Computer. http://citeseer.nj.nec.com/494954.html.
8. EIBA (1995) EIBA: Handbook for Development, v. EIB 2.21.
9. EIBA (1998) EIBA: Handbook for Development, Pre Release EIB 3.0.

10. EIBA (1999) EIBA: Handbook for Development, Release 3.0.
11. European Telecommunications Standards Institute (2000) ETSI EN 300 220–1,2, V1.3.1. http://www.etsi.org.
12. Gokhale AS, Schmidt DC (1999) Techniques for Optimizing CORBA Middleware for Distributed Embedded Systems. In: INFOCOM (2), pp. 513–521 http://citeseer.nj.nec.com/article/gokhale99techniques.html.
13. Gupta R, Talwar S, Agrawal DP (2002) Jini Home Networking: A Step toward Pervasive Computing. In: IEEE Computer.
14. HAVi Organization (2001) HAVi, the A/V digital network revolution. http://www.havi.org/techinfo/whitepaper.html.
15. Heinzelman WR, Sinha A, Wang A, Chandrakasan AP (2000) Energy-Scalable Algorithms and Protocols for Wireless Microsensor Networks. In: ICASSP.
16. HP HP JetSend. http://www.jetsend.hp.com.
17. Infineon Technologies AG (1999) C161 PI Microcontrollers. Data Sheet, Preliminary, Infineon Technologies AG, Munich.
18. Infineon Technologies AG (2002) TDA5255 E1 ASK/FSK 434 MHz Wireless Transceiver. Data Sheet, Preliminary Specification, Version 1.1, Infineon Technologies AG, Munich.
19. Jagiello J, Tay N, Biddington B, Dacray R (2000) Mobile functionality in a pervasive world. In: IEEE International Symposium on Performance Analysis of Systems and Software, pp. 178–183.
20. Kahn JM, Katz RH, Pister KSJ (1999) Next Century Challenges: Mobile Networking for "Smart Dust". In: MOBICOM, pp. 271–278.
21. Konnex Konnex Association http://www.konnex.org.
22. LAN/MAN Standards Committee of the IEEE Computer Society (2003) IEEE Standard for Part 15.4: Wireless Medium Access Control (MAC) and Physical Layer (PHY) Specifications for Low-Rate Wireless Personal Area Networks (LR-WPANs), P802.15.4.
23. Microsoft Corporation (2000) Universal Plug and Play Device Architecture 1.0. http://www.upnp.org, Microsoft Corporation, Redmond, USA.
24. Min R, Bhardwaj M, Cho S, Sinha A, Shih E, Wang A, Chandrakasan A (2000) An architecture for a power-aware distributed microsensor node. In: IEEE Workshop on Signal Processing Systems (SiPS '00) http://ciiteseer.nj.nec.com/min00architecture.html.
25. Monks JP, Bharghavan V, Hwu WE (2001) A Power Controlled Multiple Access Protocol for Wireless Packet Networks. In: INFOCOM.
26. Motorola (1995) LONWorks Technology Device Data.
27. Mowbray TJ, Malveau RC (1997) CORBA Design Patterns. Ed. Katherine Schowalter.
28. Neff AJ, Chung T (2001) iAppliances 2001: You've Got the Whole World in Your Hand. Bear Stearns Equity Research, Technology.
29. Object Management Group (2002) Minimum Corba Specification http://www.omg.org.
30. Open Service Gateway Initiative (2000) Open Service Gateway Initiative: Specification Overview. http://www.osgi.org.
31. Project JXTA (2002) JXTA. http://www.jxta.org].
32. Rabaey JM, Ammer MJ, da Silva JL, Patel A, Roundry S (2000) PicoRadio Supports Ad Hoc Ultra-Low Power Wireless Networking. IEEE Computer, vol. 33(7), pp. 42–48.

33. Salutation Consortium (1999) Salutation Architecture Specification – v2.0c. http://www.salutation.org/ordrspec.htm.
34. Satyanarayanan M (2001) Pervasive computing: vision and challenges. IEEE Personal Communications, vol. 8 (4), pp. 10–17.
35. Schurgers C, Kulkarni G, Srivastava MB (2001) Distributed Assignment of Encoded MAC Addresses in Sensor Networks. In: MobiHoc.
36. Sun Microsystems (2000) Connected, Limited Device Configuration. http://java.sun.com/aboutJava/communityprocess/final/jsr030/index.html.
37. Sun Microsystems (2000) Java 2 Platform Micro Edition Technology for Creating Mobile Devices. http://java.sun.com/products/cldc/wp/KVMwp.pdf.
38. Sun Microsystems (2002) Jini. http://www.jini.org.
39. Vrana G (2002) Pervasive computing: a computer in every pot. EDN http://www.ednmag.com.
40. W3C (2001) SOAP Version 1.2 W3C Working Draft. http://www.w3c.org/TR/2001/WD-soap12-20010709.
41. Warneke B, Last M, Leibowitz B, Pister KSJ (2001) Smart Dust: Communicating with a Cubic-Millimeter Computer. IEEE Computer, vol. 34 (1), pp. 44–51.
42. Weber W (2000) The Konnex Technology. In:EIB Event 2000 http://hotletter.emt.e-technik.tu-muenchen.de/eiba/Conference_00/Proceedings00.html.
43. Yoshimi B, Bolam GB, Sukaviriya N, Elliott J, Carmeli B, Morgan J, Derby H (2002) Bradio: a wireless infrastructure for pervasive computing environments. In: Performance, Computing, and Communications Conference, pp. 309–316.
44. ZigBee^{TM} Alliance (2003) ZigBee^{TM} Alliance http://www.zigbee.org:

Part III

Components and Technologies

Ambient Intelligence Technology: An Overview

F. Snijders

1 Introduction

In the second half of the 19th century Jules Gabriel Verne, one of the science fiction's founding fathers, wrote his famous novels playing in a world full of technology most of which was unknown yet at that time, but to become reality some 100 years later. Still considered to be hall-marks of a genius, his books "From the earth to the moon" (1865), "Twenty thousand leagues under the sea" (1869), "Around the moon" (1869) and "The mysterious Island" (1873) predict – among others – technologies like electric lighting, broad spectrum medicines, submarines, large displays, rockets avant-la-lettre and a mysterious propulsion power that – in hindsight – well might be interpreted as nuclear energy. However, what the talented writer did not forecasted was the dawning of the digital era with computers, the Internet, digital encoding of audio and video and all kinds of wireless communication. This gives us, the humble technology prognosticators of the 21st century; the reassuring feeling that predicting the technological course of history remains difficult and speculative, even for men of genius.

Now at the dawn of the 21st century breathtaking advances in computer and communication technology open a window on a new and exciting world, where flocks of electronic devices become an integral and natural part of our everyday lives. A world that is full of smart objects because of embedded microcomputers with adequate communication capabilities. Where integrated sensors create a "contextual aware" environment by sensing and communicating events and personal data to the system, that combines them in meaningful responses to the end-user. With electronics integrated into clothing, furniture, and mirrors and probably even in the construction materials of our houses and offices. And where the electronic functions spontaneously organize themselves in networks to provide their human companions with communication and entertainment services and with access to rich content wherever they are. A world referred to as "Ambient Intelligence".

2 Ambient Intelligence Technologies

Predicting the technologies that will shape the future Ambient Intelligence world is no sinecure as we have learned from the example of Jules Verne. It is to be expected, however, that Ambient Intelligence technology will develop incrementally. The danger of such simple extrapolations is, that they often miss major effects from disruptive forces, but disruptive forces are – by nature – unpredictable. Technology trends certainly have momentum, but lateral forces can create rapid changes in direction. It is clear, however, that in particular the evolutionary information processing technologies: computing, communication, software, storage, displays, sensors and digitization of analog signals representing all physical modalities have paved the way for Ambient Intelligence and will shape this world further.

In accordance with Moore's law data density on integrated circuits is continuing to double every eighteen months. Other ingredients for distributed computing like storage capacity, CPU speeds and (wireless) communication speeds all show significant rates of change. Areal density of HDDs (Hard Disk Drive), for instance, has on average doubled every year in the last decade and the improvement cycle seems to continue without a hitch [1]. In the same period CPU speed grew with about a factor of 400 and wireless transfer speed with at least a factor of 20. And – most of all – the costs for those goodies are coming down year after year making them affordable for large-scale embedded use.

Battery technology is a bit of a "dark horse" in this success story of formidable and continuing growth in performance characteristics of technologies. Over the last decade the capacity (Wh/kg) of Li-ion type rechargeable batteries has improved by a factor of 4–5 compared to traditional nickel-cadmium (NiCd) types (Fig. 1). With exotic and (still) expensive electrode materials for nickel-metal hydride batteries (e.g., magnesium-scandium hydride) energy density figures up to four times those of Li-ion batteries have been reported [2]. A lot of work however, has to be done to bring down the price point and the discharge cycle characteristics of batteries using this type of electrode materials.

Besides the "classical" technologies mentioned above, the concept of Ambient Intelligence requires major technological advances in sensor and micro-actuator technology, in ultra-low power radio and in smart materials to create the adaptiveness and responsiveness of the Ambient Intelligence environment. Furthermore, massive research and progress in energy scavenging from the environment to power autonomous, unattended micro systems and in the self-organization aspects of those Ambient Intelligence devices is needed. Last but not least, all technology elements have to be integrated into the architectural framework of an Ambient Intelligence System with which the user interacts at the application level in a multi-modal way. These technologies will form a "library" of building blocks for general ambient intelligence system architectures.

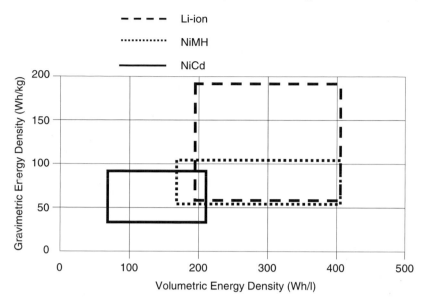

Fig. 1. Performance of present and future battery technologies

It's clear that engineering is much driven by the fashion of the moment and that "occasionally" fashion is simulated by the latest technology. The entire industry changes directions to pursue the latest gizmo. There have been more useless gadgets invented in the last few years than in any period of human history, just because it's "doable" from a technology point of view. A lot of technology driven innovations fail because they ignore social, economic and market factors. Technology should make the life of people easier instead of bombarding them with new gadgets. People must be at the heart of Ambient Intelligence environments. Technological building blocks should integrate into systems and products that are serving the people and that are logical steps in the social and cultural context of the users.

3 Remarks on Power Constraints

The common vision at the moment is that the Ambient Intelligence environment will be intelligent, responsive, connected and wireless. Despite all progress in IC technology information-processing devices that are needed to build these environments, unfortunately still require energy to function. Ambient Intelligence hardware mainly builds on four components: distributed processing, hierarchical storage, tangible interfaces and ubiquitous communication. The availability of energy and the type of energy source largely determine the architecture and functionality of the different devices and how intelligence in terms of the distribution of processing power, storage, interfaces and communication functions, is organized in the Ambient Intelligence system. Careful balancing is necessary because the energy constraints of each of these functions do not scale in the same way.

Power consumption for "processing" for instance is directly related to the complexity and operating speed of the processing circuit. Processing circuits can span orders of magnitudes in complexity ranging from a few gates doing intermittent simple tasks on sensor signals to a multi-million gates device operating at GOPS speed e.g., for processing video streams. At the moment computation requires approximately 1 nJ/OP of energy. Over time power needed for processing will roughly scale down according to Moore's law because of the shrink in transistor size and parameters that is the basis of this law (@ constant complexity). Special implementation options like asynchronous logic are useful to optimize the power constraints of specific logic circuitry further [3].

Power consumption in storage depends on the applied storage technology. Silicon storage will be used when size is a premium. Power consumption is then related to the number of read/write cycles on the memory and scales with Moore's law. Disks (either magnetic or optical) are particularly suited for bulk storage (Gbytes). Already for years the price per Mbyte for disk storage is a factor 10 lower than that of silicon memory. Because of the mechanism disk drives continuously consume (some) power when activated. However, the trend towards smaller-diameter disks improves the power efficiency of the drive because the power loss due to air shear is given by [1]:

$$P = constant \times D^{4.6} \times R^{2.8} \; . \tag{1}$$

Where P is the power loss, D the disk diameter and R the rotation rate. HDDs with 70 mm disks rotating at 15000 rpm consume approximately 100 mW per Gbyte; a value that has been dropping steeply over the past decade.

Power consumption in interfaces spans a wide dynamic range from narrow bandwidth interfaces to sensors to high-speed audio and video interfaces. The power efficiency of especially the high-speed interfaces becomes more and more important because of the trend towards high quality displays in portable, battery-powered devices.

Power consumption in communication has two facets. On the one hand we have the power consumption of the digital signal (pre) processing and encoding part, following more or less the same rules as "processing" and on the other hand of the RF and mixed-signal functions of the communication circuit. In the latter parts the energy efficiency in terms of energy/bit depends on noise and distortion parameters. Shannon's law [4] defines the theoretical upper bound for the maximum obtainable error-free bit rate in a (bandwidth limited) channel as a logarithmic function of the signal-to-noise ratio of the channel. In formula:

$$C = B \log_2(1 + s) \tag{2}$$

where C is the maximum obtainable error-free data speed in bit/s, B the bandwidth of the channel in Hz and s the signal-to-noise ratio in that channel.

In layman's terms Shannon's law implies that a certain amount of energy per bit is needed to bring it over the channel and that this energy increases as noisier the channel is. Furthermore, in wireless communication this figure strongly depends on the distance to be bridged as can be seen from the power link formula below.

$$P_{\text{link}} = \varepsilon \left(\alpha + \beta.d^{\gamma} \right) C + P_{\text{stand-by}} . \tag{3}$$

Where ε is a bit-overhead factor (e.g., for error correction purposes), α the computation energy to transceive a single bit, β the transmission cost factor per bit, d the transmission distance, γ the path loss exponent, which varies between 2 and 4 (factor 4 for indoors), C the transmission rate (bit/s) and $P_{\text{stand-by}}$ the standby power to keep the receiver enabled for signaling purposes (wake-up). As a rule of thumb one can use an energy figure of about 150 nJ/bit to bridge a distance of 10 meters; the equivalent of 150 instructions on a low power microprocessor. As a consequence, power consumption in communication does not scale down with advances in IC technologies.

These reflections are summarized in Fig. 2, published earlier in [5]. In this figure some examples of hardware technologies needed for Ambient Intelligence systems are combined in a single graph, where the X-axis represents some ball-park figures for power consumption, while the Y-axis tells something about the information processing capacity that might be bytes of

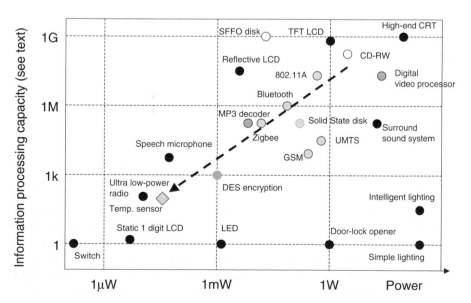

Fig. 2. Technologies mapped on power/information graph [5]

memory for storage, displayed raw bit rate for displays, captured information for sensors, effective bit rate for communication and operations per second for computing. Most technologies cluster in the upper-right part of the graph because high performance (still) goes hand in hand with (high) power consumption. To create context awareness in an Ambient Intelligence system current communication technologies like Zigbee, Bluetooth and IEEE802.11 (dotted line) have to be supplemented with "to develop" ultra-low power communication devices in the lower left corner of the graph, because current local and home area communication technologies consume too much power for remote sensor purposes.

4 Ambient Intelligence Architectures

Many devices or sub-systems working together in a network create Ambient Intelligence environments. All Ambient Intelligence devices have on the one hand communication capabilities for the interaction with other devices and on the other hand computational capabilities to implement the interaction protocols and to process the information. We have seen that energy constraints certainly are a determinative factor for the distribution of functionality over the devices in the network. It is common understanding that, because of these power constraints, ambient intelligence environments will consist of a hierarchy of devices with optimized functionality. We distinguish three generic classes based on their dimensions, power source and mobility factor viz.:

- Autonomous micro devices (so-called micro-Watt nodes)
- Portable mini devices (so-called milli-Watt nodes)
- Static maxi devices (so-called Watt nodes)

Figure 3 maps these device categories on the capacity/power graph that we used before.

4.1 Autonomous Micro Devices

The class of autonomous micro devices is meant to collect and disseminate a range of environmental data like light, temperature, sound pressure, vibrations, humidity, heart frequency, etc. Thus, these devices harvest data in the process of perceiving the environment, making them an essential element in the "awareness" part of Ambient Intelligence. Networks of these devices might implement applications like environmental control in homes and offices; person or health monitoring; identification; diagnostics; smart home; security; robot control; etc. Crucial is that the devices are lightweight, extremely small and very cheap. They have to operate autonomously without any supervision over their entire lifetime and need to organize themselves in networks without any manual interaction. For their energy supply

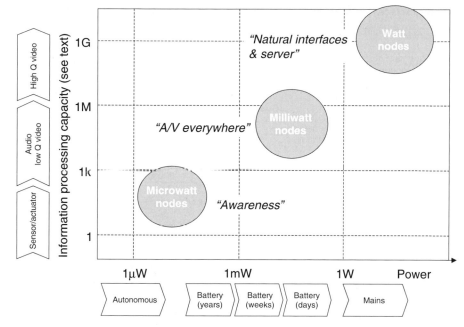

Fig. 3. Hierarchy and specialization of Ambient Intelligence nodes

they depend on energy extraction from the environment by scavenging either light or electromagnetic energy or mechanical energy from vibrations, sound pressure or temperature differences. Conflicting requirements on size and energy harvesting capabilities severely limits the power consumption of autonomous micro systems as we can learn from Fig. 4 that compares some micro-power-generators that have been reported in literature. For comparison: zinc-air batteries produce 1000–$1500\,\mathrm{mWh/cm^3}$ and lithium batteries (rechargeable) some 200–$400\,\mathrm{mWh/cm^3}$. Specifically micro-power-generators using vibrations as energy source suffer from decreasing conversion efficiency when their physical dimensions are scaled down. Smaller dimensions mean that the resonance frequency goes up, which detunes the device from the normally low-frequency vibrations in a building.

Figure 5 gives a schematic of an autonomous micro device. Besides the micro power generator (μPG) it includes one or more sensors, some computational element (ASIC), a radio (RF) and last but not least some element for energy storage (battery or capacitors). Current prototypes of such micro systems pack in volumes of approximately $5\,\mathrm{cm^3}$. The Ferdinand-Braun Institute in Berlin (Germany) [11] prospects a shrink towards a volume of $1\,\mathrm{cm^3}$ in 2005 and of 1–$5\,\mathrm{mm^3}$ in 2010. So within a decade autonomous micro devices can become the "electronic dust" or "e-grain" that is often mentioned in Ambient Intelligence vision documents.

262   F. Snijders

Energy source	Conversion technique	Institute	Size MPG	Harvested power
Fuel	Combustion engine	Berkeley [6]	1 cm³	> 30 mW
Thermal	thermo electric	DTS [6]	225 mm³	15 µW
Light (outdoors)	photovoltaic	-	-	0.15-15 mW/cm²
Light (indoors)	photovoltaic	-	-	0.01 mW/cm²
Acoustic noise	piezo	-	-	3 nW/cm² @75dB 960 nW/cm² @100dB
Vibration	electro-magnetic	- [7]	240 mm³	1 mW@320Hz
Vibration	electro-magnetic	AML [8]	1 cm³	830µW@60Hz
Vibration	electro-magnetic	Sheffield [9]	25 mm³	1µW@70Hz 0.1 mW@300Hz
Vibration	electro-magnetic	Southampton [10]	4 cm³	4.9 mW@99Hz
Vibration	piezo	Southampton [10]	529 mm²	3 µW@95Hz

Fig. 4. Comparison of Micro Power Generator principles

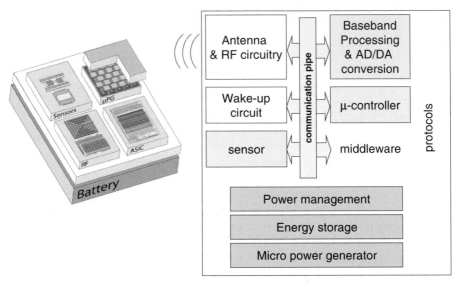

Fig. 5. Schematic of an autonomous micor system with micro power generator

Autonomous micro devices with physical dimensions of a few mm³ can harvest only a very limited amount of energy from their environment. Consequently they have to operate in a burst mode, sending their information in a short burst to the receiver when sufficient energy has been scavenged. Afterwards they have to go in a sleep mode until they have collected sufficient energy to emit the next burst. Because of the limited resources the communication protocol of autonomous micro devices must be very simple; no complicated handshake is possible.

Passive cavity resonators are special instantiations of autonomous micro devices. They consist of a cavity of which the RF reflection coefficient is influenced by the event to be measured. This event e.g., temperature, vibrations or acoustical noise influences the dimensions of the cavity. As a result the device backscatters a "modulated" signal that can be detected at some distance when excited by an outside RF source. The cavity adds a unique "signature" to the reflected signal for identification purposes. Passive cavities are particularly suited for integration in the construction of the building.

An early example of this passive resonator principle was discovered in 1952 in the American Embassy in Moscow as a spying device. A metal cavity resonating at 330 MHz that included an acoustic diaphragm was found integrated in the Great Seal of the United States hanging above the ambassador's desk. Thus when excited by an RF signal the device did backscatter the conversation in the room without any batteries, wires or active components. Modern versions of passive cavities can be made in silicon technology.

4.2 Portable Mini Devices

Portable mini devices will be the workhorses of the Ambient Intelligence environment. They are wireless infotainment pocket consoles; battery or micro fuel cell powered and small enough to be carried on the body. In this category one will find the current and future GSM and UMTS phones with built-in cameras and Audio/Video (A/V) storage, wireless video tablets, portable access devices to audio and video information, personal digital assistants and so on. In terms of functionality they are the "A/V everywhere" part of the Ambient Intelligence environment.

Battery capacity is not expected to increase dramatically in the coming years, which sets some limits on the processing, storage and communication capabilities of the portable mini devices. We expect that these portable devices will have processing powers of up to a few hundreds of MOPS/s, a handful of megabytes of storage and a communication speed of some tens of Mbit/s. They will be equipped with power efficient bright displays. PolyLED (polymer light emitting diode) or OLED (organic light emitting diode) are promising display technologies for this purpose, because they are lightweight, flexible and power efficient since they do not require backlighting and can work with low drive voltages. With FOLED (flexible OLED) technology roll-up displays will become feasible in the next few years to combine large(r) displays with small device dimensions.

Figure 6 gives a functional diagram of a portable mini device or milli-Watt node. The nodes are capable of 2-way communication (transmit/receive pipe) and have sufficient processing power on-board to execute complex communication protocols both for access to wireless public networks as well as in-home networks. It is to be expected that they will switch adaptively between both network types depending upon the environment they are in. In general they will lodge embedded processor and digital signal processing cores for communication and audio/video processing functions. Depending on the functionality they contain sensors (e.g., for biometric security), microphones, speakers and/or displays with the appropriate drivers and preprocessors.

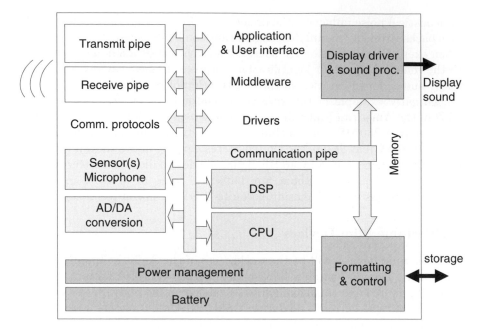

Fig. 6. Functional diagram of milli-Watt node

4.3 Static Maxi Devices

Static maxi devices or "Watt nodes" constitute the highest layer in the Ambient Intelligence device hierarchy. Watt nodes have neither power (besides heat and fan noise) nor volume constraints and consequently can lodge large amounts of processing horsepower and bulk storage. Examples of "Watt nodes" in the Ambient Intelligence environment are residential gateways, home servers, large (plasma) displays, etc. Their functional architecture (Fig. 7) resembles that of milli-Watt nodes but with additional interfaces to (wired) broadband access networks and of course much higher performance figures. Watt nodes also implement the processing-intensive multi-modal interfaces and User Interface (UI) applications that ultimately will allow the user to interact in a natural way with his/her Ambient Intelligence environment. They give access to personal audio and video databases that are managed via automatic feature extraction and retrieval. For instance the user can search the audio database via "query by humming [12]". Because of their almost unlimited processing power Watt nodes will play an important role in the control of the total Ambient Intelligence environment and in the format conversion of the audio/video material for consumption on the milli-Watt nodes with their limited internal processing power and different display technologies and resolution. Uninterruptible power supplies are, however, needed if the functionality needs to be permanently available.

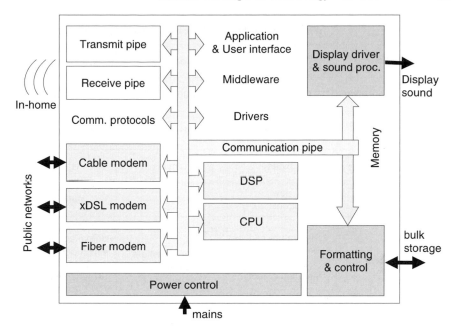

Fig. 7. Functional diagram of Watt-node

Figure 8 summarizes the main characteristics of the micro-Watt, milli-Watt and Watt-node.

4.4 Node Communication Structure

The various nodes in the Ambient Intelligence architecture organize them-selves spontaneously in networks (Fig. 9). Micro-Watt nodes are, due to their power limitations, very dependent on neighboring nodes for reliable commu-nication with the control center of the Ambient Intelligence environment they are in. We expect the milli-Watt nodes to act as a kind of relay station be-tween the micro-Watt nodes and the Watt-node that acts as control station. There are two possibilities. Micro-Watt nodes could use adjacent micro-Watt nodes as "hopping" devices to bridge some distance to the milli-Watt node or are triggered by a wake-up signal from the milli-Watt node when the latter is in reach. This requires the (ultra-low power) receiver to be on for most of the time which is unattractive from an energy point of view.

Signal hopping is very power efficient as we have seen from the power link formula (3). Breaking up the distance d in smaller hops helps a lot to save power for a single node. However, it requires some kind of store-and-forward mechanism and protocol in the micro-Watt nodes and of course the presence of sufficient micro-Watt nodes at certain distances to implement the hopping strategy efficiently. Furthermore, some extra storage capacity

	Micro-Watt node	Milli-Watt node	Watt node
Power source	energy scavenging from environment	battery	mains
Processing power	01-1 KOPS/s	1-200 MOPS/s	> 1 GOPS/s
Communication speed	0.01-1 kbit/s	0.01-10 Mbit/s	0.1-10 Gbit/s
Storage capacity	bytes	Mbytes	Gbytes
Mobility	low (electronic dust)	high	stationary
functions	temperature sensor person detection light sensor vibration detector	A/V everywhere terminal	• home server • comm.control center • high-performance A/V-terminal • human interface center
Main communication partners	micro-Watt nodes, (hopping) milli-Watt nodes	micro-Watt nodes, milli-Watt nodes & Watt nodes	• milli-Watt nodes • Watt nodes and • service providers

Fig. 8. Summary of the node characteristics according to the micro-, milli- and Watt-node model

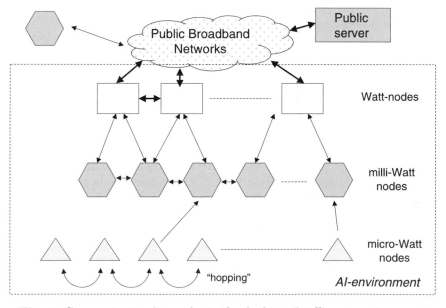

Fig. 9. Communication hierarchy in the Ambient Intelligence environment

is needed. An alternative communication strategy is that milli-Watt nodes trigger the transmission of micro-Watt nodes when they come close enough for efficient transmission and that they relay the information to the Watt-node that acts as the control center. We expect the nodes to switch automatically between those communication strategies depending on their environment to optimize the power budget of the system.

Milli-Watt nodes in their turn communicate with other milli-Watt nodes and with Watt-nodes in their Ambient Intelligence system environment. Watt-nodes serve as gateways to the external world for that Ambient Intelligence environment. Milli-Watt nodes can also communicate with external servers directly, but we expect this only to be the case for milli-Watt nodes that have left their "native" Ambient Intelligence environment and operate autonomously. An example of such operation is remote access to information in the home via a mobile milli-Watt pocket console. Secure access mechanisms are a prerequisite for this application.

5 Technology Timeframe

Looking to the technologies needed to implement Ambient Intelligence environments, the conclusion should be that the basic ingredients for on the one hand Watt-nodes and on the other hand milli-Watt nodes with somewhat limited functionality are already available. Of course a lot of improvements in terms of size, power dissipation, display resolution etc. are on the wish list, but in terms of functionality these nodes are doable (or exist) today. Micro-Watt nodes with their energy harvesting and pico-Watt radios are still a formidable technological challenge and it will take some 5–8 years to progress sufficiently to include micro-Watt nodes in the Ambient Intelligence concept. In our opinion Ambient Intelligence environments will evolve in 3 successive steps viz. "Connected Home", "A/V Anywhere" and "Context Aware Ambient Intelligence environments" as depicted in Fig. 10.

In 1–2 years time industry standards will be available to implement the first step referred to as the "connected home" or "digital home". In June 2003 seventeen leading companies[1] from the consumer electronics, mobile phone and computer industry announced the formation of the "Digital Home Working Group (DHWG)" with the objective to develop an interoperability framework of design guidelines based on open industry standards to obtain cross-industry convergence of digital home products. Examples of these products include PCs, TVs, set-top boxes, audio equipment, printers, mobile phones, PDAs, DVD players, etc., so mainly Watt-nodes and some milli-Watt nodes. Cross-industry convergence would give the end-users seamless access to digital content (music, video, photos) via wired and wireless home networks.

---

[1] Fujitsu, Gateway, Hewlett-Packard, IBM, Intel, Kenwood, Lenovo, Microsoft, NEC, Nokia, Panasonic, Philips, Samsung, Sharp, Sony, STMicroelectronics, Thomson.

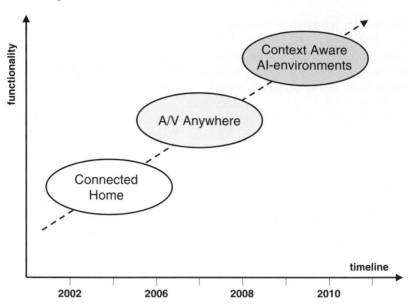

Fig. 10. Successive instantiations of Al-environments

The design guidelines will utilize well-known and established standards such as Internet Protocol (IP), Universal Plug and Play (UPnP), IEEE802.11xx and other common formats. Flaws in these standards for the audio/video environment will be patched up and over time the guidelines will evolve to incorporate emerging or subsequent versions of existing standards. Before the end of 2004 the Digital Home Working Group wants to have most of the guidelines in-place and certification and compliance testing bodies operational.

Within a couple of years in-home radio standards and (polyLED) display technology will have evolved sufficiently to implement the second phase in the creation of Ambient Intelligence environments (A/V Anywhere). In this phase a lot of portable mini devices (milli-Watt terminals) with A/V capabilities, bright high-resolution screens and miniaturized HDD drives will populate the Ambient Intelligence home. These devices communicate wirelessly at high bit rates with the stationary Watt-terminals (home servers and gateways). Video programs will migrate on demand from the stationary high-resolution displays onto the wireless pocket consoles, while the home server takes care of format and resolution conversion when needed. A few micro-devices in the form of intelligent tags for person identification and tracking will be part of the Ambient Intelligence system as a first step towards "awareness". Full-fledged context awareness (phase 3) will be introduced around 2008 when the picoradio and energy scavenging technologies have matured.

References

1. E. Grochowski and R. Halem; *Technological Impact of magnetic hard disk drives on storage systems*; IBM Systems Journal, Vol. 42, No. 2, 2003.
2. P. Notten et al.; *From materials research to battery integration;* submitted to J. Power Sources.
3. C. van Berkel, M. Josephs, and S. Nowick; *Scanning the technology: Applications of asynchronous circuits;* Proceedings of the IEEE, 87(2) pp. 223–233.
4. C. Shannon; *Communication in the presence of noise;* Proceedings of the IRE, vol. 37, Jan. 1949, pp. 10–21.
5. E. Aarts and R. Roovers; *Design, Automation and Test in Europe Conference and Exhibition (DATE)*, 2003, March 3–7, 2003.
6. O. Cugat and J. Lelamare; Energie portable: autonomie et integration dans l'environment humain; Colloque ENS de Cachan, 21–22 March 2002.
7. M. El-Hami et al.; *Design and fabrication of a new vibration-based electro-mechanical power generator*; Sensors and Actuators A, Vol. 92, pp. 335–342, 2001.
8. N. Ching et al.; *A laser-micromachined multi-modal resonating power transducer for wireless sensing systems*; Sensors and Actuators A, Vol. 97–98, pp. 685–690, 2002.
9. C. Williams and R. Yates; *Analysis of a micro-electric generator for Microsystems*; Sensors and Actuators A, Vol 52, pp. 8–11, 1996.
10. P. Glynne-Jones et al.; Towards a piezoelectric vibration-powered microgenerator; IEE Proc.-Sci Meas Technology, Vol. 148, no.2, March 2001.
11. E. Reichl (project coordinator); *Autarke Verteilte Mikrosysteme*; presentation for the Industriebeirat, Berlin, 21-05-2003.
12. S. Paauws; Cubthum: *A fully operational query by humming system*; In Fingerhut M., Proceedings ISMIR 2002, Paris, France.

Powering Ambient Intelligent Networks

S. Roundy, M. Strasser, and P.K. Wright

1 Introduction

The vision of ambient intelligence requires that there be thousands of small sensing, computation, and communication devices (or nodes) in the environment around us. Of course these devices need to have some source of power to perform their function. While power can be transmitted to them or energy stored on them, it would be highly desirable if most of the individual nodes could be completely self-sustaining. While this is not an absolute requirement for all types of devices in the network, it is hard to imagine a truly intelligent environment without at least many of the endpoint sensor devices being completely self-sustaining.

As advances in technology continue to reduce the size and cost of wireless sensor nodes, the problem of how to power them becomes more acute. At issue is the fact that the scaling down in size and cost of CMOS electronics has far outpaced the scaling of energy density in batteries, which are by far the most prevalent power supply currently used for wireless sensor devices. Therefore, the power supply is becoming the largest and most expensive component of the emerging wireless sensor nodes being proposed and designed. Furthermore, the power supply (usually a battery) is also the limiting factor on the lifetime of a sensor node. If wireless sensor networks are to become truly ubiquitous, then replacing batteries every year or two is simply cost prohibitive.

Several small low power wireless platforms on which the beginnings of ambient intelligence could be based are available commercially and are being developed in the research community. These platforms range in power consumption from hundreds of milliwatts [9] down to hundreds of microwatts [15,34]. Many approaches are used to reduce the aggregate power consumption of the network [32]. However, if some level of data throughput is to be maintained, power consumption cannot continue to fall indefinitely. While reduction of power consumption is essential, it is not in itself sufficient to solve the power problem without the development of new alternative sources.

The purpose of this chapter, then, is to review existing and potential power sources for the wireless sensor nodes that will make up an intelligent environment. Current state of the art, ongoing research, and theoretical limits for many potential power sources will be discussed. Possible methods of providing power for wireless sensor nodes may be classified into three groups: store energy on the node (i.e. a battery), distribute power to the node (i.e. a wire), scavenge available ambient power at the node (i.e. a solar cell). Power sources that fall into each of these three categories will be reviewed.

A direct comparison of vastly different types of power source technologies is difficult. For example, comparing the efficiency of a solar cell to that of a battery is not very useful. However, in an effort to provide general understanding of a wide variety of power sources, the following metrics will be used for comparison: power density, energy density (where applicable), and power density per year of use. Additional considerations include the complexity of the necessary power electronics and whether secondary energy storage is needed.

2 Energy Storage

Energy storage, in the form of electrochemical energy stored in a battery, is the predominant means of providing power to wireless devices today. In fact electrochemical batteries have been a dominant form of energy storage for the past 100 years. Batteries are probably the easiest power solution for wireless electronics because of their versatility. However, there other forms of energy storage that may be useful for wireless sensor nodes. Regardless of the form of the energy storage, the lifetime of the node will be determined by the fixed amount of energy stored on the device. While it is cost effective in some applications to repeatedly change or recharge batteries, if wireless sensor nodes are to become a ubiquitous part of the environment, it will no longer be cost effective in most situations.

The primary metric of interest for all forms of energy storage will be usable energy per unit volume (J/cm^3). Additionally, energy storage devices usually have a maximum *rate* at which energy can be drawn out (i.e. a maximum power level). Generally, this maximum power is related to the size of the device. In the case of a battery, the output voltage is more or less constant, and there is a limit on the maximum amount of current that can be drawn. Making the battery larger increases not only the amount of stored energy, but also the maximum allowable current draw. Thus a battery can be characterized by an energy density (J/cm^3), and a power density ($\mu W/cm^3$), which represents its maximum current draw (the voltage is constant) divided by its size. Another metric of interest, directly related to energy density, is the average power supplied per unit volume per unit time ($\mu W/cm^3/year$). This metric is simply the energy density divided by the time of operation (e.g., the number of seconds in a year). It does not take into account the maximum amount of instantaneous power that can be drawn from the energy storage device, but rather indicates the level of average power consumption that the device can supply for a year of operation. All three metrics will be used in the subsequent discussion.

2.1 Batteries

Macro-scale primary batteries are commonly available in the marketplace. Table 1 shows the energy density and standard operating voltage for a few common primary battery chemistries. Figure 1 shows the average power available from these battery chemistries versus lifetime. The data shown in Fig. 1 relate to the average power supplied, and not the maximum instantaneous current density that can be supplied. In the context of wireless sensor networks, maximum current is generally not an issue for macro-scale batteries. Therefore, the capacity, or energy density, of the battery dominates its consideration for use. Current limitations are an issue for micro-scale batteries and fuel cells and will be discussed below in that context. Note that while zinc-air batteries have the highest energy density (see Table 1), their lifetime is very short (see Fig. 1). They are, therefore, primarily used in applications with fairly high and relatively constant power consumption, such as hearing aids. While lithium batteries have excellent energy density and longevity, they are also the most expensive of the chemistries shown. Because of their combination of reasonably high energy density and low cost, alkaline batteries are widely used in consumer electronics.

Table 1. Energy density and voltage of three primary battery chemistries

Chemistry	Zinc-air	Lithium	Alkaline
Energy density (J/cm^3)	3780	2880	1200
Voltage (V)	1.4	3.0–4.0	1.5

Because batteries have a fairly stable voltage, electronic devices can often be run directly from the battery without any intervening power electronics. While this may not be the most robust method of powering the electronics, it

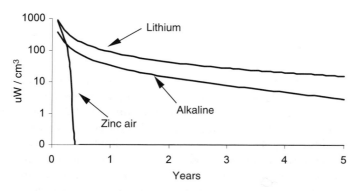

Fig. 1. Continuous power per cm^3 vs. lifetime for three primary battery chemistries

is often used and is advantageous in that it avoids the extra power consumed by power electronics.

Macro-scale secondary (rechargeable) batteries are commonly used in consumer electronic products such as cell phones, PDA's, and laptop computers. Table 2 gives the energy density and standard operating voltage of a few common rechargeable battery chemistries.

It should be remembered that rechargeable batteries are a *secondary* power source. Therefore, in the context of wireless sensor networks, another primary power source must be used to charge them. In some cases (such as cell phones or notebook computers) this is easily accomplished because the device can be periodically connected to the power grid. However, as wireless sensor devices become smaller, cheaper, and more widespread, periodically connecting the device to an energy rich source or the power grid will not be cost effective or may not even be possible. More likely, an energy scavenging source on the node itself, such as a solar cell, would be used to recharge the battery. One item to consider when using rechargeable batteries is that electronics to control the charging profile must often be used. These electronics add to the overall power dissipation of the device. However, like primary batteries, the output voltages are stable and power electronics between the battery and the load electronics can often be avoided.

Table 2. Energy density and voltage of three secondary battery chemistries

Chemistry	Lithium	NiMHd	NiCd
Energy density (J/cm^3)	1080	860	650
Voltage (V)	3.0	1.5	1.5

2.2 Micro-Batteries

The wireless sensor nodes available today tend to be about the size of a small matchbox or pager. One such sensor node, the Crossbow Mica Mote [4], is shown in Fig. 2. It is readily evident that the majority of the space is consumed by the batteries, packaging, and interconnects, not by the electronics. Therefore, in order to dramatically reduce the size of wireless sensor nodes, greater integration is necessary. A large research effort, with the ultimate goal of developing micro-batteries (or on-chip batteries), is underway at many institutions. This section will review some of the issues in the development of micro-batteries and some of the research in the field.

One of the main stumbling blocks to reducing the size of micro-batteries is power output due to surface area limitations of micro-scale devices. The maximum current output of a battery depends on the surface area of the electrodes. Because micro-batteries are so small, the electrodes have a small

Fig. 2. Mica Mote wireless sensor node from Crossbow [4]

surface area, and their maximum current output is also very small. This problem can also be alleviated to a certain degree by placing a large capacitor in parallel with the battery capable of providing short bursts of current. However, the capacitor itself consumes additional volume, and therefore may not be desirable in many applications.

The challenge of maintaining (or increasing) performance while decreasing size is being addressed on multiple fronts. Bates et al. at Oak Ridge National Laboratory have created a process by which a primary thin film lithium battery can be deposited onto a chip [1]. The thickness of the entire battery is on the order of 10's of μm, but the area is in the cm^2 range. This battery is in the form of a traditional Volta pile with alternating layers of Lithium Manganese Oxide (or Lithium Cobalt Oxide, LiCoO2), Lithium Phosphate Oxynitride and Lithium metal. Maximum potential is rated at 4.2 V with continuous maximum current output on the order of 1 mA/cm^2 and 5 mA/cm^2 for the LiCoO2 – Li based cell. A schematic of a battery fabricated with this process is shown in Fig. 3.

Work is being done on thick film batteries with a smaller surface area by Harb et al. [11], who have developed micro-batteries of Ni/Zn with an aqueous NaOH electrolyte. Thick films are on the order of 0.1 mm, but overall battery

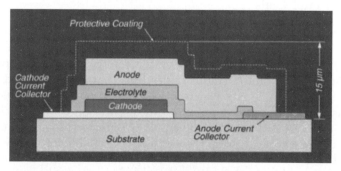

Fig. 3. Primary Lithium on chip battery proposed by Bates et al. [1]

thickness is minimized by use of three-dimensional structures, which reduce the overall thickness because the cathode and anode are side by side rather than stacked one on top of another. While each cell is only rated at 1.5 V, geometries have been duty-cycle optimized to give acceptable power outputs at small overall theoretical volumes (4 mm by 1.5 mm by 0.2 mm) with good durability demonstrated by the electrochemical components of the battery. The main challenges lie in maintaining a microfabricated structure that can contain an aqueous electrolyte.

Radical three dimensional structures are also being investigated to maximize power output. Hart et al. [12] have theorized a three dimensional battery made of series alternating cathode and anode rods suspended in a solid electrolyte matrix. Theoretical power outputs for a three dimensional microbattery of this type are shown to be many times larger than a two dimensional battery of equal size. Additionally, the three dimensional battery would have far lower ohmic ionic transport distances, and thus lower ohmic losses.

For example, a 1 cm^2 thin film with each electrode having a thickness of 22 μm and a 5 μm electrolyte, would have a maximum current density on the order of 5 mA. If the battery is restructured to have the same total volume, with square packing electrode rods (as Hart et al. have proposed) of 5 μm radius with 5 μm surface to surface distance, geometry dictates that the electrode surface area increases to 3.5 cm^2, which increases the current to 17.5 mA. Moreover, the ionic transport scale in the 2D structure is about 350% longer than the 3D case because the electrodes for the 3D case are much thinner. Therefore, decreased ohmic losses could further improve the maximum throughput to 20 mA at 4.2 volts. However, while the power density of the 3D battery would be larger, the energy density would reduce to 39% of the 2D case due to a lower volume percentage of electrolyte in the 3D battery. An additional concern is that the inherent non-uniformities in current distribution in three dimensional batteries (exacerbated by the particular complexity of this cell) may lead to difficulties with regard to device reliability on primary battery systems and cycle life in secondary battery systems.

2.3 Micro-Fuel Cells

Hydrocarbon based fuels have very high energy densities compared to batteries. For example, methanol has an energy density of 17.6 kJ/cm^3, which is about 6 times that of a lithium battery. Therefore, fuel cells are potentially very attractive for wireless sensor nodes. Like batteries, fuel cells produce electrical power from a chemical reaction. A standard fuel cell uses hydrogen atoms as fuel. A catalyst promotes the separation of the electron in the hydrogen atom from the proton. The proton diffuses through an electrolyte (often a solid membrane) while the electron is available for use by an external circuit. The protons and electrons recombine with oxygen atoms on the other side (the oxidant side) of the electrolyte to produce water molecules. This process is illustrated in Fig. 4. While pure hydrogen can be used as a

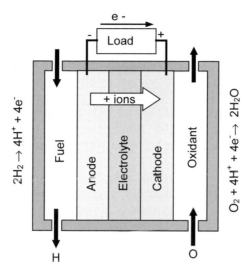

Fig. 4. Illustration of how a standard hydrogen fuel cell works [19]

fuel, other hydrocarbon fuels are often used. For example in Direct Methanol Fuel Cells (DFMC) the anode catalyst draws the hydrogen atoms out from the methanol.

Most single fuel cells tend to output open circuit voltages around 1.0–1.5 volts. Of course, like batteries, the cells can be placed in series for higher voltages. The voltage is quite stable over the operating lifetime of the cell, but it does fall off with increasing current draw. Figure 5 shows the voltage versus current load for a typical fuel cell. Notice that as the current density increases, the dominant loss mechanism also changes. Because the voltage

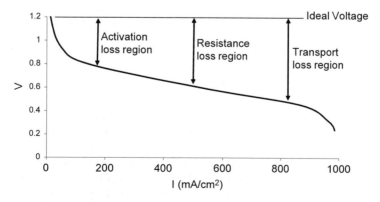

Fig. 5. Typical voltage versus current curve for a fuel cell [19]

drops with current, it is likely that some additional power electronics will be necessary if replacing a battery with a fuel cell.

Large scale fuel cells have been used as power supplies for decades. For example, the Apollo spacecraft used alkaline fuel cells for electricity. More recently, fuel cells have been developed as alternative power supplies for automobiles. Cells using a variety of fuels and electrolytes have been successfully used at the macro scale. Recently fuel cells have gained favor as a replacement for consumer batteries [14]. Small, but still macro-scale, fuel cells are likely to soon appear in the market as battery rechargers and battery replacements [45].

The research trend is toward micro-fuel cells that could possibly be closely integrated with wireless sensor nodes. Like micro-batteries, a primary metric of comparison in micro-fuel cells is power density in addition to energy density. As with micro-batteries, the maximum continuous current output is dependent on the electrode surface area. Microfabricated fuel cells offer an advantage in surface to volume ratio, thereby giving them a higher power density. Likewise microfabricated features can potentially improve gas diffusion and lower the internal resistance [19], both of which improve efficiency.

Fuel cells tend to operate better at higher temperatures, which are more difficult to maintain for micro-fuel cells. Efficiencies of large scale fuel cells have reached approximately 45% electrical conversion efficiency and nearly 90% if cogeneration is employed [21]. Efficiencies for micro-scale fuel cells will certainly be lower. The maximum obtainable efficiency for a micro-fuel cell is still uncertain. Demonstrated efficiencies are generally below 1% [17].

Many research groups are working on microfabricated partial systems that typically include an electrolyte membrane, electrodes, and channels for fuel and oxidant flow. Recent examples include the hydrogen based fuel cells developed by Hahn et al. [10] and Lee et al. [23]. Both systems implement microfabricated electrodes and channels for fuel and oxidant flow. The system by Hahn et al. produces power on the order of $100\,\mathrm{mW/cm^2}$ from a device of $0.54\,\mathrm{cm^2}$ in size. The system by Lee et al. produces $40\,\mathrm{mW/cm^2}$. It should be noted that the fundamental characteristic here is power per unit area rather than power per unit volume because the devices are fundamentally planar. Complete fuel storage systems are not part of their studies, and therefore an energy or power per unit volume metric is not appropriate. Fuel conversion efficiencies are not reported.

Hydrogen storage at small scales is a difficult problem that has not yet been solved. Primarily for this reason, methanol based micro-fuel cells are also being investigated by numerous groups. Holloday et al. [17] have demonstrated a research methanol fuel processor with a total size on the order of several $\mathrm{mm^3}$. This fuel processor has been combined with a thin fuel cell, $2\,\mathrm{cm^2}$ in area, to produce roughly $25\,\mathrm{mA}$ at 1 volt with 0.5% overall efficiency. They are targeting a 5% efficient cell. Additionally Mench et al. [27] have proposed a complete three dimensional methanol fuel cell with a volume of $1\,\text{cm}^3$. The system would contain all necessary elements except a methanol reservoir. The projected power output is $1\,\text{W}/\text{cm}^3$ at a projected efficiency of 30%, however, to the authors' knowledge, this has not been demonstrated. It should be noted that this is a stacked fuel cell and that if fuel volume were included, the power density would be lower.

Given the energy density of fuels such as methanol, fuel cells need to reach efficiencies of at least 20% in order to be more attractive than primary batteries. Nevertheless, at the micro scale, where battery efficiencies are also lower, a lower efficiency fuel cell could still be attractive. Finally, providing for sufficient fuel and oxidant flows is a very difficult task in micro-fuel cell development. The ability to microfabricate electrodes and electrolytes does not guarantee the ability to realize a micro-fuel cell. The problem of microfabricating the fuel reservoir and all of the plumbing is arguably a more difficult task than the microfabrication of electrodes. To the authors' knowledge, a self-contained on-chip fuel cell has yet to be demonstrated.

2.4 Ultra-Capacitors

Ultra-capacitors represent a compromise of sorts between rechargeable batteries and standard capacitors. Capacitors can provide significantly higher power densities (or current densities) than batteries, however their energy density is lower by about 2 to 3 orders of magnitude. Ultra-capacitors (also called super-capacitors or electrochemical capacitors) achieve significantly higher energy density than standard capacitors, but retain many of the favorable characteristics of capacitors, such as long life and short charging time.

Rather than just storing charge across a dielectric material, as capacitors do, ultra-capacitors store ionic charge in an electric double layer to increase their effective capacitance. By introducing an electrolyte researchers hope to limit ionic diffusion between plates, trading power generation for longer running times. However, this is still an area of technical difficulty. The energy density of commercially available ultra-capacitors is about 1 order of magnitude higher than standard capacitors and about 1 to 2 orders of magnitude lower than rechargeable batteries (or about 50 to $100\,\text{J}/\text{cm}^3$). Because of their increased lifetimes, short charging times, and high power densities, ultra-capacitors could be very attractive as secondary power sources in place of rechargeable batteries in some wireless sensor network applications. Corporations working on such ultra-capacitors include NEC [30] and Maxwell [26].

2.5 Micro-Heat Engines

Like fuel cells, heat engines convert the very high chemical energy density of hydrocarbon fuels to either electrical or mechanical energy. At large scales, fossil fuels are still the dominant source of power generation. Gasoline, for example, has an energy density of approximately $31\,\text{kJ}/\text{cm}^3$. In large power

280   S. Roundy et al.

generation facilities, chemical energy embedded in fossil fuels is converted
to thermal energy through combustion, then to mechanical energy through
a heat engine, and finally to electricity through a magnetic generator. The
complexity of this process, along with the many complicated parts has served
as a barrier to the miniaturization of this technology. Recent progress made
in silicon micromachining is removing this barrier by making it possible and
cost effective, to fabricate extremely small (on the order of microns) complex
mechanical and fluidic structures. As a result, there have been a number of
recent research projects with the goal of a achieving a micro-scale heat engine.

To the authors' knowledge, the first such effort was proposed and undertaken by Epstein et al. in the mid 1990's [5]. Epstein and colleagues designed and built high speed turbomachinery with bearings, a generator, and
a combustor all within a cubic centimeter using a combination of silicon deep
reactive ion etching, fusion wafer bonding, and thin film processes. An application ready power supply would additionally require auxiliary components,
such as a fuel tank, engine and fuel controller, electrical power conditioning
with short term storage, thermal management and packaging. The expected
performance of the system proposed by Epstein et al. was 10–20 Watts of
output electrical power at thermal efficiencies of 5–20%. Figure 6 shows an
example microturbine test device used for turbomachinery and air bearing
development.

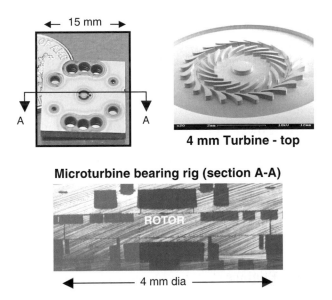

Fig. 6. Micro-turbine development devices, which consist of a 4 mm diameter single
crystal silicon rotor enclosed in a stack of five bonded wafers used for micro air
bearing development. Reproduced with permission from Luc Frechette at Universite
de Sherbrooke

Many other research groups have subsequently undertaken the development of micro-heat engines. Various approaches have been taken including micro gas turbine engines [18], Rankine steam turbines [22], rotary Wankel internal combustion engines [8], free and spring loaded piston internal combustion engines [25, 44], and thermal-expansion-actuated piezoelectric power generators [36, 48].

Most of these and similar efforts are at initial stages of development and predicted performance has not been demonstrated. However, predictions range from 0.1–10 W of electrical power output, with typical masses $\sim$1–5 g and volumes $\sim$1 cm^3. Microengines are not expected to reduce further in size due to manufacturing and efficiency constraints. At small scales, viscous drag on moving parts and heat transfer to the ambient and between components increase, which adversely impacts efficiency.

Given the relatively large power outputs of microengines (approximately 1 to 10 watts), the engine would need to operate at very low duty cycles, and secondary, short term energy storage would be needed for wireless sensor network applications. In this scenario, the engine would intermittently charge up a secondary battery or capacitor. While the low duty cycle operation would alleviate lifetime issues for the engine, it would also increase the complexity, size, and cost of the system.

The two primary potential benefits of microengines over primary batteries are their high power density and their high energy density. While high power density is an advantage in some applications, it could be detrimental in wireless sensor applications. This leaves only the higher energy density of the fuel as a significant benefit for wireless sensor networks. Future projected thermal efficiencies are approximately 20%. At this efficiency, the energy density of the system is below 6 kJ/cm^3, constituting approximately a 2X improvement over the energy density of primary lithium batteries.

2.6 Radioactive Power Sources

Radioactive materials contain extremely high energy densities. As with hydrocarbon fuels, this source of energy has been used for decades on a much larger scale. However, it has not been exploited on a small scale as would be necessary to power wireless sensor networks. The use of radioactive materials can pose a serious health hazard, and is a highly political and controversial topic. It should, therefore, be noted that the goal here is neither to promote nor discourage investigation into radioactive power sources, but to present their potential, and the research being done in the area.

The total energy emitted by radioactive decay of a material can be expressed as in 1.

$$E_t = A_c E_e T \qquad (1)$$

where E_t is the total emitted energy, A_c is the activity in Curies (Ci), E_e is the average energy of emitted particles, and T is the time period over which power is collected. Table 3 lists several potential radioisotopes, their half-lives, activity volume densities and energy and power densities based on radioactive decay. It should be noted that materials with lower activities and higher half-lives will produce lower power levels for more time than materials with comparatively short half-lives and high specific activities. The half-life of the material has been used as the time over which power would be collected. Only alpha and beta emitters have been included because of the heavy shielding needed for gamma emitters. Finally, uranium 238 is included for purposes of comparison only.

While the energy density numbers reported for radioactive materials are extremely attractive, it must be remembered that in most cases the energy is being emitted over a very long period of time. Second, efficient methods of converting this power to electricity at small scales do not exist. Therefore, efficiencies would likely be extremely low.

Table 3. Comparison of radio-isotopes

Material	Half-life (years)	Activity Volume Density (Ci/cm^3)	Energy Density (J/cm^3)	Power Density (mW/cm^3)
^{238}U	4.5×10^9	6.34×10^{-6}	2.23×10^{10}	1.6×10^{-4}
^{32}Si	172.1	151	3.3×10^8	60.8
^{63}Ni	100.2	506	1.6×10^8	50.6
^{90}Sr	28.8	350	3.7×10^8	407
^{32}P	0.04	5.2×10^5	2.7×10^9	2.14×10^6

Recently, Li and Lal [24] have used the ^{63}Ni isotope to actuate a conductive cantilever. As the beta particles (electrons) emitted from the ^{63}Ni isotope collect on the conductive cantilever, there is an electrostatic attraction. At some point, the cantilever contacts the radioisotope and discharges, causing the cantilever to oscillate. Up to this point the research has only demonstrated the actuation of a cantilever, and not electric power generation. However, electric power could be generated from an oscillating cantilever. The reported power output, defined as the change over time in the combined mechanical and electrostatic energy stored in the cantilever, is 0.4 pW from a 4 mm × 4 mm thinfilm of ^{63}Ni. This power level is equivalent to $0.52 \, \mu W/cm^3$. However, it should be noted that using $1 \, cm^3$ of ^{63}Ni is impractical. The reported efficiency of the device is 4×10^{-6}.

3 Power Distribution

In addition to storing power on a wireless node, in certain circumstances power can be distributed to the node from a nearby energy rich source. It is difficult to characterize the effectiveness of power distribution methods by the same metrics (power or energy density) because in most cases the power received at the node is more a function of how much power is transmitted rather than the size of the power receiver at the node. Nevertheless an effort is made to characterize the effectiveness of power distribution methods as they apply to wireless sensor networks.

3.1 Electromagnetic (RF) Power Distribution

The most common method (other than wires) of distributing power to embedded electronics is through the use of RF (Radio Frequency) radiation. Many passive electronic devices, such as electronic ID tags and smart cards, are powered by a nearby energy rich source that transmits RF energy to the passive device. The device then uses that energy to run its electronics [7, 16]. This solution works well, as evidenced by the wide variety of applications where it is used, if there is a high power scanner or other source in very near proximity to the wireless device. It is, however, less effective in dense ad-hoc networks where a large area must be flooded with RF radiation to power many wireless sensor nodes.

Using a very simple model neglecting any reflections or interference, the power received by a wireless node can be expressed by 2 [39].

$$P_r = \frac{P_0 \lambda^2}{4\pi R^2} \tag{2}$$

where P_0 is the transmitted power, λ is the wavelength of the signal, and R is the distance between transmitter and receiver. Assume that the maximum distance between the power transmitter and any sensor node is 5 meters, and that the power is being transmitted to the nodes in the 2.4–2.485 GHz frequency band, which is the unlicensed industrial, scientific, and medical band in the United States. Federal regulations limit ceiling mounted transmitters in this band to 1 watt or lower. Given a 1 watt transmitter, and a 5 meter maximum distance the power received at the node would be 50 μW, which is probably on the borderline of being really useful for wireless sensor nodes. However, in reality the power transmitted will fall off at a rate faster than $1/R^2$ in an indoor environment. A more likely figure is $1/R^4$. While the 1 watt limit on a transmitter is by no means general for indoor use, it is usually the case that some sort of safety limitation would need to be exceeded in order to flood a room or other area with enough RF radiation to power a dense network of wireless devices.

3.2 Wires, Acoustic, Light, Etc.

Other means of transmitting power to wireless sensor nodes might include wires, acoustic emitters, and light or lasers. Distributing power through a wired power grid may be effective in certain circumstances. For example, if a new "smart" building was being designed, then wires for distributed sensors could be included in the design. However, in most situations, sensors would be distributed in an existing environment and data communication would take place wirelessly. Installing wires for power distribution and/or data communication would be cost prohibitive in most situations.

Energy in the form of acoustic waves has a far lower power density than is sometimes assumed. A sound wave of 100 dB in sound level only has a power level of $0.96 \, \mu W/cm^2$.

One could also imagine using a laser or other focused light source to direct power to each of the nodes in the sensor network. However, to do this in a controlled way, distributing light energy directly to each node, rather than just flooding the space with light, would likely be too complex and not cost effective. If a whole space is already flooded with light, then this source of power becomes attractive. However, this situation has been classified as "power scavenging" and will be discussed in the following section.

4 Power Scavenging

Unlike power sources that are fundamentally energy reservoirs, power scavenging sources are usually characterized by their power density rather than energy density. Energy reservoirs have a characteristic energy density, and how much average *power* they can provide is then dependent on the lifetime over which they are operating. On the contrary, the *energy* provided by a power scavenging source depends on how long the source is in operation. Therefore, the primary metric for comparison of scavenged sources is power density, not energy density.

4.1 Photovoltaics (Solar Cells)

At midday on a sunny day, the incident light on the earth's surface has a power density of roughly $100 \, mW/cm^2$. Single crystal silicon solar cells exhibit efficiencies of 15%–20% [33] under high light conditions, as one would find outdoors. Common indoor lighting conditions exhibit far lower power density than outdoor lighting conditions. Common office lighting provides about $100 \, \mu W/cm^2$ at the surface of a desk. Single crystal silicon solar cells are better suited to high light conditions and the spectrum of light available outdoors [33]. Thin film amorphous silicon or cadmium telluride cells offer better efficiency indoors because their spectral response more closely matches that of artificial indoor light. Still, these thin film cells only offer about 10% efficiency. Therefore, the power available from photovoltaics ranges from about $15 \, mW/cm^2$ outdoors at midday to $10 \, \mu W/cm^2$ indoors. Table 4 shows the measured power outputs from a cadmium telluride solar cell (Panasonic BP-243318) at varying distances from a 60 watt incandescent bulb.

Table 4. Power from a cadmium telluride solar cell at various distances from a 60 watt incandescent bulb and under standard office lighting conditions

Distance	20 cm	30 cm	45 cm	Office Light
Power (μW/cm^2)	503	236	111	7.2

A single solar cell has an open circuit voltage of about 0.6 volts. Individual cells are easily placed in series, especially in the case of thin film cells, to get almost any desired voltage needed. A current vs. voltage (I–V) curve for a typical five cell array (wired in series) is shown below in Fig. 7. Unlike the voltage, current densities are directly dependent on the light intensity.

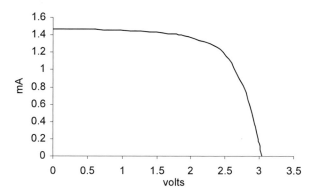

Fig. 7. Typical I–V curve from a cadmium telluride solar array (Panasonic BP-243318)

Solar cells provide a fairly stable DC voltage through much of their operating space. Therefore, they can be used to directly power electronics in cases where the current load is such that it allows the cell to operate on the high voltage side of the "knee" in the I–V curve and where the electronics can tolerate some deviation in source voltage. More commonly solar cells are used to charge a secondary battery. Solar cells can be connected directly to rechargeable batteries through a simple series diode to prevent the battery from discharging through the solar cell. This extremely simple circuit does not ensure that the solar cell will be operating at its optimal point, and so power production will be lower than the maximum possible. Secondly, rechargeable batteries will have a longer lifetime if a more controlled charging profile is employed. However, controlling the charging profile and the operating point of the solar cell both require more electronics, which use power themselves. An analysis needs to be done for each individual application to determine what level of power electronics would provide the highest net level of power to the load electronics. Longevity of the battery is another consideration to be considered in this analysis.

4.2 Temperature Gradients (Thermoelectrics)

Conversion of heat into electricity and vice versa can take place by making use of thermoelectric effects. The most important thermoelectric effects are the closely related Seebeck and Peltier effects. Certain metals and semiconductors possess thermoelectric properties and can thus be used to construct heat conversion devices, so called thermoelectric generators. Thermoelectric generators are configured in a similar way as their famous brothers, Peltier elements (see Fig. 8). However, whereas electrically powered Peltier elements are used for cooling purposes, accomplished by dissipating Peltier heat on one side and setting Peltier heat free on the other side, thermoelectric generators convert external heat into electric energy. Given a thermal gradient between the two ends of a bar made of a thermoelectric material, a voltage is generated by the thermoelectric Seebeck effect. The bigger the thermal gradient and the larger the Seebeck-coefficient of the thermoelectric material employed, the more voltage is generated. In a thermoelectric device, a large number of such bars of thermoelectric materials ("legs") are connected in series and give a total voltage that is the sum of all individual Seebeck voltages. If an electrical load is attached to an operating thermoelectric generator, an electric current flows through the load. The output power of the device is equal to the heat that is converted into electricity by the Seebeck effect.

Because of their high cost and relatively low efficiency, thermoelectric generators have traditionally only been used in niche applications, such as space

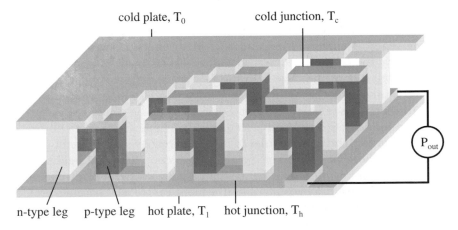

cold plate, T_0 cold junction, T_c

P_{out}

n-type leg   p-type leg   hot plate, T_1 hot junction, T_h

Fig. 8. Schematic view of a thermoelectric generator that consists of n- and p-type thermoelectric legs which are electrically connected in series and that are thermally arranged in parallel

exploration. However, recent improvement in the thermoelectric properties of semiconductor materials [46] and the greatly reduced power consumption of microelectronic circuitry have opened up new fields of application for thermoelectric generators. In particular, thermoelectric devices can be applicable in the field of ambient intelligence where robust, self-sustaining, small-size power sources are needed. Thus, the following discussion will be focused on small-size and low-power thermoelectric generators that could easily be integrated into small sensor units.

From thermodynamics, the theoretical efficiency, η, of a thermal converter based on irreversible effects like those taking place in a thermoelectric generator must hold. The theoretical efficiency is given by:

$$\eta < \frac{\Delta T}{T} \tag{3}$$

where ΔT is the temperature difference and T the temperature of the hot side in Kelvin. Thus, for $\Delta T = 5\,\mathrm{K}$ and $T = 300\,\mathrm{K}$ which could be found for many ambient applications, there is an efficiency limit of about 1%. While this low efficiency estimate may initially stifle enthusiasm for thermoelectric generators, it must be remembered that many low-power applications exist for which power consumption is in the mW or µW range. Hence, the first wrist-watches powered by body-heat have been manufactured by Seiko and Citizen. The power consumption of the Seiko watch is specified at $1\,\mu\mathrm{W}$ with a driving voltage of 1.5 V [20]. Citizen [3] employs a thermoelectric generator with a contact area of $0.7\,\mathrm{cm} \times 0.7\,\mathrm{cm}$ and a height of 1.5 mm that generates a voltage of 0.5 V and has a power output of $13.8\,\mu\mathrm{W}$ under load at a temperature difference of 1 K. Thermal environments other than the human body, such as those encountered in the automotive field or domestic environments, provide higher temperature gradients and, therefore, higher potential power output.

There is a significant difference between conventional, large-size thermoelectric generators and small-size generators. Because a silicon chip is usually on the order of hundreds of micrometers thick, its internal thermal resistance is small compared to the thermal contacts of the surrounding assembly. This means that most of the heat is dissipated in the environment rather than in the actual thermoelectric device. Clearly, small-size thermoelectric generators need to be optimized separately to make the best use of the heat available. It can be shown that if the heat current rather than the temperature difference is given, it is the output power and not the thermal efficiency that needs to be maximized [29]. The output power, P_{out}, of a thermal converter is given by:

$$P_{\mathrm{out}} = \eta q_{\mathrm{in}} < \frac{\Delta T}{T} K \Delta T \tag{4}$$

where q_{in} is the heat current entering the device and K is its thermal resistance.

Indeed, the output power of a small-size thermoelectric generator has been shown to be proportional to the squared temperature difference [42], and is given by:

$$P_{\text{out}} = \beta Z^* V_g (\Delta T)^2 \tag{5}$$

$$Z^* = \frac{\alpha^2}{\lambda^2 \rho} \tag{6}$$

where V_g is the volume of the generator, and β is a constant depending on the internal geometrical arrangement of the generator and the thermal resistors connected to the device, i.e. the assembly.

The figure of merit Z^* is used to compare thermoelectric materials suitable for small-size thermoelectric generators. As in the case of the classical, large-scale devices [42], efficient small-size thermoelectric generators should be constructed of thermoelectric materials possessing a large Seebeck coefficient α, a low electrical resistivity ρ and a low thermal conductivity λ. However, because λ determines the actual temperature drop for a given heat current, its impact on the output power is stressed in small-size devices given the squared dependence of P_{out} on ΔT.

Large Seebeck effects are found in semiconducting materials which makes silicon an interesting choice for thermoelectric devices. Even more promising are compound semiconductors such as bismuth tellurides because of their low thermal conductivity [46]. However, these V–VI-semiconductors are difficult to produce and large efforts are necessary to make these materials compatible with standard silicon chip fabrication processes [2]. Conventional approaches use thermoelectric materials that are deposited on foils and laminated to stacks to build up the generator [41,47]. Stordeur and Stark recently presented such a discrete thermoelectric device which generates a voltage of 2.5 V and an electrical power output of 20 μW under load at a temperature drop of 5 K. Its contact area is 0.9 cm × 0.52 cm and its height is 1.8 mm.

Another approach is to build silicon-based thermoelectric generators directly in CMOS technology as presented by Infineon Technologies. For such devices, a voltage of 5 V per cm^2 of chip area and an electrical power output of 1 μW per cm^2 are achieved for a matched load at a temperature drop of about 5 K [43]. The dense CMOS integration allows a large number of several thousand thermoelectric legs on a chip area of a few mm^2. This means that a comparatively high open circuit voltage can be achieved using a silicon-based thermoelectric generator. However, silicon is not a first-choice thermoelectric material because of its relatively high thermal conductivity. The advantage of a silicon-based device rather lies in the low cost CMOS fabrication and the possible monolithic on-chip integration of the microelectronic circuitry. The realized silicon chips have dimensions of 3.2 mm × 2.2 mm and a height of 150 μm. The chip area can be easily enlarged to 1 cm^2 by simple design changes. It is also possible to increase the generator's volume by stacking several chips on top of each other. The parameters of the three quoted generators are summarized in Table 5. In order to enable comparison of the data in the table, the parameters are related to a standardized generator volume of $1\,cm^3$.

Table 5. Standardized data of some small-size thermoelectric generators

Reference of Thermoelectric Generator	Voltage per Temp. Difference and Volume $[V/K/cm^3]$	Power Per Square of temp. drop and volume $[\mu W/K^2/cm^3]$	Realized Generator Size $[mm^3]$
Citizen 2003	6.8	187.8	73.5
Stordeur 1997	5.9	9.5	84.2
Strasser 2003	66.7	2.7	1.1

As a matter of course, the power output per volume is determined by the type of thermoelectric generator chosen. On the other hand, the devices can be connected in series or in parallel just like batteries to give either a high voltage or high current depending on the specific application. Stable energy supplies based on thermoelectric generators may always need some sort of power management to control varying heat currents given by the environment. If, for example, the temperature gradient changes sign (i.e. hot and cold side of the generator are exchanged) the generated voltage also changes its sign. This has to be compensated for by an appropriate circuit which should also include some kind of rechargeable storage.

4.3 Human Power

An average human body burns about $10.5\,MJ$ of energy per day. (This corresponds to an average power dissipation of $121\,W$.) Starner has proposed tapping into some of this energy to power wearable electronics [40]. For example watches are powered using both the kinetic energy of a swinging arm and the heat flow away from the surface of the skin [37].

The conclusion of studies undertaken at MIT suggests that the most energy rich and most easily exploitable source occurs at the foot during heel strike and in the bending of the ball of the foot [38]. This research has led to the development of piezoelectric shoe inserts capable of producing an average of $330\,\mu W/cm^2$ while a person is walking. The shoe inserts have been used to power a low power wireless transceiver mounted to the shoes. While this power source is of great use for wireless nodes worn on a person's foot, the problem of how to get the power from the shoe to the point of interest still remains.

The sources of power mentioned above are passive power sources in that the human doesn't need to do anything other than what he or she would normally do to generate power. There is also a class of power generators that could be classified as active human power in that they require the human to perform an action that he or she would not normally perform. For example Freeplay [6] markets a line of products that are powered by a constant force spring that the user must wind up. While these types of products are extremely useful, they are not very applicable to the concept of ambient intelligence or wireless sensor networks because it would be impractical and not cost efficient to individually wind up every node in a dense network.

4.4 Wind/Air Flow

Wind power has been used on a large scale as a power source for centuries. Large windmills are still common today. However, the authors' are unaware of any efforts to try to generate power at a very small scale (on the order of a cubic centimeter) from air flow. The potential power from moving air is quite easily calculated as shown in (7).

$$P = \frac{1}{2}\rho A v^3 \tag{7}$$

where P is the power, ρ is the density of air, A is the cross sectional area, and v is the air velocity. At standard atmospheric conditions, the density of air is approximately $1.22\,\mathrm{kg/m^3}$. Figure 9 shows the power per square centimeter versus air velocity at this density.

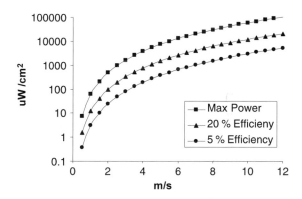

Fig. 9. Maximum power density vs. air velocity. Power density assuming 20% and 5% conversion efficiencies are also shown

Large scale windmills operate at maximum efficiencies of about 40%. Efficiency is dependent on wind velocity, and average operating efficiencies are usually about 20%. Windmills are generally designed such that maximum efficiency occurs at wind velocities around $8\,\mathrm{m/s}$ (or about 18 mph). At low air velocity, efficiency can be significantly lower than 20%. Figure 9 also shows power output assuming 20% and 5% efficiency in conversion. As can be seen from the graph, power densities from air velocity are quite promising. As there are many possible applications in which a fairly constant air flow of a few meters per second exists, it seems that research leading to the development of devices to convert air flow to electrical power at small scales is warranted.

4.5 Pressure Variations

Variations in pressure can be used to generate power. For example one could imagine a closed volume of gas that undergoes pressure variation as the daily temperature changes. Likewise, atmospheric pressure varies throughout the day. The change in energy for a fixed volume of ideal gas due to a change in pressure is simply given by

$$\Delta E = \Delta PV \tag{8}$$

where ΔE is the change in energy, ΔP is the change in pressure, and V is the volume. A quick survey of atmospheric conditions around the world reveals that an average atmospheric pressure change over 24 hours is about 0.2 inches Hg or 677 Pa, which corresponds to an energy change of 677 μJ/cm^3. If the pressure cycles through 0.2 inches Hg once per day, for a frequency of 1.16×10^{-5}, the power density would then be 7.8 nW/cm^3.

An average temperature variation over a 24 hour period would be about 10°C. The change in pressure to a fixed volume of ideal gas from a 10°C change in temperature is given by

$$\Delta P = \frac{mR\Delta T}{V} \tag{9}$$

where m is mass of the gas, R is gas constant, and ΔT is the change in temperature. If 1 cm^3 of helium gas were used, a 10°C temperature variation would result in a pressure change of 1.4 MPa. The corresponding change in energy would be 1.4 J per day, which corresponds to 17 μW/cm^3. While this is a simplistic analysis and assumes 100% conversion efficiency to electricity, it does give an idea of what might be theoretically expected from naturally occurring pressure variations.

To the authors' knowledge, there is no research underway to exploit naturally occurring pressure variations to generate electricity. Some clocks, such as the "Atmos clock", are powered by an enclosed volume of fluid that undergoes a phase change under normal daily temperature variations. The volume and pressure change corresponding to the phase change of the fluid mechanically actuates the clock. However, this is on a large scale, and no effort is made to convert the power to electricity.

4.6 Vibrations

Low level mechanical vibrations are present in many environments. Examples include HVAC ducts, exterior windows, manufacturing and assembly equipment, aircraft, automobiles, trains, and household appliances. Table 6 shows the results of measurements on several different vibration sources performed by the authors. It will be noticed that the primary frequency of most sources is between 60 and 200 Hz. Acceleration amplitudes range from about 1 to 10 m/s².

Table 6. Summary of several vibration sources

Vibration Source	Peak Acc. (m/s²)	Freq. (Hz)
Base of 3-axis machine tool	10	70
Kitchen blender casing	6.4	121
Clothes dryer	3.5	121
Door frame just as door closes	3	125
Small microwave oven	2.25	121
HVAC vents in office building	0.2–1.5	60
Wooden deck with foot traffic	1.3	385
Breadmaker	1.03	121
External windows next to street	0.7	100
Notebook computer w/CD.	0.6	75
Washing machine	0.5	109
Second story floor of a wood frame office building	0.2	100
Refrigerator	0.1	240

A simple general model for power conversion from vibrations has been presented by Williams and Yates [49]. The final equation for power output from this model is shown here as equation (10).

$$P = \frac{m\zeta_e^2 A^2}{4\omega \left(\zeta_e + \zeta_m\right)^2} \qquad (10)$$

where P is the power output, m is the oscillating proof mass, A is the acceleration magnitude of the input vibrations, ω is the frequency of the driving vibrations, ζ_m is the mechanical damping ratio, and ζ_e is an electrically induced damping ratio. In the derivation of this equation, it was assumed that the natural frequency of the oscillating system matched the frequency of the driving vibrations. While this model is oversimplified for many implementations, it is useful to get a quick estimate on potential power output from a given source. Three interesting relationships are evident from this model.

1. Power output is proportional to the oscillating mass of the system.

2. Power output is proportional to the square of the acceleration amplitude of the input vibrations.
3. Power is inversely proportional to frequency assuming a constant input acceleration magnitude across all frequencies.

Point three indicates that the generator should be designed to resonate at the lowest frequency peak in the vibrations spectrum provided that higher frequency peaks do not have a higher acceleration magnitude. Many spectra measured by Roundy et al. [35] verify that in general the lowest frequency peak has the highest acceleration magnitude.

Figures 10 through 12 provide a range of power densities that can be expected from vibrations similar to those listed above in Table 6. The data shown in the figures are based on calculations from the model of Williams and Yates and do not consider the technology that is used to convert the mechanical kinetic energy to electrical energy.

Several researchers have developed devices to scavenge power from vibrations [28, 31, 35, 49]. Devices include electromagnetic, electrostatic, and piezoelectric methods to convert mechanical motion into electricity. Theory, simulations, and experiments performed by the authors suggest that for devices on the order of $1\,\mathrm{cm}^3$ in size, piezoelectric generators will offer the most attractive method of power conversion. Piezoelectric conversion offers higher potential power density from a given input, and produces voltage levels on the right order of magnitude. Roundy et al. [34] have demonstrated a piezoelectric power converter of $1\,\mathrm{cm}^3$ in size that produces $200\,\mu\mathrm{W}$ from input vibrations of $2.25\,\mathrm{m/s^2}$ at $120\,\mathrm{Hz}$. Both Roundy et al. and Ottman et al. have demonstrated wireless transceivers powered from vibrations.

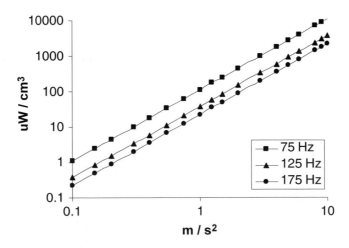

Fig. 10. Power density vs. vibration amplitude for three frequencies

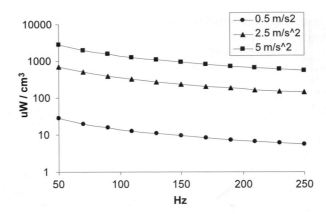

Fig. 11. Power density vs. frequency of vibration input

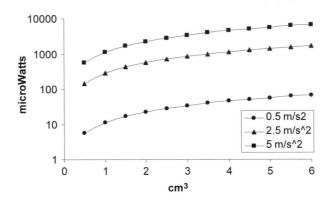

Fig. 12. Total power vs. device size. Frequency of input vibrations is 125 Hz

Figure 13 shows the generator, power circuit, and transceiver developed by Roundy et al.

Because vibration based power generators almost always have fairly low damping (Q ∼ 30), it is essential that the natural frequency of the converter match the dominant frequency of the input vibrations. In many applications the vibration spectrum is known beforehand, and the system can be designed to resonate at the appropriate frequency. However, in other applications the frequency of the input vibrations is either unknown or changes with time. Therefore, self-tuning generators would be necessary in these situations.

The power signal generated from vibration generators needs a significant amount of conditioning to be useful to wireless electronics. The converter produces an AC voltage that needs to be rectified. Additionally the magnitude of the AC voltage depends on the magnitude of the input vibrations, and so is not very stable. Typically, some sort of energy reservoir is needed along with a voltage regulator or DC-DC converter. However, the energy reservoir

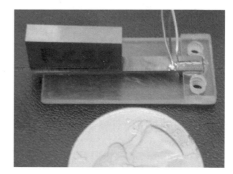

Fig. 13. Piezoelectric generator, power circuit, and radio powered from vibrations of $2.25\,\mathrm{m/s^2}$ at $120\,\mathrm{Hz}$

could be as small as a capacitor of several microfarads depending on the application. Although more power electronics are needed compared with some other sources, commonly occurring vibrations can provide power on the order of hundreds of microwatts per cubic centimeter, which is quite competitive compared to other power scavenging sources.

5 Summary

An effort has been made to give an overview of the many potential power sources for wireless sensor networks. Well established sources, such as batteries, have been considered along with potential sources on which little or no work has been done. Because some sources are fundamentally characterized by energy density (such as batteries) while others or characterized by power density (such as solar cells) a direct comparison with a single metric is difficult. Adding to this difficulty is the fact that some power sources do not make much use of the third dimension (such as solar cells), so their fundamental metric is power per square centimeter rather than power per cubic centimeter. Nevertheless, in an effort to compare all possible sources, a summary table is shown below as Table 7. Note that power density is listed as $\mu\mathrm{W/cm^3}$, however, it is understood that in certain instances the number reported really represents $\mu\mathrm{W/cm^2}$. Such values are marked with a "*". Note also that with only two exceptions, values listed are numbers that have been demonstrated or are based on experiments rather than theoretical optimal

Table 7. Comparison of various potential power sources for wireless sensor networks. Values shown are actual demonstrated numbers except in two cases which have been italicized

Power Source	P/cm^3 ($\mu W/cm^3$)	E/cm^3 (J/cm^3)	$P/cm^3/yr$ ($\mu W/cm^3/Yr$)	Secondary Storage Needed	Voltage Regulation	Comm. Available
Primary Battery	–	2880	90	No	No	Yes
Secondary Battery	–	1080	34	–	No	Yes
Micro-Fuel Cell	–	3500	110	Maybe	Maybe	No
Ultra-capacitor	–	50–100	1.6−3.2	No	Yes	Yes
Heat engine	–	3346	106	Yes	Yes	No
Radioactive(^{63}Ni)	0.52	1640	0.52	Yes	Yes	No
Solar (outside)	15000*	–	–	Usually	Maybe	Yes
Solar (inside)	10*	–	–	Usually	Maybe	Yes
Temperature	40*†	–	–	Usually	Maybe	Soon
Human Power	330	–	–	Yes	Yes	No
Air flow	*380***	–	–	Yes	Yes	No
Pressure Variation	*17*††	–	–	Yes	Yes	No
Vibrations	200	–	–	Yes	Yes	No

* Denotes sources whose fundamental metric is power per *square* centimeter rather than power per *cubic* centimeter.

† Demonstrated from a 5°C temperature differential.

**Assumes air velocity of 5 m/s and 5% conversion efficiency.

††Based on a $1\,cm^3$ closed volume of helium undergoing a 10°C change once per day.

values. The two cases in which theoretical numbers are listed have been italicized. In many cases the theoretical best values are explained in the text above.

Almost all wireless sensor nodes are presently powered by batteries. This situation presents a substantial roadblock to the widespread development of ambient intelligence because the replacement of batteries is cost prohibitive. Furthermore, a battery that is large enough to last the lifetime of the device would dominate the overall system size and cost, and thus is not very attractive. It is therefore essential that alternative power sources be considered and developed.

This chapter has attempted to characterize a wide variety of such sources. It is the authors' opinion that no single alternative power source will solve the problem for all, or even a large majority of cases. However, many attractive and creative solutions do exist that can be considered on an application-by-application basis.

Acknowledgements

The authors especially thank Brian Otis and Professor Jan Rabaey at the Berkeley Wireless Research Center (BWRC) for their collaboration and development of the radio transceiver shown in Fig. 13. The authors gratefully acknowledge the Intel Corporation for support under the Noyce Fellowship, the support of DARPA under grant # F33615-02-2-4005, and the support of the California Energy Commission under contract # 500-01-043.

References

1. Bates JB, Dudney NJ, Neudecker B, Ueda A, Evans CD (2000) Thin-film lithium and lithium-ion batteries. Solid State Ionics 135: 33–45.
2. Boettner H, et al. (2001) New Thermoelectric Components Using Micro-System-Technologies, Proc. Sixth European Workshop on Thermoelectrics, Germany, Sept. 20th to 21st (2001).
3. Citizen Eco-Drive Thermo (2003) Citizen press release, Basel, March 2003.
4. Crossbow (2003) http://www.xbow.com/Products/Wireless_Sensor_Networks. htm, 2003.
5. Epstein AH, et al (1997) Micro-Heat Engine, Gas Turbine, and Rocket Engines – The MIT Microengine Project. AIAA 97-1773, 28th AIAA Fluid Dynamics Conf., Snowmass Village, CO, June 1997.
6. Freeplay (2003) http://www.freeplay.net, 2003.
7. Friedman D, Heinrich H, Duan D-W (1997) A Low-Power CMOS Integrated Circuit for Field-Powered Radio Frequency Identification. Proceedings of the 1997 IEEE Solid-State Circuits Conference, pp. 294–295.
8. Fu K, Knobloch AJ, Martinez FC, Walther DC, Fernandez-Pello C, Pisano AP, Liepmann D, (2001) Design and Fabrication of a Silicon-Based MEMS Rotary Engine, ASME IMECE, New York, November 11–16, 2001.
9. Haartsen JC, Mattison S (2000) Bluetooth – A New Low-Power Radio Interface Providing Short-Range Connectivity. Proceedings of the IEEE, 88(10):1651–1661
10. Hahn et al. (2003) www.pb.izm.fhg.de/hdi/040_groups/group4/fuelcell_micro. html.
11. Harb J, LaFollete RM, Selfridge RH, Howell LL (2002) Microbatteries for self-sustained hybrid micropower supplies. Journal of Power Sources 104: 46–51.
12. Hart RW, White HS, Dunn B., Rolison DR (2003) 3-D Microbatteries. Electrochemistry Communications 5:120–123.
13. Heikes RR, Ure RW (1961) Thermoelectricity: Science and Engineering, Interscience Publishers, 1961.
14. Heinzel A, Hebling C, Muller M, Zedda M, Muller C (2002) Fuel cells for low power applications. Journal of Power Sources 105:250–255.
15. Hill J, Culler D (2002) Mica: A Wireless Platform for Deeply Embedded Networks. IEEE Micro 22(6):12–24
16. Hitachi (2003) http://www.hitachi.co.jp/Prod/mu-chip/, 2003.
17. Holloday JD, Jones EO, Phelps M, Hu J (2002) Microfuel processor for use in a miniature power supply. Journal of Power Sources 108:21–27.
18. Isomura K, Murayama M, Yamaguchi H, Ijichi N, Asakura H, Saji N, Shiga O, Takahashi K, Tanaka S, Genda T, Esashi M (2002) Development of Microturbocharger and Microcombustor for a Three-Dimensional Gas Turbine at Microscale. ASME IGTI 2002 TURBO EXPO, Paper GT-2002-30580, Amsterdam, Netherlands, June 6, 2002.

19. Kang S, Lee S-JJ, Prinz FB (2001) Size does matter: the pros and cons of miniaturization. ABB Review 2:54–62.
20. Kishi M, et al. (1999) IEEE 18th Int. Conf. on Thermoelectrics (1999) 301–307.
21. Kordesh K, Simader G (2001) Fuel cells and their applications. VCH Publishers, New York.
22. Lee C, Arslan S, Liu Y-C, Fréchette LG, (2003) Design of a Microfabricated Rankine Cycle Steam Turbine for Power Generation. ASME IMECE, Washington, D.C., Nov. 16–21, 2003.
23. Lee SJ, Chang-Chien A, Cha SW, O'Hayre R, Park YI, Saito Y, Prinz FB (2002) Design and fabrication of a micro fuel cell array with "flip-flop" interconnection. Journal of Power Sources 112: 410–418.
24. Li H, Lal M (2002) Self-reciprocating radio-isotope powered cantilever, Journal of Applied Physics 92(2):1122–1127.
25. Matta LM, Nan M, Davis SP, McAllister DV, Zinn BT, Allen MG (2001) Miniature Excess Enthalpy Combustor for Microscale Power Generation. AIAA Paper 2001-0978, 39th Aerospace Sciences Meeting and Exhibit, Reno, NV, January 8–11, 2001.
26. Maxwell (2003) http://www.maxwell.com/ultra-capacitors/, 2003.
27. Mench MM, Wang ZH, Bhatia K, Wang CY (2001) Design of a micro direct methanol fuel cell (.DMFC). ASME IMECE, New York, November 11–16, 2001.
28. Meninger S, Mur-Miranda JO, Amirtharajah R, Chandrakasan AP, Lang, JH (2001) Vibration-to-Electric Energy Conversion. IEEE Trans. VLSI Syst., 9:64–76.
29. Min G, Rowe DM (1995) Peltier Devices as Generators, CRC Handbook of Thermoelectrics, CRC Press, 1995.
30. NEC (2003) http://www.nec-tokin.net/now/english/product/hypercapacitor/outline02.html, 2003.
31. Ottman GK, Hofmann HF, Lesieutre GA (2003) Optimized piezoelectric energy harvesting circuit using step-down converter in discontinuous conduction mode. IEEE Transactions on Power Electronics. 18(2):696–703.
32. Rabaey JM, Ammer MJ, da Silva JL, Patel D, Roundy S (2000) PicoRadio Supports Ad Hoc Ultra-Low Power Wireless Networking. IEEE Computer, 33(7):42–48.
33. Randall JF (2003) On ambient energy sources for powering indoor electronic devices, Ph.D Thesis, Ecole Polytechnique Federale de Lausanne.
34. Roundy S, Otis B, Chee, Y-H, Rabaey J, Wright PK (2003) A 1.9 GHz Transmit Beacon using Environmentally Scavenged Energy. ISPLED 2003, August 25–27, 2003, Seoul Korea.
35. Roundy S, Wright PK, Rabaey J (2003) A Study of Low Level Vibrations as a Power Source for Wireless Sensor Nodes. Computer Communications, 26 (11):1131–1144.
36. Santavicca D, Sharp K, Hemmer J, Mayrides B, Taylor D, Weiss J, (2003) A Solid Piston Micro-engine for Portable Power Generation. ASME IMECE, Washington, D.C., Nov. 16–21, 2003.
37. Seiko (2003) http://www.seikowatches.com, 2003.
38. Shenck NS, Paradiso JA (2001) Energy Scavenging with Shoe-Mounted Piezoelectrics. IEEE Micro, 21 (2001) 30–41.
39. Smith AA, (1998) Radio frequency principles and applications: the generation, propagation, and reception of signals and noise, IEEE Press, New York.

40. Starner T (1996) Human-powered wearable computing. IBM Systems Journal, 35 (3):618–629.
41. Stordeur M, Stark I (1997) Low Power Thermoelectric Generator – Self-sufficient Energy Supply for Micro Systems, IEEE 16th Int. Conf. on Thermoelectrics, S. 575–577 (1997).
42. Strasser M, et al. (2002) Sens. Act. A, Vol. 97-98C, 528–535.
43. Strasser M, et al. (2003) Micromachined CMOS Thermoelectric Generators as On-Chip Power Supply, Tech. Dig. Transducers '03, Boston, USA (2003).
44. Toriyama T, Hashimoto K, Sugiyama S (2003) Design of a Resonant Micro Reciprocating Engine for Power Generation. Transducers'03, Boston, MA, June 2003.
45. Toshiba (2003) http://www.toshiba.co.jp/about/press/2003_03/pr0501.htm, 2003.
46. Venkatasubramanian R, et al. (2001) Thin-film Thermoelectric Devices with High Room-Temperature Figures of Merit, Nature, Vol. 413, Oct. 2001.
47. Wetzig K, Schneider CM (2003) Metal Based Thin Films for Electronics, Wiley-VCH, 2003.
48. Whalen S, Thompson M, Bahr D, Richards C, Richards R (2003) Design, Fabrication and Testing of the P3 Micro Heat Engine. Sensors and Actuators 104(3):200–208.
49. Williams CB, Yates RB (1995) Analysis of a micro-electric generator for Microsystems. Transducers 95/Eurosensors IX, 369–372.

Ultra-Low Power Integrated Wireless Nodes for Sensor and Actuator Networks

J. Ammer, F. Burghardt, E. Lin, B. Otis, R. Shah, M. Sheets, and J.M. Rabaey

1 Introduction and Motivation

In earlier chapters, it was described how the capability to build and deploy dense wireless networks of hetcrogcncous nodes collecting and disseminating wide ranges of environmental data enables a multiplicity of exciting scenarios. To mention just a few: smart homes with optimized environmental control and energy management, coordinated media delivery, integrated security, identification and personalization, robot control and guidance in automatic manufacturing environments, warehouse inventory, integrated patient monitoring, diagnostics and drug administration in hospitals, and interactive toys and museums. The mind-boggling opportunities emerging from these technologies indeed give rise to new definitions to the terms "ubiquitous computing" and "user interface". Regardless of the specific application, however, they all rely on a network of ubiquitously-distributed sensor, compute and actuation nodes, which are integrated and embedded into the fabrics of our daily living environment. This explains why the name "ambient intelligence" is often attributed to such environments.[1]

Widespread deployment of these ubiquitous networks requires that some economic and physical realities are met. More precisely, the physical implementation of an individual network node is constrained by three important metrics: **cost, size and power**.

A true ubiquitous deployment is only economically feasible if the cost of the individual elements is ignorable, or, in other words, the **electronics have become disposable**. Depending upon the intended market and the number of nodes required for a single deployment, this translates to price points per node ranging from $10 to substantially below $1. Achieving a node cost this low requires a minimal number of components, a high level of integration, simple and cheap packaging and assembly, and avoidance of any expensive components and/or technologies. In addition, the cost for the deployment and the maintenance of the network should be ignorable.

---

[1] In this chapter, we focus solely on the sensing-and-actuation aspects of an ambient intelligence environment. The networking requirements of the other components of such an environment, such as multimedia acquisition, transport, and display, are well understood and have been covered extensively in existing literature.

Embedding the components into the daily environment (walls, furniture, clothing, etc) further requires that the form factor of the entire sensor node must be **very small**. Typically, sizes smaller than $1\,\mathrm{cm}^3$ are necessary. Again, a very high level of integration is mandatory if such small dimensions are to be achieved.

However, of the three implementation constraints, power (or energy, depending on how the node is powered) turns out to be the most fundamental one. To keep cost down and to allow for a flexible deployment, most or even all nodes must be un-tethered. Cost considerations also dictate that frequent replacement of the energy source of the node (especially in a ubiquitous deployment scenario) is out of the question. This leads to the general guideline that a network node must be **self-sufficient from an energy perspective for the lifetime of the product**. This could be multiple years for applications such as smart homes. The energy storage capability of a node is limited by the size constraints. While a single-time charge could work for applications with life cycles below one year, replenishment of the energy supply using energy-scavenging is often a necessity. The table below illustrates the finite power density of state-of-the-art energy sources. (A more detailed discussion of the power sources of ambient intelligence networks can be found in another chapter in this book.)

Table 1. Power Density of Energy Scavenging Sources [14]

Power Source	Power Density ($\mu\mathrm{W/cm}^3$/year)	Lifetime
Primary batteries	90	1 year
Capacitors	2–4	Needs replenishment
Photovoltaic	10–15,000 (in $\mu\mathrm{W/cm}^2$)	∞
Vibrational Converter	200 (in $\mu\mathrm{W/cm}^3$)	∞

From the table, we learn that the average power consumption that a node is allowed to consume mainly depends upon the *volume* of the node (assuming that the electronic components of the node are small compared to the power supply which proves to be a reasonable assumption), the *intended lifetime*, and the *rate of replenishment*. An interesting design guideline is that an **average power dissipation of about $100\,\mu\mathrm{W/cm}^3$** is what can be comfortably sustained by a host of non-replenishable and replenishable power sources (or a combination thereof). This number represents quite a challenge, as the power consumption of most state-of-the-art integrated wireless processing nodes is approximately three orders-of-magnitude higher. Hence, power considerations restrict the amount of processing that can be performed within a node, and further determine the type of wireless connectivity that can be obtained between the nodes.

In the design of these nodes, we have experienced that the wireless interface takes up a large fraction of the power and size budget of the node. While the demands of the sensing and digital processing components cannot be ignored, their duty cycle is typically very low. As is explained later, the exploitation of advanced sleep and power-down techniques makes it possible to make their average power dissipation virtually ignorable. Based on these considerations, a disproportional section of the chapter is devoted to the design of ultra-low power wireless links (for sensor networks). While optical communication approaches offer the potential of very low power and small size, line-of-sight and directivity considerations make them less attractive. We therefore limit our discussion to radio-frequency (RF) interfaces.

Integrated wireless sensor nodes occupy a unique corner of the semiconductor and embedded system design space, and push against many traditional design boundaries. This chapter gives an insight on what techniques and technologies may help to lead to the successful implementation of these ultra-low {cost, size, power} nodes. It starts with an overview of the state-of-the-art in wireless sensor nodes constructed from commercial- of-the-shelf components. Next, techniques to dramatically reduce the power consumption of the wireless node are analyzed. The chapter concludes with a discussion on how to accomplish full integration and a roadmap for the future.

2 State-of-the-Art in Wireless Sensor Nodes

Already in the late 80s and early 90s, a number of visionaries pointed out how a combination of technology trends – that is, cheaper and lower power electronics resulting from the continuation of Moore's law, major progress in ad-hoc wireless connectivity, and the availability of mesoscale peripheral devices such as sensors, energy sources and antennas, would lead to a so-called "third wave of computing" (with main-frames and personal computers the first and second ones [2, 21]. This third wave is now commonly called **pervasive computing and/or ambient intelligence**. Over the past decade(s), we have witnessed a number of experiments trying to make such an environment a reality. Examples are the Apple Newton, the Berkeley InfoPad [3], and a variety of wireless palmtops. None of these approaches however led to true ubiquity. It is only with the advent of the concept of the ad-hoc wireless sensor and actuator network (AWSAN) that we got a first glimpse of the true nature of an ambient intelligence environment.

AWSANs were conceived by a number of researchers working on DARPA-related projects. In the defense context, ad-hoc networks of distributed wireless sensor nodes are seen as a means to collect detailed and up-to-date information about the topology, composition and dynamics of a complex scene such as a battlefield in situations where centralized data gathering does not work or apply. The latter may be due to a lack of installed infrastructure, line-of-sight observation, or blockage. While most of the community focused on the data gathering, routing, collection and processing challenges, a number of researchers set out to build prototype test-beds. These would allow testing the developed algorithms under real world conditions. One of the best known of these early prototypes is the WINS node, developed jointly by UCLA and Rockwell [22]. A derivative node, developed by Sensoria [17], was the cornerstone of the default test bed of the DARPA Sense IT program – which was one of the driving programs for sensor network developments in the early days. A picture of the WINS node and a table with its key parameters are shown in Fig. 1. A 900 MHz ISM-band DSSS radio from Rockwell is used for the wireless link. The node features plenty of computational power (using a full-fledged 32 bit RISC processor) and memory. However, while the node supports all the functionality needed for sensor networking, its dimensions and its power consumption are respectively 2 and 3 orders of magnitude above the goals set for a truly embedded sensor network.

- StrongARM SA 1100 32-bit RISC processor, 1MB SRAM, 4MB flash
- 900MHz spread spectrum radio, with dedicated microcontroller: 32KB RAM, 1MB bootable flash
- 3.5"x3.5"x3" package size
- 0.5 to 1 W operational power (on)

Fig. 1. Rockwell/UCLA WINS node

Similar nodes with comparable performance numbers were constructed at a range of research institutes (just to mention a few, the Berkeley "PicoNodeI" [11] and the MIT "µAmps" [19]. See Fig. 2), and have been used extensively to validate some of the ground truths and falsities in distributed

PicoNodeI   µAmps-I

Fig. 2. Pogrammable wireless sensor nodes (using the stack of cards model). Both these nodes have a StrongArm processor at their center. The PicoNodeI further has an FPGA and uses a Bluetooth radio. The µAMPS node uses a ChipCon radio for the wireless link

sensor networks that had emerged over the years. Both these nodes consume up to 0.5 W when in full operation (they do support sophisticated power management techniques, however) and have a volume of 10's of cubic centimeters.

The UC Berkeley "Smartdust" project [20] was the first to explicitly target the size and power issue. In its initial incarnation, SmartDust envisioned the usage of MEMS-based optical links to provide communication between mm^3 nodes. Various prototypes were constructed to validate the basic concepts (Fig. 3). While it was established that optical links would indeed result in the lowest possible power consumption, line-of-sight and focusing issues limit their use in general.

Fig. 3. Prototypes of optical smart dust "micro" motes [20]

Interestingly enough, it was a sideline of the original Dust project that ultimately had the most impact. To create an experimental testbed and to demonstrate that small size and low-power was indeed possible, Prof. K. Pister and his students at Berkeley designed the first RF-based "mote" solely using off-the-shelf components. It used an Atmel 4 MHz processor, and an RFM 916 MHz 10 kbps cordless phone radio (Fig. 4). The small size and the low power dissipation proved to be quite attractive – in fact, they made a true deployment of large networks possible for the first time.

Fig. 4. Two examples of the UC Berkeley "COTS" motes [6]

Further, it was instrumental that two companies, Crossbow and Intel, engaged in a large-scale manufacturing of derivatives of the original motes, hence making them available to a large community at a reasonable cost (Fig. 5a). The emergence of an easy-to-use and light-weight software environment called TinyOS[2] further added to the appeal of the mote implementation platform. Without a doubt, the Mote concept has had a tremendous impact on the increasing world-wide interest in sensor networks and ambient intelligence over the last few years. This success has inspired industrial entrepreneurs and academic researchers from all over the world to develop similar platforms. Just to name a few: the EmberNet (Fig. 5b), the BTnode from ETH (Fig. 5c), and nodes from Infineon and Kaist (Korea).

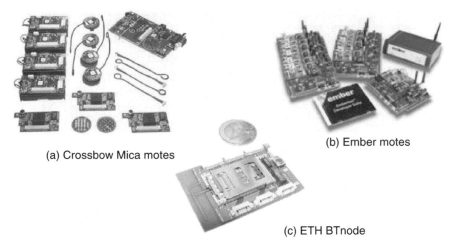

(a) Crossbow Mica motes

(b) Ember motes

(c) ETH BTnode

Fig. 5. Recent wireless sensor node incarnations, as made available by industry and academia. For more information, see [4, 5, 8]

However, while these environments have gone a long way in making ubiquitous deployment plausible, they do not yet meet by a long shot the requirements we put forward for truly embedded or "disappearing" computation and communication. Sizable reductions in cost, size, and foremost power are an absolute necessity for a wide-spread deployment of ambient-intelligence [12]. For instance, a node that on the average uses $10\,\mathrm{mW}$ of power would need $260\,\mathrm{cm}^3$ of alkaline battery to run for a year.

3 The Road to Low-Cost and Low-Power Electronics

Before examining techniques to reduce the power and cost of the wireless nodes, it is essential to examine the crucial metrics and design goals. The

---

[2] Described in more detail in another chapter in this book. See also (TinyOS).

operation mode of a sensor node is so fundamentally dissimilar from most other wireless transceivers that completely different optimization criteria apply. This results mostly from the **traffic patterns** that impact the power dissipation profile of a node. Data packets in sensor networks tend to be relatively **rare** and **unpredictable** events. In most application scenarios, each node in the network sees at most a couple of packets/sec. In addition, the packets are relatively short (typically less than 200 bits/packet), as the payloads normally represent data measurements, which typically require a resolution of less than 16 bit/measurement. Combined, this means that the average data rate of a single node rarely exceeds 1 kbit/sec. These observations are of foremost importance when designing the wireless transceiver, as we will highlight in the following sections.

Average Power Dissipation

Traditional quality metrics for radios used in wireless LANs are the data throughput (bit/sec) and spectral efficiency (bit/sec/Hz). None of these is truly important for a wireless sensor node, as the required average data rates are very low and the traffic per unit volume is not such that it creates a major bottleneck. Energy efficiency (in nJ/bit) is often quoted to be the most important measure of quality. Since the bits are few and the nodes are closely spaced, it turns out not to be a dominant metric either. In fact, the power used for the actual data transmission and reception is only a fraction of the total power dissipated in the front-end. As is shown in a subsequent section, the average power turns out to be dominated by **the power of having just the RF receiver turned on**, independent of the data activity (as long as the latter is low). A more relevant measure hence is the power consumed by the transceiver, averaged over all the operation modes under typical traffic conditions.

To put things in perspective, let us assume a total average power budget of 100 μW and that the RF module is allotted 20% of this power budget. At a 1% radio duty cycle, this means an on-state power consumption goal of 2 mW for the entire RF transceiver.

Agility

In an environment where the radio is in idle or off mode most of the time, and where data communications are rare and packets short, it is essential that the radio can start up very quickly. For instance, a 1 Mbps radio with 500 μs turn-on time would be poorly suited for the transmission of short packets. The on-time to send a 200 bit packet would be only 200 μs. Start-up and acquisition hence represent an overhead which is larger than the actual payload cost, and may very well dominate the power budget (given that channel acquisition is typically the most power-hungry operation).

Hence, fast start-up and acquisition is essential. An **agile** radio architecture that allows for quick and efficient channel acquisition and synchronization is therefore desirable. Complex wireless transceivers tend to use sophisticated algorithms such as interference cancellation and complex modulation schemes to improve bandwidth efficiency. These techniques translate into complex and lengthy synchronization procedures and may require accurate channel estimations. Packets are spaced almost seconds apart, which is beyond the coherence time of the channel. This means that these procedures have to be repeated for every packet, resulting into major overhead. **Simple modulation and communication schemes** are hence the desirable solution if agility is a prime requirement.

Size (Integration) and Cost

In RF circuit design, the term "fully integrated" typically refers to a transceiver that still requires an off-chip quartz crystal and a few assorted passive components. To meet the cost and form-factor requirements of this application, a true fully integrated transceiver is mandatory. In addition to increasing the size, off-chip passives add to the complexity and cost of the board manufacturing and package design.

One method that can be used to achieve a high level of integration is the use of a **relatively high carrier frequency**. Currently available simple low-power radios, as used in control applications, typically operate at low carrier frequencies between 100 and 800 MHz. A high carrier frequency has the distinct advantage of reducing the required values of the passive components, making integration easier. For example, a 2.53 µH inductance is needed to tune out a 1 pF capacitor in a narrow-band system at 100 MHz, requiring a surface-mount inductor. For a 2 GHz carrier frequency, the inductance needed is only 6.33 nH, which can easily be integrated on-chip using interconnect metallization layers. In addition, the antenna form-factor is very dependent upon carrier frequency. For a given antenna gain, a higher carrier frequency allows for a much smaller antenna. A quarter-wavelength monopole antenna at 100 MHz would be 0.75 m long. At 2 GHz, the size shrinks to 37.5 mm, making board-level integration or use of small chip-antennas possible. The drive to higher carrier frequencies to achieve high integration is in direct conflict with the need for low power consumption. As the carrier frequency increases, the active devices in the RF signal path must be biased at higher cutoff frequencies, increasing the bias current and decreasing the transconductance-to-current (g_m/I_d) ratio. This results in increased power dissipation at higher carrier frequencies. Thus, there exists an inherent integration/power consumption tradeoff that must be dealt with through architectural decisions and the use of new technologies.

In conclusion, transceivers for wireless sensor networks should be simple, have a minimum number of off-chip components, consume a minimum amount of on-current, and operate at higher carrier frequencies. Based on these observations, we introduce **three major guidelines** that we believe to be essential to reach the goal of truly embedded, self-contained electronics.

3.1 Simplicity Rules

While the common trend in the semiconductor industry has been towards ever more complex integrated circuits with higher transistor counts and higher clock frequencies, the interesting trend in sensor networks is that complexity of the application lies not in the individual nodes, but in the collaborative effort of a large number of distributed elements, each of which can (or actually should) be simple by itself. Only by keeping the complexity of wireless sensor node to a minimum can the aggressive cost and power constraints be met. Rather than using super-GHz clock frequencies and wireless data-rates of many Mbit/sec, individual nodes get away with a rather restricted amount of computational power, and aggregate data rates between nodes that rarely supersede 10 kbit/sec. This observation is quite crucial, and in some sense goes against the dominant trends in both the worlds of computation and wireless communication. For example, if one examines a wireless data link such as used in 802.11(abg) WLAN applications, one discovers that a large amount of complexity (and power) is put into optimizing the capacity (that is, the use of the spectrum) of the radio link. This leads to a large overhead, which is a waste for most ambient intelligence applications where spectrum utilization is not really an issue.

   To quantify these statements, it is worthwhile to model the behavior of the wireless node as a function of (a) the physical parameters of the transceiver; (b) the prevailing traffic patterns; and (c) the link and media-access operational properties [7]. This behavior is best expressed as a statistical state machine as is shown in Fig. 6. At any point in time, the transceiver is in one of the following states:

- Transmitting state (TX) during transmission of data
- Receiving state (RX) during reception of data
- Acquiring state (AQ) while acquiring synchronization at the start of the packet. It precedes the receive state, and represents the compute and power intensive timing, phase and/or frequency acquisition phase as represented by the preamble.
- Monitoring state (MN) when the transceiver is monitoring the channel (carrier sense)
- The idle state (IL) when the majority of the transceiver is turned off and it is considered to be sleeping. It may be assumed that the power dissipation in this state is zero.

The average power dissipation of the transceiver is then expressed as

$$P_{av} = p_{TX}P_{TX} + p_{RX}P_{RX} + p_{AQ}P_{AQ} + p_{MN}P_{MN} + p_{IL}P_{IL} \qquad (1)$$

where P_x is the average power dissipation in state x and p_x the probability the transceiver is in that particular state. The power dissipation in the TX state is determined mostly by the dissipation in the power amplifier. The average power dissipation of the RF front-end in the three other modes (RX, AQ, and MN) is approximately equal, although P_{AQ} may dominate slightly.

Given the low data rates and duty cycles, the transceiver **should be in the idle state** for most of the time, assuming that proper sleep disciplines are adopted.[3] Amongst the four other states, the monitoring state (MN) is the most probable, as became apparent from simulations based on this model. Assuming a dense network and sparse traffic, the average power of the transceiver is well approximated as

$$P_{av} = p_{TX}P_{TX} + (p_{RX} + p_{AQ} + p_{MN})P_{RXon} \cong p_{MN}P_{RXon} \qquad (2)$$

where P_{RXon} is the dissipation when the receiver is on.

Simple Wireless Transceiver

It is hence fair to state that the average power is dominated by **the power of having just the RF receiver turned on** (LNA, down-converter, synthesizer, etc), independent of the data activity. Minimizing the average power then translates into minimizing the active current draw of the RF front end, and, obviously, the time that the transceiver is turned on. This leads to the important conclusion that **simple RF transceivers with a minimal number of active components** are the preferred option for use in wireless sensor networks.

To address these stringent requirements, it is worth examining alternative technologies and radio architectures. In particular, mostly passive transceivers built around emerging RF-MEMS resonators are quite attractive. Recently, much progress has been made on GHz-range MEMS resonators, typically for use in bandpass filters and duplexers [1, 16]. Since these resonators exhibit quality (Q) factors greater than 1000, these devices have the potential to facilitate the design of low power RF transceivers. An example of a successful RF-MEMS resonator is the Agilent FBAR (Field Bulk Acoustic Resonator). It employs a metal-piezo-metal sandwich to achieve a high frequency, tightly controlled second-order resonance with an unloaded Q of 1000s at frequencies up to 5 GHz. (Fig. 6).

The presence of an RF frequency reference can eliminate the need for quartz crystals in the system, greatly increasing the level of integration and decreasing the cost. In addition, the availability of highly-selective RF filters makes it possible to drastically simplify the architecture of the transceiver.

---

[3] This is not the case in many existing wireless network protocols, in which the radio is in the monitoring state when not transmitting or receiving. In most Wireless LAN protocols, the radio only goes into the idle mode when explicitly instructed by a basestation.

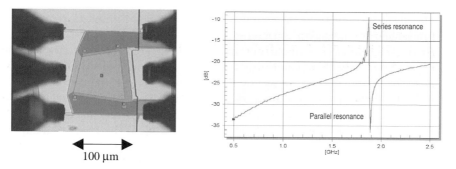

Fig. 6. Agilent FBAR RF-MEMS resonator and its frequency response [16]. Observe the small size and the high selectivity

Filtering at RF makes it possible to down-convert the signal directly to the baseband using a simple non-linear element, eliminating the need for an Intermediate Frequency (IF) stage (such as used in Heterodyne Receivers). Architectures of old such as the *tuned radio frequency (TRF)* and the *regenerative* architectures did just that. While these were well known for their effectiveness and simplicity, they suffered from the large size of the passive and the need for extensive tuning – two problems that are overcome through the use of RF-MEMS [9].

- The **Tuned Radio Frequency** architecture, one of the simplest receiver architectures, eliminates the RF frequency synthesizer and mixers by filtering in the RF domain and directly detecting the RF signal. This architecture relies on sharp RF filters with high frequency stability, two requirements which are met by RF MEMS components. The filtered signal can be detected and down-converted in a variety of ways, including envelope detection and sub-sampling. For example, the envelope detection, also referred to as diode detection, performs a self-mixing operation on the signal. The RF signal drives a non-linear element such as a diode or envelope detector, providing a DC component containing the signal spectrum of interest, as shown in Fig. 7. After down-conversion via envelope detection, the signal is simply low-pass filtered to remove the fundamental and higher harmonics, leaving only the baseband signal. After down-conversion

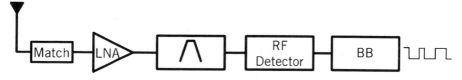

Fig. 7. TRF receiver architecture with envelope detector

via envelope detection, the signal is simply low-pass filtered to remove the
fundamental and higher harmonics, leaving only the baseband signal.

- **Super-regenerative architecture.** The TRF architecture takes advantage of RF MEMS technologies to perform channel selection without a
  need for mixers and/or frequency synthesizers. The RF gain needed, however, is high due to the noisy detection circuitry. A super-regenerative
  front-end provides extremely high RF amplification and narrowband filtering at low bias-current levels. As shown in Fig. 8, the heart of a super-
  regenerative detector is an RF oscillator with a time-varying loop gain.
  The time-varying nature of the loop-gain is designed such that the oscillator trans-conductance periodically exceeds the critical g_m necessary to
  induce instability. Consequently, the oscillator periodically starts up and
  shuts off. The start-up time of the oscillator is exponentially dependent
  upon the initial voltage in the oscillator tank. This dependency translates
  into an achievable gain, which is quite substantial. The potential of the
  super-regenerative receiver to generate large signal gain at very low bias
  currents makes it one of the preferred architectures for integrated ultra-low
  power wireless receivers.

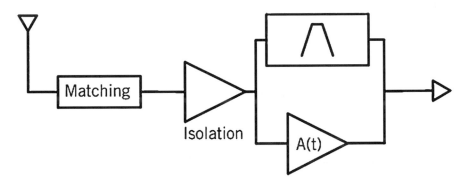

Fig. 8. Super-regenerative detection

The simplicity concept proves to be quite effective. Figure 9 shows an
example of a radio implemented following the tuned-radio concept outlined
above [10]. The overall power dissipation of the receiver (when on) is less
than 3 mW. The transmitter, which transmits at an output level of 0 dBm,
consumes approximately 5 mW when on. Fortunately, the latter is a relatively
rare event. The two-channel transceiver has been demonstrated to work reliably over distances of 20 m and more. The most striking properties of the
radio are its very small size (less than 5 mm^2 of active area) and the very rapid
turn-on time of the transceiver, which is in line with the agility requirement
phrased earlier.

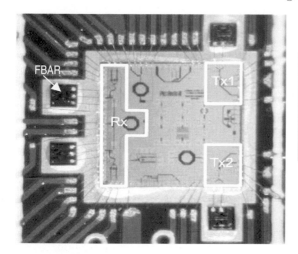

- 130 nm CMOS
- Carrier frequency: 1.9 GHz
- 0 dBm OOK
- Channel Spacing ~ 50MHz
- 10-160 kbps/channel
- 10 µs start-up time
- Total area < 5 mm²

Fig. 9. Implementation of tuned-radio OOK transceiver. The FBAR and silicon components are mounted using a Chip-on-Board packaging technology, and then wire-bonded together

Simple Digital Processor

While the radio takes the bulk of the power of the wireless node, the digital processor should not be ignored. The tasks of the processor are manifold: implementation of the wireless protocol stack (including physical, link, media-access and network layers) and application processing, interfacing to sensors and actuators and performing the necessary signal conditioning, and the realization of miscellaneous other functions such as conditioning and time synchronization (some of which can be quite compute intensive). Again, to reduce the power of the processor, simplicity is the essential message. This means choosing the simplest micro-controller that is up to the computational task, minimizing the amount of memory used, use of dedicated accelerators for fixed compute-intensive functions such as the locationing engine and the link-layer processor, and most importantly, minimizing the supply voltage and the clock frequency. As is typical for ambient intelligence applications, the power of the system does not lie in the individual components but in the ensemble of components.

A block-diagram of the protocol processor of the PicoRadio wireless sensor node is shown in Fig. 10. At the core is an 8051 microcontroller, running at 16 MHz. It implements the top (network) layers of the protocol stack as well as the application tasks and various signal conditioning functions. It is surrounded by a number of hardwired accelerator units implementing the lower layers of the protocol stack, localization hardware, and a variety of peripheral interfaces. Observe that the total memory capacity of the node is small in line with the concept that the system functionality lies in the ensemble of simple nodes, not in the individual nodes.

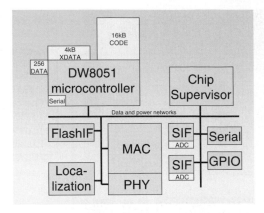

Fig. 10. Architecture of PicoNode protocol processor [13]

3.2 Standby Power Dominates

In current-day electronic components, power dissipation is dominated by far by the active power, this is the power that is consumed when the device is computing or communicating (even though the high leakage levels of the most advanced semiconductor technologies are gradually changing this perspective). The opposite is true in most ambient intelligence nodes. The duty cycle of an individual node is, in general, quite low. A typical sensor might sample the data only a couple of times per second, or even much slower than that. Hence the node is in standby mode most of the time, and standby power (due to leakage and channel monitoring) is by far the dominant source of power consumption in typical sensor networks. Clever techniques can go a long way in alleviating that cost. The primary option is to turn off the node completely when non-active. While this helps to eliminate the leakage, it makes communication between neighboring nodes non-trivial: How does one create a reliable link between two communicating entities that are asleep most of the time, and that only wake up at irregular intervals?

The answer to this is to establish a *rendezvous scheme*, which arranges simultaneous on-time of nodes wishing to communicate [Lin04]. There are multiple ways of doing so. They can be grouped in three major categories:

1. **Purely synchronous.** Nodes are synchronized in time and agree on specific time slots for communication. This is hard to accomplish in a fully ad-hoc multi-hop scenario with 100's of nodes, and tends to translate into a major overhead in power consumption.
2. **Purely asynchronous.** Nodes have the capability of waking up one another on demand. While this approach has the potential to lead to the lowest power dissipation, it requires extra hardware that is typically not available on standard parts (such as a wakeup radio, as introduced in

Rabaey 2002). It also leads to some small standby power, which might become dominant for networks with very low activity levels.

3. **Pseudo-asynchronous (or cycled receiver).** Nodes establish rendez-vous on demand, but underlying is a periodic wakeup scheme.

The *cycled receiver* scheme is a popular approach, versions of which have been proposed by a variety of authors. Nodes are powered on and off period-ically, and a beaconing approach is used to express the desire or willingness to communicate. Obviously, the pattern of periodic on/off powering has to satisfy some performance requirements, like throughput and delay. An exam-ple of such a scheme called TICER (Transmitter Initiated Cycled Receiver) is shown in Fig. 11. When a sensor node has no data packet to transmit, it periodically wakes up to monitor the channel with a period T, and if nothing is observed it goes back to sleep after a wakeup duration T_{on}. As soon as a node has a data packet to transmit either generated from the upper layers of the protocol stack or forwarded by another node it wakes up and monitors the channel for duration of T_{on}. If it does not hear any ongoing transmissions on the channel, it starts transmitting request-to-send (RTS) signals to the destination node and monitors the channel for a time T_1 for responses after each RTS transmission. The destination node, upon waking up according to its regular wakeup schedule, immediately acquires and receives the RTS's, upon which it responds with a clear-to-send (CTS) signal to the source node. After reception of the CTS signal, the source node transmits the data packet. The session ends with an acknowledgement (ACK) signal transmitted from the destination node to the source node, after correctly receiving the data packet.

Using schemes of this nature goes a long way in reducing the standby monitoring power in the network. However, to get reasonable results it is essential that the parameters of the scheme, most importantly the wake-up period T, adjust dynamically to the operational system requirements such as the traffic volume and the latency. For example, Fig. 12 shows the impact of choosing a non-optimal wake-up period T on the overall power dissipation.

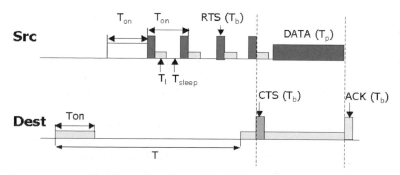

Fig. 11. Transmitter Initiated CyclEd Receiver (TICER) Rendez-vous scheme

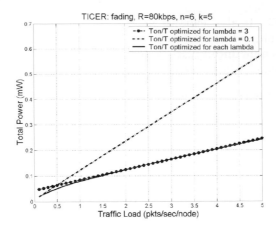

Fig. 12. Impact of choosing non-optimal wake-up period on power. Lambda stands for the traffic in packets/sec/node

The penalty can be quite large indeed. One observation however is that it is better to err towards the overestimation of traffic.

While monitoring power is an important part of the standby power of the system, other sources should be ignored. With the power levels this low, other ambient power consumption caused by the off-current (leakage) of the components starts to take a large bite out of the power budget. An estimate of the leakage power of the processor of Fig. 10 implemented in a 130 nm CMOS technology and operating at 1V showed that its value superseded the overall allotted budget of 100 μW, especially due to the embedded SRAM. Active leakage reduction in the off-mode is hence essential. The most effective means of doing so is by actively turning down the supply rail of the modules that are not in operation (to 0 V for modules without state, 300 mV for memory modules and components with state). This leads to the concept of "power domains" – modules that are in the off-mode of a scheduling event. To make this power-management process more efficient – turning modules on and off comes with some overhead – some centralized decision making is essential. This explains the "system supervisor" module in Fig. 10, which acts as a global "scheduler", helping to decide which module to turn on (or off) and when, based on external events, scheduled events (such as timers) and understanding or learning of sequences of events. An example block-diagram of such a power supervision unit is shown in Fig. 13. While such a module is absolutely essential in the ultra-low power low-activity application space such as wireless sensor nodes, they will rapidly become an important part in the high-performance high-activity area as well as leakage is gaining dominance and effective power management is crucial. The impact of using this active power management is huge. For the PicoNode protocol processor, it reduced the standby current from over 100 μW to less than 1 μW.

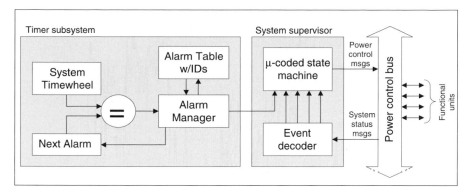

Fig. 13. Block diagram of system supervisor module. Through the "power control bus", the module manages the turning on and off of the system components. It can be extended to include such functions as dynamic voltage scaling and dynamic body biasing

3.3 Redundancy and Randomness Provide Reliability

For ambient intelligence applications really to become acceptable, it is crucial that the systems perform in an absolutely reliable manner over a long period of time. Repeated network failures can lead to life-threatening situations, or will quickly turn off the user.

While simplicity is a virtue from a power and cost perspective, it has a negative impact on the reliability of an individual link or node. For example, a major part of the complexity in a high-speed wireless point-to-point link is devoted to ensure that the link performs well under a wide variety of channel conditions. To mitigate the impact of fading, transceivers use diversity in the time, frequency and/or space dimensions. Simplicity does away with it, and hence results in rather unreliable individual links. This is illustrated in Fig. 14, which plots the measured link quality of a pair of the simple radios discussed earlier over the course of a day under varying occupation conditions. The plot clearly demonstrates that issues such as blockage of line-of-sight can result in total loss of the channel.

The beauty of sensor networks again lays in the fact it offers substantial redundancy, and that the reliability of the system does only marginally depend upon the robustness of the individual nodes or links. For instance, fading might cause a particular link to temporary fail, but that this is most probably not an issue since many other links are available to forward a packet to its destination. The impact of redundancy on channel quality is demonstrated in Fig. 15.

Curve 1 (2 nodes) shows the measured quality of a single link while moving the sender and receiver gradually apart. At certain distances (for instance, 90 cm), multi-path results in deep fades and no communication is possible. By adding a third node, which may present an alternative route between

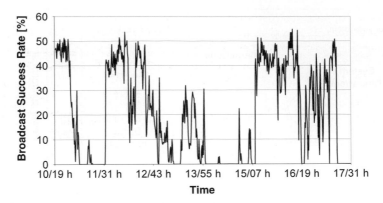

Fig. 14. Measured link quality of narrow band radios over the course of a day under actual operational conditions

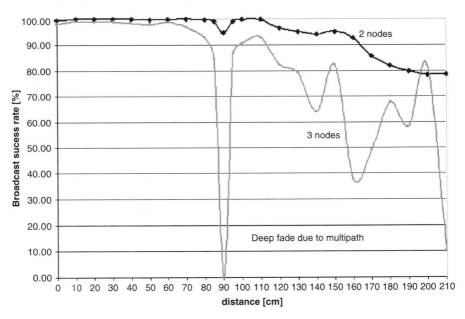

Fig. 15. Impact of adding extra node to link quality narrow-band radios

nodes 1 and 2 using multi-hop, the chance of successful data delivery increases substantially (as shown in the second graph – 3 nodes). In fact, the channel response starts to look almost flat. In a sense, an ad-hoc wireless sensor network offers spatial diversity through its redundancy. The art in designing an efficient wireless node and network is to exploit the opportunities offered by these system-level characteristics. This requires that the protocol stack (routing and media-access control) is designed such that no end-to-end connection ever relies on the presence of a specific node. We call such an approach "opportunistic routing" [23]. While forwarding a packet through the network, the selection of the next destination node in a chain is not based on routing tables or pre-defined paths, but on the presence of a reachable node that is awake and is located in the direction of the final destination. The concept is illustrated in Fig. 16.

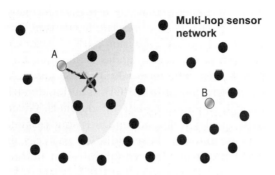

Fig. 16. Opportunistic routing in ad-hoc networks creates reliability in the presence of fading channels and failing nodes

Assume that node A has a packet that has to be delivered to node B. Rather than selecting a specific node in its immediate neighborhood for the next hop, A broadcasts a beacon in the TICER style. The first node that responds and is in the right direction, as indicated by the angle in the Figure, gets the packet. This approach improves both the reliability – transmission is ensured even in the presence of fading and failure conditions – and the energy dissipation as re-transmissions are minimized. Observe that the next node selection has now become a random process. Adding a little bit more cleverness (such as waiting a little while so that even better positioned nodes can emerge), it is even possible to out-perform the average wireless channel, which is one of the holy grails in wireless communications. This example shows how the co-design of hardware and system can lead to substantial energy reductions. The concept of having applications running on "unreliable" hardware platforms while still behaving reliably is one of central hallmarks of ambient intelligence. Another important premise that emerges from this discussion is that co-optimization of the different layers of the protocol stack – from physical to application – is absolutely essential.

4 System Integration

In the previous section, we have described a number of basic concepts that may help to reduce cost, power and size of the wireless nodes and their components. One aspect that further deserves some attention is that the packaging and physical integration strategy of the transceiver is crucial as it dramatically impacts the performance, cost, and form-factor of the completed system. The size of the node is determined by the number and size of the components and the assembly technology. As mentioned above, careful architectural decisions can help to minimize the part-count. Some components that deserve special attention are the antenna and the energy-supply chain.

The antenna size can be minimized by using a relatively high carrier frequency. Three main parameters that determine the type of antenna suitable for wireless sensor network are the radiation pattern, size, and bandwidth. As the deployment pattern of the sensor nodes is random, an omni-directional radiation pattern is desired. The form factor of the antenna should be small and allow for easy integration onto a PCB. The bandwidth of the antenna should be large enough to accommodate all channels. Given those considerations, commercially available chip-antennas (using ceramic dielectrics) are good candidates. These antennas have radiation patterns similar to a dipole and measure only about 30 mm by 5 mm. An even better and more cost-effective solution is offered by the so-called PIFA (Printed Inverted F) antennas, which are realized by patterns printed on the circuit board, as shown in Fig. 17. The advantage of the PIFA is that its characteristic impedance and center frequency is easily controlled by design. More importantly, it comes almost for free as it just needs a limited amount of board real estate, while no extra components are required. Finally, the radiation pattern is almost omni-directional, which is quite desirable in ad-hoc networks. A number of recent node implementations have hence adopted this approach.

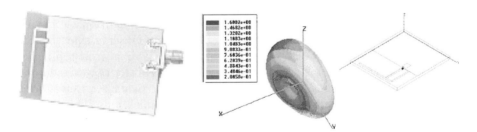

Fig. 17. Prototype of PIFA antenna and its radiaton pattern

The energy-scavenging and the energy-storage (battery or capacitance) devices are the most volume-consuming components of the completed node. As shown in Table 1, their respective volumes are directly proportional to the average and peak power dissipation levels of the node. Thus, decreasing the power dissipation of the transceiver – the most power-hungry component of the sensor node ultimately results in a linear decrease in the node size.

These guidelines are best illustrated with the aid of an implementation example, which also serves to demonstrate how the techniques introduced

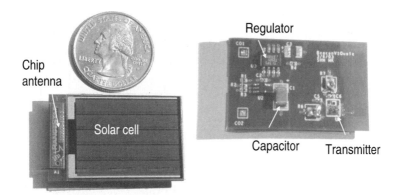

Fig. 18. PicoBeacon, an integrated energy-scavenging wireless transmitter (*front* and *back*)

in this chapter effectively lead to ultra low-power and small size. Figure 18 shows a picture of an integrated node, called the PicoBeacon, which uses a $3\,\text{cm}^2$ solar cell to drive a 10 m wireless transmitter (RF only) [15].

This system contains an ultra low power 1.9 GHz transmitter,[4] a 1.9 GHz chip antenna, a solar panel, an energy storage capacitor, and voltage regulation circuitry. Due to the high carrier frequency used, the antenna is small and takes up little board space. It is mounted on the back of the circuit board, which also acts as the ground plane of the antenna. Glued on top of that ground plane is the $3\,\text{cm}^2$ solar cell, which is able to provide all the energy necessary for transmitter operation. Figure 19 shows a conceptual diagram of the PicoBeacon power-train, which consists of the solar cell, a capacitor for temporary energy storage, shutdown control logic, and a linear voltage regulator, which supplies the RF circuitry with a stable 1.2 V supply. The operation of the beacon is quite simple. The solar cell charges the capacitor. Once a threshold voltage, guaranteeing reliable operation, is obtained, the transmitter is turned on and starts broadcasting a stream of bits. This continues until the voltage on the capacitor drops to a minimum voltage, at which point the transmitter is powered down.

The transmitter, which is based on the tuned radio approach described earlier, is fully integrated, requiring only one MEMS resonator. It does not use any crystals or external inductors, and takes up approximately 2 mm × 3 mm of board space. The transmitter die and the FBAR resonators are mounted on the board using a chip-on-board technology, as is illustrated in Fig. 20. It delivers 0 dBm (1 mW) to the $50\,\Omega$ chip antenna.

---

[4] The choice of the 1.9 GHz carrier frequency is solely due to the availability of the corresponding FBAR resonators. For more realistic deployments, 2.4 GHz ISM band resonators would be employed.

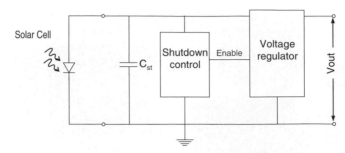

Fig. 19. PicoBeacon power-train

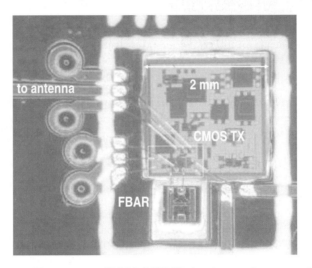

Fig. 20. Transmitter CMOS/MEMS circuitry and antenna feed

The transmitter operates at approximately a 1% duty cycle in low light conditions to a 100% duty cycle in sunlight. Figure 21 shows pertinent waveforms during the operation of the transmit beacon. The top waveform is the voltage on the storage capacitor. Once the voltage regulator is enabled (bottom waveform), the RF transmitter turns on (middle waveform). During transmission, the energy in the storage capacitor is dissipated, and the voltage on the storage capacitor decreases. Once the energy on the storage capacitor is depleted, the transmitter is disabled and the charge process starts again. The board operates indefinitely with no batteries or power supplies.

The PicoBeacon is just one example of how the combination of low-power techniques, system-level optimization and integration can lead to ultra-low {cost, size, power} wireless nodes. We have demonstrated that average power dissipation levels below $400\,\mu W$ are currently attainable for traffic patterns and network densities that are typical for smart-home applications. These numbers include RF, base-band, digital protocol stack, networking, and

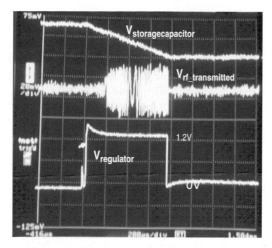

Fig. 21. Transmit beacon output waveforms

application level processing. Further improvements in power management can lead to another factor of at least 5 to 10, clearly enabling energy-scavenging.

5 Summary

In summary, self-contained low-cost and ultra low-power computation and communication nodes for ambient intelligence are clearly attainable and realizable. The secret to their realization is (1) to move in directions that are somewhat orthogonal to trends that dominate in the wireless and semiconductor communities, and (2) to maximally exploit the ubiquitous property that is at the core of the ambient intelligence idea.

Yet a large number of challenges still remain. Especially, getting the cost down to numbers that make true redundancy and ubiquitous embedding possible will still take 5 to 10 years. In the mean time, the technology will gradually but surely start to penetrate our daily lives, having potentially an even more profound impact than the computer and networking revolutions of the previous decades.

Acknowledgements

The authors would like to thank Agilent Technologies for the resonator fabrication and ST Microelectronics for the CMOS fabrication. The generous support of DARPA (under the PACC and the IMT programs) is gratefully acknowledged.

References

1. B. Bircumshaw, G. Liu, H. Takeuchi, T.-J. King, R. Howe, O. O'Reilly, A. Pisano, "The Radial Bulk Annular Resonator: Towards a 50 Ohm RF MEMS Filter," *Technical Digest, 12th Int. Conf. on Solid State Sensors, Actuators and Microsystems*, Boston, pp. 875–878, June 8–12, 2003.
2. Birnbaum, J., *http://www.hpl.hp.com/speeches/pervasive.html*.
3. R. Brodersen et al, "Research Challenges in Wireless Multimedia," *IEEE Personal, Indoor and Mobile Radio Communications (PIMRC) Workshop*, pp. 1–5, Sept. 1994.
4. Btnode, *http://www.btnode.ethz.ch/*.
5. Ember - Embedded Wireless Networking, *http://www.ember.com/*.
6. Jason Hill, Robert Szewczyk, Alec Woo, Seth Hollar, David Culler, and Kristofer Pister, "System architecture directions for network sensors," *ASPLOS 2000*.
7. E.A. Lin, A. Wolitsz, J. Rabaey, "Power Efficiency Analysis of Two Rendezvous Schemes For Dense Wireless Sensor Networks" *IEEE Int. Conf. on Communications*, Paris, June 2004.
8. Mica, *http://www.xbow.com/Products/productsdetails.aspx?sid=3*.
9. B.P. Otis, Y.H. Chee, R. Lu, N.M. Pletcher, S. Gambini, and J.M. Rabaey, *Highly Integrated Ultra-Low Power RF Transceivers for Wireless Sensor Networks*, in Low-Power Electronics, C. Piguet, Ed., CRC Press, 2004.
10. B. Otis, Y.H. Chee, R. Lu, N. Pletcher, J. Rabaey, "An Ultra-Low Power MEMS-Based Two-Channel Transceiver for Wireless Sensor Networks," *Proceedings VLSI Circuits Symposium*, Honolulu, June 2004.
11. J. Rabaey, J. Ammer, J.L. da Silva Jr., D. Patel, "PicoRadio: Ad-hoc Wireless Networking of Ubiquitous Low-Energy Sensor/Monitor Nodes," *Proceedings of the WVLSI*, Orlando, Fl, USA, April 2000.
12. J. Rabaey, J. Ammer, J. da Silva, D. Patel, S. Roundy, "Picoradio Supports Ad-hoc Ultra-low Power Wireless Networking", Cover Article, *IEEE Computer Magazine*, July 2000.
13. Rabaey JM, Ammer J, Karalar T, Suetfei Li, Otis B, Sheets M, Tuan T. "PicoRadios for wireless sensor networks: the next challenge in ultra-low power design," *2002 IEEE International Solid-State Circuits Conference Digest of Technical Papers*, pp. 200–1, San Francsisco, February 2002.
14. S. Roundy, P. Wright, and J. Rabaey, *Energy Scavenging for Wireless Sensor Networks with Special Focus on Vibrations*, Kluwer Academic Press, 2003.
15. Roundy, S., Otis, B. P, Chee, Y-H., Rabaey, J. M., Wright, P., "A 1.9 GHz RF Transceiver Beacon using Environmentally Scavenged Energy," *ISPLED '03*, Seoul, Korea, Aug. 25–27, 2003.
16. R. Ruby, P. Bradley, J. Larson III, Y. Oshmyansky, D. Figueredo, "Ultraminiature high-Q filters and duplexers using FBAR technology", *IEEE ISSCC Digest of Technical Papers*, pp. 120–1, Feb. 2001.
17. Sensoria, *http://www.sensoria.com/*.
18. TinyOS, *http://www.tinyos.net/*.
19. uAmps: Adaptive Multi-Domain Power Aware Sensor, *http://www-mtl.mit.edu/research/icsystems/uamps/*.
20. B. Warneke, M. Last, B. Leibowitz, and K. S. J. Pister, "Smart dust: Communicating with a cubic-millimeter computer," *IEEE Computer*, 32(1):43–51, January 2001.

21. M. Weiser, "The Computer for the 21st Century," *Scientific American*, Volume 165 Number 3, pp 94-104, September 1991. Reprinted in IEEE Pervasive Computing March 2003, pp. 19–25. Also see Weiser's essay [http://www.ubiq.com/hypertext/weiser/UbiHome.html].
22. WINS, *http://wins.rsc.rockwell.com/*.
23. M. Zorzi and R. R. Rao, "Geographic random forwarding (GeRaF) for ad hoc and sensor networks: energy and latency performance," *IEEE Transactions on Mobile Computing*, 2(4), Oct-Dec 2003.

Packaging Challenges in Miniaturization

C. Kallmayer, M. Niedermayer, S. Guttowski, and H. Reichl

Classically, packaging consists of assembly, interconnection and passivation. The technological advances in miniaturization, however, have changed all three aspects and have moved the focus onto system integration. Instead of developing hardware, software and technology separately, the whole system has to be considered and optimized for a further size reduction. The following chapter discusses the design and realization of tiny, highly integrated devices. It will show interdependenc es between the miniaturization techniques as well as the design of hardware and software. Initially several system aspects will be mentioned. Thereafter the integration technologies will be reflected in more detail.

1 System Aspects

In order to realize highly integrated devices in large quantities, several development phases have to be passed. Initially the envisioned functionality has to be specified according to the requirements of the application. This results in a concept of the general architecture. Then the components are defined followed by the selection of matcrials and technologies. The network structure has to be determined and the geometric arrangement of the components has to be set up. The physical couplings between tightly positioned devices require verifying the design by system simulations. The accordant optimization has to be done by simulation and measurement of particular subsystems. The concept, the design and the verification phases may require several iterations dependent on the results (Fig. 1).

1.1 System Conception

At the beginning the application requirements have to be analyzed in terms of boundary conditions such as size limitations, target costs, time restrictions, and desired features. Simple problems allow the use of word processors or spreadsheets for a system description. Higher complexities may necessitate dedicated tools at algorithmic levels for concept modeling and evaluation. This will result in a determination as to which design parameters are really

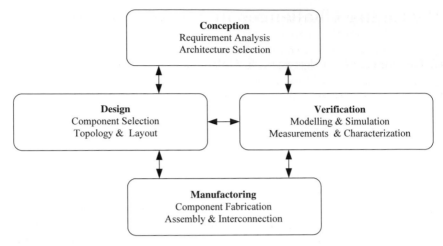

Fig. 1. Simplified concept of a development process

bounded by the requirements of the application. These constraints will evolve during the design refinement.

In the field of system miniaturization the question is which parts of the whole system mainly determine the final dimensions. The optimization focus is to be put on the design of the analog circuitry for instance in measurement applications, because the size of analog components is substantially determined by the desired accuracy. The size of digital circuitry, on the other hand, can be more critical in applications like tracking fast movements, in which case the processing of huge amounts of data requires higher clock rates with extra space for heat reduction. In many applications a self-sufficient operation is required during a long time, as a consequence of which the energy supply can strongly influence the final size. Also, most systems for ambient intelligence need some interface elements like sensors, actuators, antennas, displays or some signal generating or receiving parts, which may be size dominating as well.

After the system requirements have been determined, alternative architectures have to be evaluated. Currently, there is no automatic procedure available that can map a desired functionality and the relevant constraints to a target architecture. Likewise there is no algorithm to verify, that the architecture really matches a required behavior.

This situation reminds one of the problems in complex software design where it is equally difficult to prove "correctness". Concepts like learning or adaptation processes, redundancies or self-repair may be carried over to systems design. In fact, the distinction between hardw are and software design becomes blurred. It may be possible to realize a specific function either through hardware or through software modules. Thus, a holistic point of view (co-design) has to be adopted here, too.

The architecture selection also depends on the decision, whether the envisioned functionality can be realized by standard components or whether special custom-designed components are more advantageous. Thereby, some components can be realized on the same substrate by monolithic integration. Not all components will be processed by the same manufacturing technology because of different material requirements and functional principles. Hence, the application of hybrid technologies is often necessary. This allows an integration of several components from different functional units.

The requirement of utmost miniaturization makes it desirable to assign several functions to the same component. A hermetically sealed metal package of a sensor can be used as an antenna or simultaneously serve as a substrate for other components. Similarly, properties that were usually regarded as unwanted or "parasitic" may be turned into useful ones. For example, the inductance of a transmission element can be used to make up an inductor.

It is thus desirable to characterize components as well as packaging elements like bond wires and archive their properties in a form suited for the use in a design environment (e.g., as equivalent circuits or as behavioral models).

Within the system description it is useful to partition the functionality across multiple blocks. Distributed micro systems can be divided into four subsystems namely communication interface, environmental interface, data handling and energy supply (Fig. 2). These functional units are to be considered as to their influences on the general architecture.

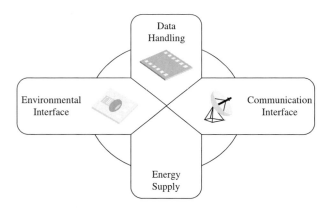

Fig. 2. Constituents of a distributed micro system

1.1.1 Communication Interface

The ability of communication is an elementary property for the task coordination among distributed systems. Not only the limitation of volume and energy influence the hardware design, but also the kind of communication has to be taken into account. For example, the task of detecting rare events recommends communication hardware with an event driven wake-up functionality, whereas a communication at defined time slots is often the better choice for monitoring frequent changes. The use of very narrow time intervals often increases the energy efficiency but requires a higher effort for synchronization.

The communication functionality is dictated by the surrounding infrastructure in many cases. Otherwise the choice of a simplified modulation scheme and access technique allows limiting power consumption and data memory.

While some applications allow an effective communication via ultrasonic signals, magnet field or light, radio transmission will be the most universal choice. The optimization of the radio transceiver is relatively complex. The selection of the transmission frequency is influenced by several parameters such as data rate and network topology. The radio front-end dominates the power consumption at low data rates. In contrast for the same data amount, higher data rates allow reduced duty cycles at the expense of temporarily higher power consumption especially in the digital back-end.

Designing the antenna, not only the limited space but also the shielding effects of the electrical circuitry and the power supply have to be considered. The options for the realization of antennas are manifold. Besides self-contained components there are on-chip or substrate-integrated antenna solutions. The antenna efficiency depends on the size relative to the carrier wavelength. Higher transmission frequencies allow smaller antennas and cause higher dielectric losses in the transceiver front-end. This will result in different optima for slow and fast architectures. The selection of the transmission frequency has to be carried out according to the official allocation of the frequency spectrum.

1.1.2 Environmental Interface

The communication unit can also act as interface to the environment in some applications such as identification labels or location trackers. To interact with the environment, a multitude of devices for ambient intelligence will possess additional circuitry including other input and output elements such as sensors, actuators. Depending on the purpose and the physical category, the requirements can be very different in terms of costs, bandwidth and resolution.

The envisioned flexibility considerably determines the architecture selection. This includes the decision, whether features like noise suppression or drift compensation should be adapted during operation. The additional effort for an increased flexibility can lower costs, because a broader range of applications allows larger quantities.

The organization of sensors and actuators in a network implies further design trade-offs. Cooperation between the single devices can offer longer life times or accelerated response times. Thereby the data amount as well as synchronization and calibration requirements influence the necessary bandwidth of the network. An improvement of the hardware specification allows lowering the network activities by measures like more accurate crystals, more precise sensors, additional compensation circuitry, and more complex data reduction. Correspondingly cost savings can be accomplished, if additional band width of the network is available.

1.1.3 Data Handling

Different architectures have to be evaluated for processing sensor data and controlling the information flow, respectively. Algorithms can be realized directly in terms of hardware by using application specific integrated circuits (ASIC). This is often highly efficient in terms of performance, power consumption and volume.

A higher flexibility can be obtained by implementing a micro processor. According to the required bandwidth, low power microcontrollers or fast digital signal processors can be used. The software should take care of limited resources such as energy and data memory.

The component selection for the generation of clock rates comprises the decision, which frequency tolerance should be permitted. This influences considerably the general architecture. Thus, larger and thus more accurate crystals can reduce the space of the energy supply, because the lower synchronization effort allows narrow time slots. This measure facilitates a faster architecture and shorter duty cycles.

1.1.4 Energy Supply

The amount of energy depends on the available volume. Currently the supply for small electronic objects is mainly based on batteries due to their availability. The selection of appropriate batteries is a complicated topic by itself. Not only high energy densities for small volumes, but also the nominal voltage, the self discharge current, and the maximal load have to be considered for the choice of material. For instance, zinc air cells with a very high energy density can be used only for a few weeks due to their high self-discharge current. On the other hand, tiny lithium cells can only drive relatively low currents. Furthermore the battery life highly depends on how the battery is discharged.

Not all applications allow a replacement of batteries. Nevertheless, the conversion of energy from the environment by movement, heat, radiation or biochemistry will become more and more important. The size of photovoltaic cells, variable capacitances, piezoelectric devices or thermoelectric sources can be very different in dependence of the energy demand and the available intensity.

Another option is the use of wireless charging techniques to enhance the system life time. The choice of the physical principle depends on application limitations. Besides electrical charging via radio antenna or inductive coupling several alternatives can be taken into account such as charging via light, heat, vibration. Shielding, directivity and coupling factors have to be considered at the system design.

An efficient power adaptation between the source and the several functional units allow smaller energy supplies at the cost of additional circuitry. For instance in a static CMOS-based integrated circuit, a reduced operating voltage implies quadratic energy savings at the expense of an additional propagation delay through static logic. Hence, a dynamic scaling of clock rate and voltage increases energy efficiency during the operation.

1.2 System Design

After consideration of the functional units, the next step is to specify the architecture by designing the circuitry and looking for bottlenecks of the complete system. Behavior models of the whole system regarding costs, energy and size help to chose materials, components and technological processes. The selection of materials and technologies is also influenced by environmental conditions. Ambient parameters such as force magnitudes, temperature maxima, and humidity result in a very different focus of the system design. Additional design restrictions follow from the process flow. Thus, the position of several components is not arbitrarily due to effects such as field coupling and shielding, if the functional layers are folded or stacked afterwards. According to the resulting design rules, the layout is generated by defining the component positions and routing their interconnections (Fig. 3).

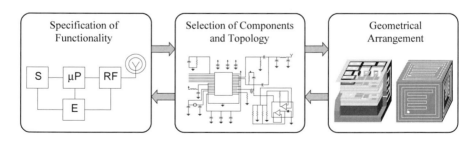

Fig. 3. Abstraction levels of system design

The interdependencies between different functional units have to be considered in case of higher component densities. For instance, the transmission frequency has to be defined for designing the communication hardware. Sole consideration of the communication circuitry would only include parameters such as antenna efficiency, antenna directivity, RF-losses, polarization and mismatch. Resulting in a changed structure of transmission data or a modification of the data rate, other modules influence the hardware optimization of the communication unit. For instance, technological improvements of an applied sensor could allow an accelerated data acquisition. This recommends a faster architecture with shorter duty cycles. Thus, the optimal transmission frequency regarding power consumption will move to higher values. The suitable solution for the whole system, however, has also to reflect environmental conditions such as absorptions, reflections, noise, and RF coverage.

Depending on the requirements of reliability, a monitoring of some internal and external conditions guarantees a safer functioning in security sensitive cases. Therefore, appropriate algorithms calculate the probability of a malfunction. As a result, an observation and evaluation of operating voltage gradients, temperature variations and vibrations can significantly improve the failure prediction.

A very compact layout with dense interconnections is the result of the system design. The debugging of high integrated prototypes can become very challenging and requires system verification at the different levels of the design process.

1.3 System Verification

An integrated treatment of physical couplings is especially important for small systems with tightly positioned components. Apart from the realization of prototypes and their characterization by measurements, the whole system or adequate subsystems have to be represented in computer memory for verification. Distributed micro systems are often heterogeneous systems and possess both spatially distributed and concentrated elements. Properties stemming from different physical domains have to be simulated to describe the behavior of the whole system. As a consequence, the modeling and simulation of these devices is more complex. Regarding the limited computation time for simulations, a generation of models often requires to neglect details, which are of less significance for the system behavior. This is hard to automize and depends considerably on the experience of the development engineer.

Besides the verification of the analog, digital and mixed-signal circuitry, the parasitics have to be considered for the components, package elements and interconnections. This includes a quantification of reflections, cross talk and radiation. Furthermore parasitic charging of insulation layers can substantially influence the system behavior.

Small distances among several components are often critical regarding heat spreading, conductance, convection, and radiation. Thus, the heating of a microprocessor can cause a thermal drift in an adjacent sensor. These thermal effects can be reduced by solutions such as thermal vias, an efficient energy management, or a sensor data processing with drift awareness. To achieve the latter, macro models of the sensors are helpful, which describe the coupled thermo-mechanical and electrical behavior. Often, for the electrical behavior white (or glass) box models are used, which are transparent and represent the "physics", while for the thermal behavior black box models are preferred, which describe the I/O behavior "merely" mathematically

Diverse computations from other domains such as acoustics or fluidics may also be required in dependence of the application. Mechanical simulations addressing shock and stress are often relevant for reliability estimates of packaging elements like solder bumps. Some verification examinations can result in coupled simulations of different domains. This is required, for instance, for a characterization of forces generated by electrostatic actuators.

Testing of highly integrated systems by measurement is certainly necessary, but faces two main problems. The first arises from the high degree of complexity, the second one stems from the extreme miniaturization, which makes interfacing to standard measurement equipment difficult. For example, even the capacitance of an RF probe may cause erroneous results.

It would thus be desirable to perform the verification of the complete system by virtual prototyping, in which the entire design is represented in computer memory and verified in the virtual world before being implemented in hardware. Software testing is not, however, immune against problems arising from complexity either, as has already been pointed out.

The more precisely the effects can be characterized, the more correctly the counter measures can be dimensioned without high design overheads. Therefore, efficient system verification allows increased components densities and smaller energy supplies due to optimized power consumption. Those energy savings can be achieved by means such as reducing timing slack or lowering voltage levels. Additionally, the increased effort for system verification can result in cheaper products by permitting higher component tolerances.

The errors, detected during the verification process, can necessitate different efforts. At best, only small changes have to be made in the layout. More serious errors require a replacement or addition of components. In the worst case, the design must be restarted at the concept level, demanding a completely different architecture. Therefore, prototyping can require several iterations for discovering and fixing problems. Furthermore, a problem solution can introduce new errors. If the verified prototypes meet the application requirements, it is often necessary to implement some adaptations for manufacturing large quantities. The next paragraphs discuss the options of assembly and interconnection from the technological point of view.

2 Packaging Technology

With the growing number of applications for disappearing electronics in objects various technologies have to be developed. They need to meet the needs of products ranging from multifunctional smart cards, electronic textiles to medical implants. Especially flexible modules have become increasingly important. They allow higher density than most rigid substrates together with minimal thickness and weight. Their mechanical flexibility allows integration in objects of various shapes by bending or even folding into 3D stacks.

Soldering techniques as well as adhesive joining have been optimized for flexible miniaturized modules. Simultaneously the development of thin flexible silicon dice provided the key technology for flexible systems.

For the realization of autonomous 3D systems (Smart Dust, eGrain, ...) thin functional layers have to be realized – ideally in wafer level processes – and connected by ultra thin electrical contacts. Thin film technologies as well as screen and stencil printing or even lamination will be applied.

In case of the integration of electronics in textiles another level of interconnection has to be taken into account: the interconnection between electronics and conductive fibers or yarn. New technologies and materials are developed and investigated to this purpose.

Table 1 shows an overview over applications for ambient intelligence together with the corresponding packaging technologies which will be described in this chapter.

Table 1. Applications for ambient intelligence and corresponding packaging technologies

Application	Technologies
Smart Dust, eGrain	Thin flexible Silicon
	Chip in polymer (CHIP)
	Thin film technology
	Screen printing
	Lamination
	Diffusion bonding
	Adhesive interconnection
Wearable electronics	Flip-Chip on flex
	Flex-to-yarn interconnection
	(adhesive, soldering,
	riveting, sewing)
	Washable plastic
	encapsulation
Body area networks	Flip-Chip on flex
	Sensor packaging
	Ultra thin Silicon
Smart labels	Flip-Chip on Flex
	Ultra thin Silicon
Smart cards	Flip-Chip on Flex
	Ultra thin Silicon
	Interconnection of foil batteries,
	flexible displays
	Lamination

2.1 Ultra Thin IC's

The capability to thin IC's down to $10\ldots30\,\mu$m is the enabling technology for flat and flexible systems. In this thickness range silicon dice show a high mechanical flexibility comparable to polymeric foils and are therefore ideally suited for assembly on polymer tapes. The dice are also more tolerant versus mechanical stress, thereby allowing better reliability. The availability of thin silicon will make it possible to produce low-cost products like smart labels on polymers or on paper in reel to reel processes and on the other hand it provides access to 3D packaging technologies either by direct stacking, embedding the dice or by folding the assembled flexible substrates.

Thinning can be performed by a combination of different processes which are already used in conventional silicon processing. A typical process starts with backside grinding. Removal rates up to $300\,\mu$m per minute are used in production. The resulting thickness variation is in the range of $1\ldots2\,\mu$m for 150 mm wafers. During the process, the backside of the silicon wafer is damaged to a certain extent. This increases the risk of wafer breakage and also of cracks in the dice after singulation. This problem is overcome by adding a stress relief process after grinding [1]. As further possibilities for stress relief, dry etching or chemical mechanical polishing (CMP) are successfully applied. CMP is especially suitable for final planarization, e.g., as a prerequisite for wafer-stacking technology. An overview on current thinning and stress relief processes is given in Table 2.

Table 2. Overview on current wafer thinning techniques

	Grinding	Spin-etching	Dry-etching	Polishing
Process type	Mechanical abrasion	Wet-chemical etching	Plasma, RIE	Chemical and mechanical
Process medium	Diamonds in ceramic wheel	$HF+HNO_3$ + additives	SF_6, NF_3, XeF_2	Slurry: SiO_2 grains in soft etchant
Removal rate	$300\,\mu$m/min	$10\ldots40\,\mu$m/min	$3\ldots30\,\mu$m/min	$<3\,\mu$m/min
Total thickness variation	$0.5\ldots3\,\mu$m	$5\ldots10\%$	n.a.	$<1\mu$m
Temperature	cool	$30\ldots40°$C	$50\ldots300°$C	$30\ldots40°$C
Application	Thinning	Stress relief	Stress relief, MEMS thinning	Surface finish, planarization

Various thinning techniques are available for reducing wafer thickness down to a few micrometers. However, at thicknesses below $100\,\mu$m the risk of wafer breakage increases dramatically so that exact process control is required. Furthermore the dicing process has a crucial impact on the quality

Fig. 4. Thin Si die on Al die compared to die in standard thickness

of the thin dice and their reliability. In Fig. 4 a thinned die assembled on a transponder antenna is shown in comparison to a conventional dice of $\sim 600\,\mu\mathrm{m}$.

2.2 Flip Chip Technologies

2.2.1 Thermode Bonding

The assembly of thin dice in thin flexible substrates requires thin contact structures. The conventional 90–100 μm high solder bumps cannot be applied here. Besides using conventional technologies like electroplating, small solder volumes can be applied by a new technology – immersion soldering [2]. This maskless process uses the wettability of a UBM (e.g., Ni/Au) by a liquid solder alloy through which the wafer is moved. The amount of solder deposited on the UBM is determined by the surface tension of the material and the pad size, typically in the range of a few micrometers. Using very small solder volumes deposited by immersion soldering together with extremely fast heating ramps (pulse heating) during soldering has been shown to be a promising technology for flip chip assembly on flex, down to finest pitches of 40 μm [3]. These thin interconnections are an important step forward regarding thin modules as well as 3-dimensional assemblies (see Fig. 5).

Due to the thin solder layer between chip metallization and conductor line the understanding of metallurgical interactions and the characteristics of dominating intermetallics has become even more important than for conventional flip chip assemblies. The intermetallic compounds determine the aging behavior and the mechanical stability of such contacts.

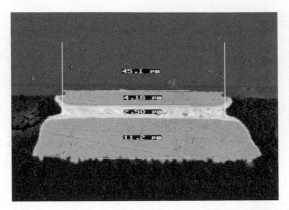

Fig. 5. Flip-chip contact after thermode bonding

2.2.2 Adhesive Bonding

The typical technology for mounting thin IC's on flexible substrates e.g., for smart labels is flip chip bonding using anisotropic conductive adhesive (ACA) or non conductive adhesive (NCA). But also for rigid substrates with metallizations which are not wettable by solder alloys (e.g., ITO on glass), adhesive bonding is applied successfully. The process requires a pad metallization on the die typically of electroplated Au, electroless Ni/Au or electroless Pd. The deposition of the adhesive can be performed by dispensing, stencil printing or dipping depending on the positioning accuracy and volume control required. Typical dispensing will already lead to good results but for chips thinner than 50 μm either special dispensers for ultra small volumes or stencil printing have to be chosen.

Bonding is performed using a flip chip bonder holding pressure and temperature over the curing time of the adhesive (5 . . . 20 s) [4]. These short cycle times which are possible with the new generation of materials allow that adhesive bonding is well suited for reel to reel assembly of low cost products. For ACA the process provides the electrical contact by clamping the conductive filler particles of the adhesive between bumps and conductor line. The resulting contact resistance for 100 μm contacts is ≤20 mΩ as there are only few particles on the contact area. A typical contact with Ni/Au bumps and Au filler particles is shown in Fig. 6. The application for a medical implant is shown in Fig. 7. The NCA process provides a metallic contact directly between bump and conductor line which leads to a lower contact resistance of ≤10 mΩ for the same pad size. Bumps with topography are preferable as a structured surface increases the process window.

Adhesive bonding allows a low stand off together with good reliability. It is ideally suited for ultra thin assemblies e.g., for smart labels or even medical implants. Figure 7 shows a module with two flip chips assembled using this technology which is implanted in the eye as an artificial retina.

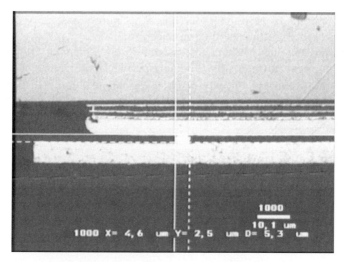

Fig. 6. Typical contact with anisotropic conductive adhesive (ACA)

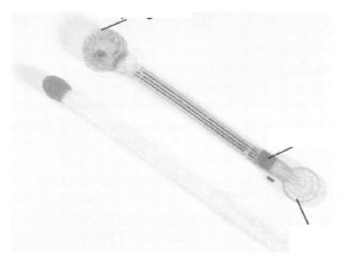

Fig. 7. Medical implant on 10 µm Polyimide foil in ACA technology

2.3 Self Assembly

Transponder chips have already reached a typical size of less than 1 mm 2. For even smaller dice it becomes increasingly difficult to achieve cost effective interconnection. Positioning and accurate placement require too long process times and too expensive equipment. A solution is seen in so-called self assembly technologies. The basic concept is that the dice are forced to place themselves on specific positions on a substrate either out of a fluid [5] or out of a large amount of chips in vibration. This allows assembly of very high volumes without high precision equipment.

Possible mechanisms which are used for self assembly are biochemical reactions, physical mechanisms (magnetism, electrostatics, ...) or just geometrical compatibility of chip and substrate. One principle which is currently discussed and investigated for self assembly of very small dice out of a liquid phase is shown in Fig. 8.

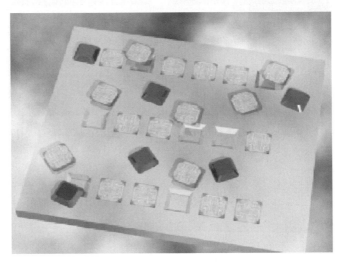

Fig. 8. Principle of fluidic self assembly of small dice (5) by Alien Technology

2.4 3D-Packaging Concepts

The assembly technologies presented offer already a high degree of miniaturization. However, there is an increasing number of applications which require even smaller modules. Often this can only be achieved by 3-D packaging. The advanced assembly technologies are also prerequisites for these 3-dimensional packaging developments.

There are different approaches to achieve 3D modules. The direct and most miniaturized approach is direct stacking of wafers. Other technologies are often based on flexible substrates which allow folding. Folded packages are especially interesting for memory modules, medical products (e.g., implants) and miniaturized microsystems (e.g., smart dust). Stackable modules are used as a matchbox approach to flexible products including sensors and actuators. On the other hand 3D integration can also be realized by integrating the IC's in the circuit boards using typical processes from PCB technology. This new technology is currently investigated and is suitable e.g., for mobile phones.

2.4.1 Folded Packages

Folded flexible packages are especially interesting for memory modules, medical products (e.g., implants) and miniaturized microsystems (e.g., smart dust). The first demands for such a technology resulted from the necessity to integrate a constantly increasing amount of memory in smaller and smaller products. As the memory dice also increased in size the only way to obtain reasonably small footprints for memory packages is 3D assembly. Thin Polyimide substrates realized by modified thin film processes have been used to create this type of folded package with one IC type. Any flip chip technology can be applied on this substrate type. As a low stand off is an advantage adhesive bonding and thermode bonding with ISB bumps are recommended. If required the interconnection with the board can be achieved by an array of BGA balls. In Fig. 9 an example for a folded structure is shown.

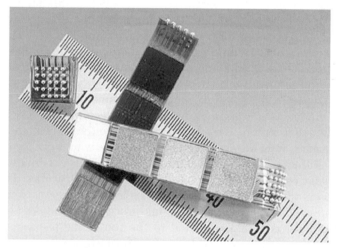

Fig. 9. Folded package in Flip Chip on flex technology (FCOF) by Hightec SA

If systems have to be realized in 3D the concept in Fig. 10 can be more suited since most passive components sensors and microsystem components are not thin enough. Here a rigid flex substrate is assembled with all components and then folded in the final shape.

2.4.2 Stacked Packages

Stackable modules are used as a matchbox approach to miniaturized flexible products including sensors and actuators. Different concepts are followed regarding interposer materials and assembly technologies. Stackable packages are available in flip-chip and wirebond technology on ceramic, FR4 and flex substrates depending on the area of application.

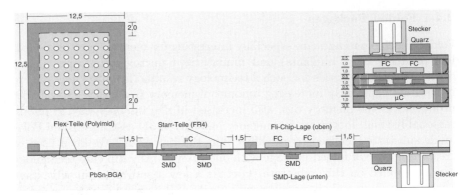

Fig. 10. Concept for folded SIP on rigid flex

The mechatronic system using StackPac is a logical development towards high integration, low volume and high functionality mechatronic solutions. This package offers also a high rate of manufacturability as it is possible to manufacture even Multi Chip Packages, including discrete components using reel to reel processes. StackPac processing starts with a polyimide tape with a photostructured metallization. The footprint is standardized, allowing fast redesigns by simply changing one single interconnection layer. The IC's and discrete components are assembled using SMD compatible reflow soldering or also adhesive assembly. The flexible tape is no longer just the carrier for the components, it is an an integral part of the whole package, providing shape and stiffness. The module is then molded with a thermosetting resin with high heat conductivity, to achieve good thermal features. The material system chosen offers all advanced properties for a package used under harsh environment conditions. After the molding-process the modules are separated and the top layer is folded and fixed to the top of the package. The result is a package of very low size and low volume, which is assembled in low cost standard IC manufacturing processes (Fig. 11).

Using StackPac it is possible to mount different types of modules in stacks together and connecting it to a standard bus system as shown in the mechatronic solution using ceramic TB-BGA's. The package allows also full testability of the assembled device and eliminates several interconnection layers. This leads to an overall reduction of interconnections of approximately 35–40% providing a drastic increase in package reliability.

2.5 Electrical Interconnection on Textiles

Woven fabric shows tolerances and mechanical deformation which are more than one order of magnitudes higher than those of conventional organic substrates. Therefore dice can not be assembled directly on textiles. The gap between the silicon die and the antenna has to be overcome e.g., by an interposer

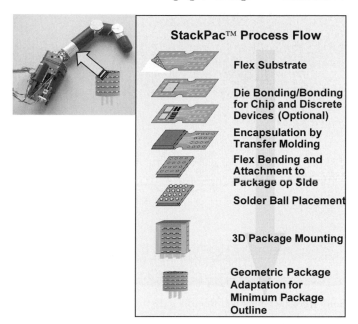

Fig. 11. Process flow and application for StackPac

or a module which provides the small contact pads for IC's and passive components and large contact areas to be assembled to the fabric.

New technologies are currently developed for this interconnection and the encapsulation of the assembly. The possible processes are derived from packaging technologies or from textile processes [6]:

- Soldering with low melting solders
- Adhesive joining
- Crimping
- Sewing
- Riveting

While these technologies provide mechanically stable permanent contacts there will also be the need for detachable contacts for those modules which are not washable and have to be attached to every garment. These interconnections are far more difficult to achieve reliably. Possible solutions – which still have to be qualified – include:

- Conductive Velcro
- Push buttons
- Zippers
- Miniaturized connectors in pockets

It could be shown that soldering of electronic modules to conductive yarn leads to reliable interconnections which even withstand washing. Figure 12

Fig. 12. Solder contact between flexible substrate and conductive yarm in fabric

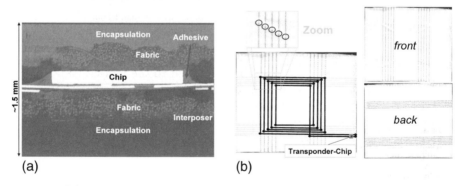

Fig. 13. (a) Transponder module between fabric layers; (b) textile transponder (antenna woven by TITV Greiz)

shows a typical contact with good wetting of the coated fibers. This technology was used to produce first transponders with textile antennas. Cross section and overview in Fig. 13 illustrate the concept of this module.

These new interconnection technologies – most of which are not yet fully developed – will be the basis for textile based electronic modules required for ubiquitous computing.

2.6 Assembly of Thin Layers

Miniaturized systems can be realized by stacking and interconnecting of functional layers. There are different approaches based on fabrication of multiple device layers using epitaxial growth of Si, wafer-level 3D packaging or stacking of wafers. Wafer-level 3D packaging can be realized in its most simple form by mounting dice to a substrate wafer, embedding and interconnecting the thinned dice with a modified multilayer thin film wiring. Wafer stacking is realized by wafer bonding of a thinned top wafer on a bottom wafer and interconnected by inter-chip vias and micro bumps. For the realization of small autonomous functional blocks (eGrains, Smart Dust, ...), wafer level assembly of functional layers is required using e.g., printing, laminating but also ultra thin solder joints and diffusion bonding. Besides the integration of the active Si devices a key issue will be the integration of passive components by thin film technologies, electroless deposition or printing. The thermal management for such highly integrated modules will require new strategies and predictive engineering. The vision for the wafer level processing of eGrains is shown in Fig. 14.

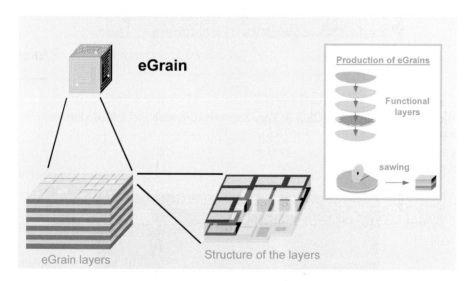

Fig. 14. Concept of building eGrains by depositing and structuring of functional layers on wafer level

2.7 Chip in Polymer

A number of approaches for integration of components have been presented lately. One of the first was the integration of power devices, shown by General Electric [7]. The Technical University of Berlin presented later an embedding of chips into ceramic substrates for a signal processing application [8]. Recently Intel announced its Bumpless Build-Up Layer (BBUL) as packaging technology for their future microprocessors [9]. At Helsinki University of Technology the so-called Integrated Module Board (IMB) technology has been developed, which embeds chips into holes in an organic substrate core [10, 11].

At the joined institute of Fraunhofer IZM and the Technical University of Berlin the Chip in Polymer (CIP) technology is under development [12, 13]. Its main feature is the emdedding of very thin chips (50 μm thickness or less) into build-up layers of printed circuit boards (PCB's), which does not sacrifice any space in the core substrate. The embedded chips can be combined with integrated passive components (see Fig. 15 and Fig. 16). A substantial advantage of the CIP approach is the embedding of components, using mainly processes and equipment from advanced PCB manufacturing. The realization of the process is illustrated by the cross section in Fig. 15?

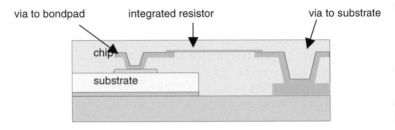

Fig. 15. Principle of the Chip in Polymer structure with embedded chip and integrated resistor

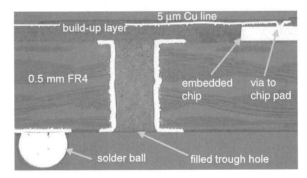

Fig. 16. Cross-section of a stackable CSP package in CIP technology

Before introducing such a technology in production, a number of challenges have to be mastered. An acceptable yield has to be obtained. Since no subsequent repair of an embedded component is possible the use of known good dice is essential. Furthermore, via formation and metallization process must be highly reliable. The reliability of the CIP package concept still has to be proven. CIP is basically a compound of materials with quite different Young's moduli and thermal expansion coefficients which might give reasons for cracking and delamination. This requires a detailed understanding by modelling and possibly an adaptation of technological parameters. The RF characteristics have to fulfill requirements of future applications, i.e. several GHz of bandwidth. This requires the use of polymers with low dielectric losses and low dielectric constant as well as very low tolerances for conductor geometries. A power management concept for components with high power dissipation has to be developed. Since no convection cooling of embedded active chips is possible, materials with low thermal resistance have to be used. Integrated resistors have to achieve sufficiently low tolerances, especially in order to be suited for RF circuits. This requires a processing technology with high geometrical resolution and tight control of deposition parameters. Finally, a new production flow has to be established in which chip assembly and testing will be part of substrate manufacturing.

However, Chip in Polymer technology is a promising approach to obtain 3-dimensional assemblies using low cost and high volume technologies. Only technologies with the potential to be applied for high volume production are suitable for microelectronic modules integrated in everyday objects and products and thereby the prerequisite for ubiquitous computing and ambient intelligence.

References

1. C. Landesberger et al. "New Dicing and Thinning Concept Improves Mechanical Reliability of Ultra Thin Silicon", Int. Symp. and Exhibition on Materials, Braselton, Georgia, March 2001.
2. S. Nieland et al., "Immersion Soldering – a New Way for Ultra Fine Pitch Bumping", Proc. of Electronic Goes Green 2000 Conference, Berlin, Germany, 2000, pp. 165–167.
3. B. Pahl et al., "A Thermode Bonding Process for Fine Pitch Flip Chip Applications Down to 40 Micron", Proc. EMAP, Korea, 2001.
4. R. Aschenbrenner et al., "Evaluation of Adhesive Flip Chip Bonding on Flexible Substrates", *Proc. Flexcon '97*, Sunnyvale, 1997.
5. Alien technology corporation: Fluidic self assembly; www.alientechnology.com/mambo_alien/library/pdf/fsa_white_paper.pdf.
6. C. Kallmayer et al., "New Assembly Technologies for Textile Transponder Systems", Proc. 53rd ECTC, New Orleans, 2003.
7. R. Filion, R. Wojnarowski, T. Gorcyzca, B. Wildi, H. Cole; "Development of a Plastic Encapsulated Multichip Technology for High Volume, Low Cost Commercial Electronics", 10. Applied Power Electronics Conf, IEEE, 1994, pp. 805–809.
8. Owzar, K. Buschik, O. Ehrmann; M. Kasper; "RF-Investigation of the electrical parameters of the embedded chip structure", Proc. VLSI Packaging Workshop, Monterey USA, October 16–18, (1995).

9. T. Waris, R. Tuominen, J. Kivilahti, "Panel-sized Integrated Module Board Manufacturing", Proc. Polytronic Conference, Oct. 21–24. 2001, Potsdam, Germany.
10. R. Tuominen, P. Palm, "Development of industrial scale manufacturing line for Integrated Module Board technology", 6th VLSI Packaging Workshop of Japan, November 12–14., 2002, Kyoto.
11. H. Braunisch, S. Towle, R. Emery, C. Hu and G. Vandentop, "Electrical Performance of Bumpless Build-Up Layer Packaging", Proc. ECTC 2002, May 28–31. 2002, San Diego, USA.
12. A. Ostmann, A. Neumann, S. Weser, E. Jung, L. Böttcher and H. Reichl, "Realization of a Stackable Package Using Chip in Polymer Technology", Polytronic Conference, June 23–26. 2002, Zalaegerszeg, Hungary.
13. A. Ostmann, A. Neumann, J. Auersperg, C. Ghahremani, G. Sommer, R. Aschenbrenner and H. Reichl, "Integration of Passive and Active Components into Build-Up Layers", EPTC 2002.

Algorithms in Ambient Intelligence

E. Aarts, J. Korst, and W.F.J. Verhaegh

Abstract. We briefly review the concept of ambient intelligence and discuss its relation with the domain of intelligent algorithms. By means of four examples of ambient intelligent systems, we argue that new computing methods and quantification measures are needed to bridge the gap between the class of existing algorithms and the ones that are needed to realize ambient intelligence. These examples include quality of experience, private recommender systems, intentional search, and multimodal user awareness. The major differences between the classical and novel approaches are formulated in terms of a number of challenges for the design and analysis of intelligent algorithms.

1 Introduction

Ambient intelligence opens a world of unprecedented experiences: the interaction of people with electronic devices will change as context awareness, natural interfaces, and ubiquitous availability of information will be realized. Distributed applications and their processing on embedded stationary and mobile platforms play a major role in the realization of ambient intelligent environments. Notions as media at your fingertips, enhanced-media experiences, and ambient atmospheres denote novel and inspiring concepts that are aimed at realizing specific user needs and benefits such as *personal expression, social presence,* and *well being* that seem quite obvious from a human perspective, but are quite hard to realize because of their intrinsic complexity and ambiguity. Obviously, the intelligence experienced from the interaction with ambient intelligent environments will be determined to a large extent by the software executed by the distributed platforms embedded in the environment, and consequently, by the algorithms that are implemented by the software [1]. Hence, the concept of ambient intelligence poses major challenges on the design and analysis of intelligent algorithms.

In this chapter we briefly review the concept of ambient intelligence and elaborate on its relation with intelligent algorithms. The emphasis of the chapter is on qualitative aspects highlighting those elements in the design and analysis of intelligent algorithms that play a role in the realization of ambient intelligence. We first review some the most profound developments in computational intelligence. As distributed computing devices and new interaction technologies are growing mature, new ways of intelligence can be pursued thus enabling more natural interaction styles among interacting devices and their users. In this way electronics can be moved into people's background allowing users to move to the foreground. We argue that new computing methods and quantification measures are needed to bridge the gap between the class of existing algorithms and the ones that are needed to realize ambient intelligence. This is illustrated by a number of typical examples brought about by the recent development of specific ambient intelligent systems. These examples include quality of experience, private recommender systems, intentional search, and multimodal user awareness. The major differences between the classical and novel approaches are formulated in terms of ten challenges for the design and analysis of intelligent algorithms.

2 Intelligence in Electronic Environments

Ambient intelligence is aimed at the realization of electronic environments that are sensitive and responsive to the presence of people [2]. The focus is on user interaction and experience, which introduces several new research issues related to natural user interaction and social intelligence. The notion ambient in ambient intelligence refers to the environment and reflects the typical system requirements such as distribution, ubiquity, and transparency. Distribution refers to non-central systems control and computation, ubiquity means that the embedding is overly present, and transparency indicates that the surrounding systems are invisible and non-obtrusive. The notion intelligence reflects that the digital surroundings exhibit specific forms of social interaction, i.e., the environments should be able to recognize the people that live in it, adapt themselves to them, learn from their behavior, and possibly show emotion.

2.1 Key Features of Ambient Intelligence

Aarts & Marzano have defined the following five key technology features of ambient intelligence [3]:

- *embedded*: many networked devices that are integrated into the environment,
- *context aware*: that can recognize you and your situational context,
- *personalized*: that can be tailored towards your needs,
- *adaptive*: that can change in response to you, and
- *anticipatory*: that anticipate your desires without conscious mediation.

The first two elements relate to the integration of hardware devices into the environment, and refer to embedded systems in general. Embedded systems play an important role in the realization of ambient intelligence because they account for the embedding of electronic devices into people's surroundings.

An extensive overview of recent developments and challenges in embedded systems is given by the research agenda compiled by the National Research Council [4]. MIT's Oxygen project [5] and IBM's effort on pervasive computing [6] are examples of networked embedded system approaches addressing the issue of integration of networked devices into people's backgrounds. The three other key elements of ambient intelligence are concerned with the adjustment of electronic systems in response to the user. They all can be viewed as system adjustments, but done at different time scales. Personalization refers to adjustments at a short time scale, for instance to install personal settings. Adaptation involves adjustments resulting from changing user behavior detected by monitoring the user's history over a longer period of time. Ultimately, when the system gets to know the user so well that it can detect behavioral patterns, adjustments can be made that relate to future user intentions.

2.2 New Developments

From a technological point of view, ambient intelligence builds on early innovation concepts such as ubiquitous computing [7] and pervasive computing [6]. The major new thing in ambient intelligence is the involvement of the user. Ubiquitous computing and pervasive computing are primarily positioned as technologies that support and improve productivity in business environments. It goes without saying that these novel concepts have played a major role in the development of new generations of mobile computing such as i-mode and UMTS, as well as in novel services such as e-business on demand. The next step is to augment ubiquity and ervasiveness with intelligence, and introduce it into everyday life situations. This is not simply a matter of introducing the productivity concepts into consumer environments. It is far more than that, because a totally new interaction paradigm is needed to make ambient intelligence work. The productivity concept is to a large extent still based on the graphical user interface known as the desktop metaphor that was developed by Xerox PARC in the 1970s, and which has become a world standard since [8]. What we need is a new metaphor with the same impact as the desktop metaphor that enables natural and social interaction within ambient intelligent environments, and this poses a major challenge on algorithm design. Especially, the requirements that ambient intelligent applications pose on the mechanisms that support users in their interaction with media and environments call for paradigms that will be substantially different from contemporary interaction concepts. This calls for social interaction paradigms build on social intelligence and this marks the difference between the classical field of artificial intelligence and the novel field of ambient intelligence. It complies with the difference between cognitive and social intelligence, where artificial intelligence is more aimed at cognitive intelligence while ambient intelligence is more aimed at social intelligence. As an example, in ambient intelligence, context-aware electronic doors that open up at a speed and duration depending on the number of people approaching them and their speed of walking are considered "more intelligent" than the ultimate chess computer Deep Blue that exhibits a level of cognitive intelligence that is way above that of average people and enables the machine to compete with the world's best chess players but does not exhibit any form of social intelligence.

2.3 Design for Intelligence

Intelligence in ambient intelligent environments can be related to two aspects: the social nature of the user interface that is used, and the extent to which the system can adapt to the user and the environment. The social character of the user interface will be determined by the extent to which the system complies with the intuition and habits of its users. The self-adaptability is determined by the capability of the system to learn through interaction with the user. The combination of human specific communication modalities such as speech, handwriting, and gesture, as well as the possibility to personalize to user needs, play a major role in the design of novel applications and services. Furthermore, ubiquitous computing environments should exhibit some form of emotion to make them truly ambient intelligent. To this end, the self-adaptive capabilities of the system should be used to detect user moods and react accordingly. This issue has led to the development of a novel research area that is called affective computing, and which again is characterized by a multidisciplinary approach, for instance, combining approaches from psychology and computer science [9]. The design of functions and services in ambient intelligence is often based on the concept of user-centered design [10, 11]. Within this design concept the user is placed in the center of the design activity, and through a number of design cycles in which the designer iterates over concept design, realization, and user evaluation, the final interaction design is created. Many interaction designers follow the *media equation* introduced by Reeves & Nass who argued that the interaction between man and machine should be based on the very same concepts as the interaction between humans is based, i.e., it should be intuitive, multimodal, and based on emotion [12]. Clearly, this conjecture is simple in its nature but at the same time it turns out to be very hard to realize. The design of ambient intelligent environments poses several challenging research questions with respect to the way intelligence is embedded and configured in such systems. Below, we discuss three elements that are related to system aspects, user aspects, and integration aspects, respectively.

Firstly, we distinguish the class of algorithms that exhibit some form of intelligent user-perceived behavior upon execution. This predominantly refers to algorithms that enable natural interaction between users and electronic systems. Classical examples are algorithms for speech processing tasks such as speech recognition, speech dialog, and speech synthesis. Other examples are algorithms for vision such as tracking, object recognition, and image segmentation techniques. More recently, algorithms have gained attraction that enable users to have a personalized access to content, e.g., through collaborative filtering and other recommendation techniques. Also sophisticated data browsing techniques that allow search based on intentions rather than on formal queries are attracting attention recently. Furthermore, techniques that can analyze and augment media by applying sophisticated algorithms that can generate meta data or can determine media context from raw data have gained much attention.

Secondly, we distinguish the class of algorithms that can be applied to control systems. As ambient intelligent environments will consist of many distributed devices, there is a need for intelligent algorithms that can manage data and computations within a networked system. This field is called resource management if the emphasis is on the distribution of activities over the available resources and quality of service if the emphasis is on the perceived quality of the tasks that are carried out. Both fields are concerned with the design and analysis of adaptive algorithms that can control and change system functionalities. Especially, requirements that are related to real-time and on-line execution of tasks are of prime interest. Also adaptation through learning is considered an important issue in this domain.

Thirdly, we mention the challenge of determining where the intelligence should reside within ambient intelligent systems. As pointed out, ambient intelligent systems are not standalone devices; they will appear as integrated environments consisting of networked embedded devices. Ambient intelligent environments can be seen as clusters of computing, storage, and input-output devices that are connected through ubiquitous broadband communication networks allowing data rates up to a few gigabytes per second. The ubiquitous communication network contains servers that act as routers and have internal storage devices. This implies that one can trade off the location of the software that provides the system intelligence. Roughly speaking there are two extremes, which are referred to as "ambient intelligence inside" versus "ambient intelligence outside". In the inside case the system intelligence resides as embedded software in the terminals. They are big footprint devices that can efficiently process large software stacks that implement sophisticated computational intelligence algorithms. The network that connects the devices is rather straightforward from a functional point view just allowing data communication, possibly at high bit rates. In the opposed view of ambient intelligence outside, the terminals may be small footprint devices just allowing for the data communication that is required to generate output or to take new input. The system intelligence is residing at the powerful servers of the ubiquitous communication network where it can be accesses by the terminals. Clearly the trade-off between ambient intelligence inside and outside relates to complex design decisions that need to be evaluated at a system level. The resulting design problem is referred to as "inside-outside co-design", and poses several new challenges for algorithm design.

3 Computational Intelligence

The algorithmic techniques and methods that apply to design for intelligence in ambient intelligent systems are rooted in the field of computational intelligence, which is the scientific and technological pursuit that aims at designing and analyzing algorithms that upon execution give electronic systems intelligent behavior. Engelbrecht gives a scholarly introduction into the field [13]. Sinčák & Vaščák present an overview of more recent advances in the field [14]. For the purpose of our presentation we distinguish the following features to characterize intelligent behavior in electronic systems, i.e., the ability to solve problems, to predict and adapt, and to reason. Below we discuss a number of algorithmic approaches in computational intelligence that can be structured along the lines of these three characteristic features. The domain of problem solving contains a collection of well-known search methods. Prediction and adaptation is covered by machine learning techniques. Finally, the domain of reasoning is treated by expert systems. For all three domains we present some applications in ambient intelligence. We also briefly address the topic of computational complexity, which plays a major role in computational intelligence, because many problems in this field are intrinsically hard to solve. Computational intelligence can be viewed as a subfield of artificial intelligence and for that reason we start with a short exposition if this quite interesting field in computer science.

3.1 Artificial Intelligence

According to Minsky, artificial intelligence is the science of making machines do things that require intelligence if done by man [15]. The origin of the subject dates back to the early days of computing science, and the classical subjects of investigation are vision, natural language and speech processing, robotics, knowledge representation, problem solving, machine learning, expert systems, man-machine interaction, and artificial life [16, 17]. Several major achievements were obtained over the years among which we mention the following ones. Expert systems have been developed that use speech and duologue technology to support hotel reservation booking and travel planning. Two other well-known examples are MYCIN, the first computer program to assist physicians in making diagnosis [18], and PROSPECTOR a program that could aid geologists to explore oil fields [19]. Robots have been constructed that can play baseball or climb stairs, and IBM's Deep Blue was the first chess computer that could compete at world masters level. For an overview of the developments in robotics or animats we refer to the work of Meyer & Wilson [20] and Brooks [21]. Neural networks were introduced as artificial computing models based on an analogy with the human brain. NETtalk was the first operational neural network that could produce speech from written text [22].

In all these cases it can be argued that the major achievements were realized through the use of sophisticated algorithmic methods that make excellent use of the vast growth in computational power. More recently, the field has given birth to several new exciting developments. One example is the development resulting from the strong involvement of elements from cognitive science. For instance, the general belief that human intelligence does not follow a specific predetermined path to accomplish a task, but merely responds to a current situation, gave rise to a novel concept that is referred to as situated action. This concept has been applied with great success in robotics by replacing classical path planning by approaches that are governed by environmental response mechanisms. Another new development is the use of case-based reasoning, which follows the idea of solving problems by making use of common sense understanding of the world based on the use of ontologies. These ontologies are embedded by agents that carry information about the meaning of fundamental everyday life concepts such as time, objects, spaces, materials, events and processes [23].

3.2 Problem Solving

Problem solving concerns finding a solution from a large set of alternatives that minimizes or maximizes a certain objective function [24]. Usually, the set of possible solutions is not given explicitly, but implicitly by a number of variables that have to be assigned numerical values and a number of constraints that have to be obeyed by this assignment. Often, this results in a compact description of an exponentially large solution set. One classical method to search such a solution space for an optimum is called branch and bound. This method recursively splits the solution space into subspaces by e.g., fixing the value of a variable. This is called the branching part, and it results in a tree-like search. The bounding part is used to prune as many parts from the search tree as possible, by comparing bounds on what can be achieved in a subspace corresponding to a subtree, with an existing solution, e.g., a feasible solution that has already been found. A second classical method is called dynamic programming. This method, which is generally more efficient, is applicable if the problem contains a certain structure, based on solution states. Basically, if one has to assign values to a number of variables, one considers one variable at a time, and checks for each solution state what happens if one chooses any of the possible values of this variable, in the sense of its direct (cost) effect, and the state in which one ends up. Consequently, the number of variables is decreased by one, and this can be repeated until all variables have been assigned a value. A third classical method, called linear programming, is applicable if all constraints and the objective function can be expressed linearly in the variables, which are continuous. In this situation, an optimal solution can be found in a running time that is polynomial in the size of the problem instance. For an overview of the theory of combinatorial optimization we refer the reader to the books by Korte & Vygen [25] and Wolsey [26].

Many problems cannot be solved to optimality in a reasonable time. In this situation one may resort to the use of heuristics, which drop the requirement of finding an optimal solution at the benefit of substantially shorter running times [27]. Local search is an example of such an approach [28]. Its working is based on making small alterations to a solution, i.e., one has a solution, and iteratively this solution is changed a bit, and the effect of this change on the objective function is evaluated. If the effect is favorable, the new solution is accepted, and used for the next iteration. If the solution deteriorates, the new solution may be rejected unconditionally (iterative improvement), or a mechanism may be used to accept it with a certain probability, which decreases with the amount of deterioration and over the course of the algorithm (simulated annealing). Local search is easily applicable in the sense that it requires (nearly) no specific problem knowledge, and it has given good results for various well-known problems. A next step is not to iterate with one solution, but a population of solutions at the same time, as is done in genetic algorithms [29]. Then, one can define crossover schemes between different solutions, to generate offspring from two parent solutions. Survival of the fittest reduces the increased population to only keep the best ones, which are the parents in the next iteration. In this way, good solutions are found by a kind of evolution process. Genetic algorithms play a major role in computational intelligence because of their flexibility and power to find good solutions in complex problem solving situations. Search problems do not always come with an objective function, but sometimes only a feasible solution is required. A common method that can be used in this situation is called constraint satisfaction [30]. This method applies a mechanism that systematically reduces the value domains of the involved variables by combining several constraints. This process is called constraint propagation. When domains cannot further be reduced, a variable is chosen, and it is assigned a value from its remaining domain. Then again constraint propagation is applied and a new variable is assigned a value, etc. The strength of this method depends on the possibility to delete infeasible values by combining constraints, and quite some research has been spent on the types of constraints that lend themselves best for constraint propagation. The field of constraint satisfaction can be positioned at the intersection of mathematical programming and artificial intelligence, and the fact that it contributed to the merger of these two fields is probably one of its major contributions in addition to the fact that it is quite a powerful method that can be applied to a large range of problems.

3.3 Machine Learning

Machine learning is a key element to predict and adapt in ambient intelligent systems. It concerns learning a certain input-output behavior from training examples, which makes it adaptive, and generalizing this behavior beyond the observed situations, to make predictions [31]. The best-known method in machine learning may be given by neural networks, which are built in analogy to the human brain [32]. In this, we have neurons with inputs and outputs, and outputs of neurons are coupled to inputs of other neurons. Each neuron takes a weighted sum of its inputs and sends a signal to its output depending on the result. For the communication with the outside world there are also input neurons and output neurons. In the training phase, the input pattern of a training example is fed to the input neurons, and the weights of the connections are adjusted such that the required output pattern is achieved. In the application phase, the input pattern of a new situation is fed to the input neurons, and the output neurons are read for the predicted response.

When the output function is binary, we speak about classification. Apart from neural networks, support vector machines have recently been introduced for this task [33]. They perform classification by generating a separating hyper-plane in a multidimensional space. This space is usually not the space determined by the input variables, but a higher-dimensional space that is able to obtain classifications that are non-linear in the input variables. A strong element of support vector machines is that they perform structural risk minimization, i.e. they make a trade-off between the error in the training examples and the generalization error. Bayesian classifiers use a probabilistic approach [34]. Here, the assumption is that classification rules are based on probability theory, and the classifier uses probabilistic reasoning to produce the most probable classification on a new example. To this end, Bayes' rule of conditional probabilities is used to come from prior probabilities to posterior probabilities. An advantage of Bayesian classifiers is that prior knowledge can be taken into account. The combination of analytic knowledge about a problem area and inferred knowledge is an important subject in current research in machine learning in general.

3.4 Expert Systems

An expert system is a computer program that provides solutions to search problems or gives advice on intricate matters making use of reasoning mechanisms within specific knowledge domains [35]. The characteristic features of such systems can be summarized as follows. It can emulate human reasoning using appropriate knowledge representations. It also can learn from past experiences by adjusting the reasoning process to follow promising tracks that were discovered on earlier occasions. Furthermore, it applies rules of thumb often called heuristics to exhibit guessing behavior. The behavior of expert systems should be transparent in the sense that they can explain the way in which solutions are obtained. The execution of expert systems often enables the use in on-line decision situations, which requires real-time behavior. The criteria used to evaluate the performance of expert systems are typically deduced from human task-performance measures, such as consistency, regularity, typicality, and adaptability, which all relate to the basic requirement that repetition and small input changes should produce results that make sense. Early expert systems used reasoning mechanisms based on logical rules. More recent approaches apply probabilistic reasoning and reasoning mechanisms that can handle uncertainty. Most recently, the concept of semantic engineering has been applied based on important aspects such as belief, goals, intentions, events, and situations, which are often expanded to the use of ontologies. Especially the use of the semantic Web is a new intriguing development that can further stimulate the development of expert systems with a truly convincing performance within a broad application domain [36, 37].

Recently, much attention has been devoted to agent systems, which are collections of small intelligent software programs that handle certain tasks in a collective way [38]. Their salient features can be described as follows. Agents execute continuously and they exhibit intelligent behavior such as adaptability and automated search. In addition they exhibit anthropomorphic properties in their interaction with users through expression of believe and obligation, and even through their appearance. They act autonomously, i.e., independent of human control or supervision. They are also context aware in the sense that they are sensitive and responsive to environmental changes. They can interact with other agents and migrate through a network. Agents have the ability to develop themselves to become reactive, intentional, and eventually social. The development of agent technology is a quickly growing field of computational intelligence. Since their introduction about a decade ago, intelligent agents have been widely recognized as a promising approach to embed intelligence in interactive networked devices and for that reason it is of special interest to ambient intelligence.

4 Examples of Intelligent Algorithms

Below we discuss four examples of intelligent algorithms used in recently developed ambient intelligent systems.

4.1 Quality of Experience

Consumer terminals, such as set-top boxes and digital TV sets, currently apply dedicated hardware components to process video. In the near future, programmable hardware with video processing in software is expected to take over for reasons of flexibility. The problem with this, however, is that software video processing shows highly fluctuating processing requirements, resulting in a worst-case load that is significantly higher than the average-case load. A safe (worst-case) resource allocation is however too costly, so one has to resort to a lower, close to average-case resource allocation.

When multiple video processing tasks are executed in parallel on a programmable component with lower than worst-case resource allocations, one has to take measures to ensure stability of such a system. For instance, a task that encounters a peak in its load should not interfere with other tasks. To achieve this, tasks are allocated individual resource budgets, which are both guaranteed and enforced. In other words, each task gets its promised share of the processor, even if other tasks misbehave, and it has to get by with this share.

To allow a task to get by with its share while still producing timely output at regular intervals, we introduce two means. The first one is working ahead, for which some buffering of input and output data is required. This allows the task to spread load peaks over time. The second means is scalable processing: if a task risks missing a deadline, it can choose to reduce the quality of the processing of some video frames, in order to save time and meet its deadline. The problem we address is now how to choose the quality level for each frame such that the experienced quality is as high as possible. This quality of experience is determined by three components: the quality of processing (which should be as high as possible), deadline misses (which should be avoided), and quality changes (which should also be minimized to avoid jumping up and down). To discuss describe this appraoch in more detail we describe below the work of Wüst & Verhaegh [39].

4.1.1 A Markov Decision Process

Wüst & Verhaegh model the problem described above as a Markov decision process (MDP) [40]. The state of the decision process is naturally given by the amount that the task has worked ahead, which we call *progress*, together with the previously chosen quality level. The latter is needed because we want to take quality changes into account. Hence we get states $s = (p, q)$, where p gives the progress, and q the previous quality level. Next, at each decision moment, which is the beginning of the processing of a frame, an action a has to be taken, which is the choice of the quality level for processing the upcoming frame. Thirdly, MDPs require transition probabilities $\Pr(s, a, s')$, which give the probability of going from state s to state s' when action a is chosen. These probabilities can be determined from processing statistics on example video streams. Finally, an MDP requires a revenue $r(s, a)$ that reflects the value of processing the upcoming frame in the quality level given by a, given state s. This revenue is chosen to be a weighted sum of a reward for the chosen quality level, a (high) penalty for the expected number of deadline misses until the next decision moment, and a (smaller) penalty for changing the quality level.

Given the MDP, an optimal strategy can be computed off-line by maximizing the average revenue per frame. There are a number of solution algorithms for this, such as *successive approximation, policy iteration,* and *linear programming.* The resulting strategy indicates for each state what the best action is, i.e., what the best choice of quality level is. This strategy can be stored in a look-up table, that can simply be consulted on-line to control the video processing algorithm.

4.1.2 Reinforcement Learning

The above approach requires processing statistics beforehand, which is used for the off-line computation of an optimal strategy. As it is difficult to find video streams that are representative for the streams that will be encountered on-line, a reinforcement learning (RL) approach [41] was developed, which is basically an on-line variant of MDP. RL learns an optimal strategy on-line, by constructing and maintaining dynamic *state-action values* $v(s, a)$, which indicate how good it is to choose action a in state s. The general idea is that, given the state s, the action a is chosen that maximizes $v(s, a)$. After such an action, the resulting revenue is measured, and this is used to update $v(s, a)$. An often used algorithm for this is the so-called SARSA algorithm.

A big issue in reinforcement learning in general is that in each iteration only the value for one state-action pair (s, a) is updated, which makes the learning very slow if the state-action space is large. Furthermore, one has to choose random actions from time to time in order to explore their value too. This can be improved in the following way. If an action a is taken in state s, the specifics of the model allow us to also update $v(s', a)$ for all other states s'. Furthermore, one can also estimate the effect for other actions a'. As a result, one can update *all* state-action values $v(s', a')$ in each iteration. Experiments have shown that this gives a drastic improvement, and that RL even outperforms MDP, even though it starts with no knowledge at all.

4.1.3 Budget Scaling

The MDP and RL control strategies implicitly assume that the processing time of successive frames are unrelated. In practice, however, load fluctuations can happen not only on a short time scale, which fit the assumption, but also on a longer time scale. For instance, some scenes are inherently more difficult to process than others. As a result, there are periods with a significantly higher load or with a lower load, which violates the independence assumption. To overcome this, an additional control algorithm is used, which basically tries to estimate the difficulty of the scene, and uses a resulting *scaled budget* to look up the action from the MDP or RL strategy.

4.1.4 Conclusion

The approaches have been tested on an MPEG-2 decoding application, for a number of video streams, of which processing time data has been used for simulating the algorithms. The approaches indeed achieved the goal to balance quality of processing, deadline misses, and quality changes. Comparing them to an off-line computed optimum by means of dynamic programming (when the entire video stream is known) revealed that the optimality gap is quite small [42].

The approach discussed above is a nice example of the "humanized cost measures" challenge mentioned in Sect. 5, reflected by the way the quality of the experience is modelled. One of the future steps is to validate the way the user-perceived value is modelled which can be done by means of real user tests. The reinforcement learning algorithm is also a nice example of the "small-footprint algorithms" challenge, as it has to run on the same resources that it tries to deploy optimally.

4.2 Private Recommender Systems

The explosive growth of the Internet has led to a situation where people are confronted with an overload of information. One of the approaches to help people to make a good selection of content is given by recommender systems. These systems estimate to what extent a user will like the available content, based on the user's likes and dislikes for previously encountered content. A well-known technique to do so is *collaborative filtering*, which uses preferences of a community of users to predict the preference of a particular user for a particular piece of content. These kind of systems are found, for example, on the Internet at music sites to recommend new music, and at book sites to recommend new books.

The most common way of collaborative filtering is that the preferences (in the form of ratings for content) of a community of users are collected at a web server. Then, a similarity measure is computed between each pair of users based on the content they jointly rated. Next, recommendations for a particular user can be made by considering users that are similar to him, and checking for content that they liked but has not yet been rated by the user.

One of the issues with the above system is that it requires the preferences of a community of users for comparison. If this kind of information is really personal (e.g., in case of medical information), users may not be willing to give this information, because of a lack of trust in the server. We will show how we can make collaborative filtering private, by introducing encryption techniques.

4.2.1 Collaborative Filtering Algorithm

First, we explain how collaborative filtering works, by giving the formulas underlying it. We will give the most commonly used approach and formulas;

other approaches and formulas are given by Van Duijnhoven [43]. The two main steps in collaborative filtering are the determination of user similarities and the prediction of scores.

User Similarities. A common similarity measure used in literature is the so-called *Pearson correlation coefficient*, given by

$$s(u,v) = \frac{\sum_{i \in I_u \cap I_v} (r_{ui} - \bar{r}_u)(r_{vi} - \bar{r}_v)}{\sqrt{\sum_{i \in I_u \cap I_v} (r_{ui} - \bar{r}_u)^2 \sum_{i \in I_u \cap I_v} (r_{vi} - \bar{r}_v)^2}} \tag{1}$$

where u and v are users, i denotes an item, I_u and I_v denote the sets of items that have been rated by user u and v, respectively, r_{ui} denotes the rating of user u for item i, and $\bar{r}_u$ denotes the average rating of user u. The numerator in this equation gets a positive contribution for each item that is either rated above average by both users or rated below average by both. If one user has rated an item above average and the other user below average, we get a negative contribution. The denominator in the equation normalizes the similarity, to fall in the interval $[-1, 1]$, where a value 1 indicates complete correspondence and -1 indicates completely opposite tastes.

Predictions. The second step in collaborative filtering is to use the user similarity to interpolate between other users' ratings to determine the rating of a new user-item combination. Using the similarities as weights, a commonly used prediction formula is given by

$$\hat{r}_{ui} = \bar{r}_u + \frac{\sum_{v \in U_i} s(u,v)(r_{vi} - \bar{r}_v)}{\sum_{v \in U_i} |s(u,v)|} \tag{2}$$

where U_i is the set of users that have rated item i. So, the prediction is the average rating of user u plus a weighted sum of deviations from the averages.

4.2.2 Encryption

Next, we show how we can encrypt the above collaborative filtering formulas, using the cryptosystem of Paillier [44]. This encryption starts with two large primes p and q, which form the private key, and their product $n = pq$ and a *generator* g, which form the public key. Then, a message $m \in \mathbb{Z}_n$ is encrypted by

$$\varepsilon(m) = g^m r^n \bmod n^2 \ ,$$

where r is a random number that disables decryption by simply encrypting all possible values of m and comparing. A nice property of this encryption system is that

$$\varepsilon(m_1)\varepsilon(m_2) \equiv g^{m_1} r_1^n g^{m_2} r_2^n \equiv g^{(m_1+m_2)} (r_1 r_2)^n \pmod{n^2},$$

which is an encryption of the sum $m_1 + m_2$. Furthermore,

$$\varepsilon(m_1)^{m_2} \equiv (g^{m_1} r_1^n)^{m_2} \equiv g^{m_1 m_2}(r_1^{m_2})^n \pmod{n^2} \, ,$$

which is an encryption of the product $m_1 m_2$. In other words, the encryption scheme allows us to perform additions and multiplications on encrypted data. We can use this to calculate the sums of products in (1) and (2), using

$$\prod_j \varepsilon(a_j)^{b_j} \equiv \prod_j \varepsilon(a_j b_j) \equiv \varepsilon\left(\sum_j a_j b_j \right) \pmod{n^2} \, .$$

The details how to do this are given by Van Duijnhoven [43].

4.2.3 Protocol

The protocol for encrypted collaborative filtering now runs as follows. Let user u be the user for which we want to give recommendations. First, user u encrypts his ratings and sends them to the server, who distributes them further to all other users v. The latter calculate the three sums in (1) in the encrypted domain, using their ratings, and send the resulting encrypted sums back via the server to u, who can decrypt the sums and compute the similarities. Next, user u encrypts these similarities again and sends them via the server to the other users v together with the index j of a candidate item. Each user v computes his contribution to the numerator and denominator of (2) in the encrypted domain, and sends the results back to the server, who multiplies the encrypted contributions to get an encryption of the sums. This is sent back to u, who can now decrypt the sums, and calculate the resulting prediction for item j.

4.2.4 Conclusion

We have shown that collaborative filtering systems can be made private, by encrypting profiles and performing the computations on encrypted data. Compared to the original set-up of collaborative filtering, the new set-up requires a more active role of the users' devices. This means that instead of a (single) server that runs an algorithm, we now have a system running a distributed algorithm, where all the nodes are actively involved in parts of the algorithm.

4.3 Intentional Search

As mentioned before, one of the basic assumptions in the development of ambient intelligent systems is that a person will be able to interact with his environment in a similar way as he interacts with other persons [12]. Such a form of natural interaction requires different capabilities of the environment involving different modalities. To capture and understand spoken commands, speech recognition and natural language processing techniques are required. To capture the context of use and understand the activities of the user, computer vision techniques how to be important.

In this section we focus our attention on the concept of *query by humming*, that can be useful in interacting with large music collections. Usually, people listen to music separately from gaining knowledge about the names of the performer or composer or the title of the song. Recalling a song title from a given melody or vice versa is notoriously difficult [45]. Query by humming allows the user to sing any melodic passage of a song, while the system seeks the song containing that melody.

The sung or hummed input is compared with the known melodies from a potentially large music collection. A query-by-humming system tries to detect the pitches in the sung input and compares these pitches with symbolic representations of the known melodies. Broadly speaking, query by humming consists of two major steps. The first step deals with the conversion of an audio signal into a symbolic representation, which is predominantly in the audio signal processing domain. The second step deals with the matching of the resulting symbolic representation to the known melodies. In this section, which is based on the work of Pauws [46], we focus on the second step. For an overview of the techniques that can be used in the first step, we refer to [46].

4.3.1 Melody Representation

A complication to be taken into account is that the input usually contains errors. One problem is that people, without any assistance, start singing in any preferred key. Therefore, melodies have to be encoded in a representation that is invariant to key. This invariance can be obtained by representing the interval between successive notes, expressed in semitones. In particular, a melody sequence $S = s_1 s_2 \ldots s_N$ comprising absolute pitches is converted to the sequence $S' = (s_2 - s_1)(s_3 - s_2) \ldots (s_N - s_{N-1})$. Temporal information is kept by storing the real value of the inter-onset-interval (IOI) of each note to each corresponding interval category. These data are used to make melody comparison invariant to global tempo. The melody distance is defined in terms of ratios of consecutive IOIs.

4.3.2 Melody Comparison

While singing, people forget or add notes. Also, intervals may be sung too flat or too sharp, or notes may be sung too long or too short. To account for these errors, a distance measure must be defined that expresses to what extent the sung input deviates from a given melody. A distance measure that is often used in the comparison of strings is the so-called *edit* distance, although also other distance measures can be used that are more tuned to query by humming. The edit distance between two strings Q and S is defined as the minimum number of deletions, insertions, and replacements of single elements that are necessary to transform Q into S or a substring of S. With each type of these elementary transformations we can associate a given cost, and the cost of transforming Q to a substring of S is given by the sum of its constituent elementary transformations. The problem of finding the known song that best matches the sung input can consequently be formulated as finding an approximate match of Q in S that has a minimum edit distance, where S can be considered as the concatenation of all the known songs in the collection.

4.3.3 Approximate String Matching Algorithms

Computing the edit distance can be solved by using classical dynamic programming for sequences of the same length [47] and for sequences of different lengths [48]. Let m denote the length of the query string Q and let n denote the length of the concatenation of all songs in the collection. Then, by using dynamic programming the approximate string matching problem can be solved in $O(mn)$ time. If the music collection is large, this might result in response times that are not acceptable in practice. Fortunately, various indexing methods have been proposed that show sub-linear computation times. For examples, we refer to [49–52].

4.3.4 Conclusion

Intentional search allows users to focus on their intentions in browsing and selecting audio and video content. In this section we focus on query by humming, which is just one of the possible ways to intuitively interact with audio and video content. An alternative way to interact with audio content is to ask for songs that are similar to a given song, where similarity is based on meta data such as genre and artist and on features that have been extracted from the content itself. Intentional search algorithms usually involve large databases while at the same time require small response times. Consequently, they require extensive computational resources. At the same time, optimization objectives and search criteria that are used in intentional search algorithms should reflect psychological measures that reflect user perceived values.

4.4 Multimodal User Awareness

Ambient intelligent environments are expected to adapt sensibly to the context of use. One important aspect is that they adapt to the users that are present. In addition, users should be able to issue commands to the environment by speech.

   In this section we discuss an example of multimodal integration. Sensor data from different modalities are combined to increase the performance of the system, especially to make it more robust and reliable. We discuss an example where speech recognition is enhanced with computer vision.

4.4.1 Voice Control Using a Microphone Array

The context is a speech recognition system that can be used in the interaction between an ambient intelligent environment and its users, where spoken commands can be used to control different devices in the environment. In a real-life setting, high-quality voice control is hampered by the presence of other sources of sound, such as back-ground music. A well-known strategy to improve the reception of specific sound sources, is the use of microphone arrays [53,54]. A microphone array consists of multiple microphones that are placed at different locations. If the location of the human speaker is known, one can tune the microphone array (beamforming), such that sound coming from other locations is filtered out. In this way, a microphone array enables the capture of spoken commands under adverse acoustic conditions. In addition, a microphone array allows, to a certain extent, the tracking of moving human speakers without requiring the use of a cumbersome tethered microphone. In general, however, persons will not be talking continuously, and multiple persons may interact with the environment.

4.4.2 Speaker Detection Using Computer Vision

To increase the performance of such a microphone array one can use computer vision to identify potential sound sources and improve the tracking of human speakers, as follows. By using one or more cameras, computer vision can be used to detect the persons that are present in a given room. Additional face detection enables to determine the position of the person that is talking at a given moment in time. Several algorithms have been proposed to realize visual tracking, ranging from algorithms that use skin color detection via algorithms that search for facial features, such as pupils, lip corners, and nostrils, to algorithms that use gaze tracking [55]. The latter algorithms are based on a 3D model of the head. They try to estimate the head pose by finding correspondences between a number of head model points and their locations in the camera images. This information from a visual tracking system can be used as input for the beamforming in the microphone array.

4.4.3 Sensor Fusion Algorithms

Algorithms for audio tracking and visual tracking can be combined in a same way as verbal and non-verbal channels are being used in human-human communication. Algorithms that can be used for tracking human speakers may use the inter-aural delay between two microphone signals, allowing an estimation of the direction of sound. In addition, vision algorithms may use skin tone and motion detection, allowing visual sensing of faces and body movement. Information from these sensors can be combined at the pixel level in order to detect "noisy face pixels" and track the position of the person speaking.

4.4.4 Conclusion

It is generally acknowledged that a bimodal tracking system can deliver more robust and reliable results than either of the two single modalities. Various projects pursue research in this area; as examples we mention the MIT's Oxygen project [56] and IDIAP's M4 project [57]. By combining multiple modalities, one creates algorithms that provide awareness in a robust way.

5 Challenges for Intelligent Algorithms

An interesting frame of reference for our discussion on intelligent algorithms is provided by the developments in semiconductor industry. It is generally known and accepted that developments in this domain follow Moore's law [58], which states that the integration density of systems on silicon doubles every eighteen months. This law seems to hold a self-fulfilling prophecy because the computer industry follows this trend for already four decades. Moreover, other characteristic quantities of information processing systems, such as communication bandwidth, storage capacity, and cost per bit of input-output communication seem to follow similar rules. These technological developments have a great impact on intelligent algorithms and we argue that the future developments in algorithms can be directly related to the developments according Moore's law. To discuss these trends we distinguish between two lines of thought being *more Moore* and *more than Moore*. These lines of thought comply with the opportunities that originate on the one hand from the improvements in efficiency of classical computing platforms that result from Moore's law, and on the other hand from the extension from classical stand-alone computing platforms to distributed networks of computing platforms that are augmented with robust context-aware sensor networks. Below we explain some of these aspects in more detail.

More Moore. According to the trends implied by Moore's law we will have within a few years from now computing platforms at our disposal that may either be small footprint systems with a performance comparable with current high-end computing systems or big footprint systems with substantially improved performance characteristics. Small footprint devices will become cheap and abundantly available. They will have compute powers of a few giga-operations per second with on chip storage of a few gigabytes, and they can wirelessly connect to other devices with a communication bandwidth of a few megabytes per seconds. In other words they have a performance of a current high-end desktop PC. Big footprint devices will have a performance that improves with a factor of two every eighteen months in all dimensions that are relevant. However, one may question whether this performance improvement will be obtained for wireless communication bandwidth and for energy density. Nevertheless, it will be definitely obtained for computing and storage capacity yielding a new generation of high-end computing platforms with substantially improved performance over the currently available platforms.

More than Moore. Developments in connectivity, nano-technology, and micro-systems design introduce two characteristic features of novel computing systems that extend beyond the classical developments implied by Moore's law and therefore can be identified as *more than Moore*. The first one is given by the fact that large scale distributed networks can be designed consisting of many interconnected devices that have substantial on-board computing and storage capacities. These so-called grids have been recently launched as the computing infrastructure that is going to succeed the Internet [59]. Indeed, massively distributed networks may introduce new emergent computing features similar to those that have already been identified at a small scale by intelligent agents and neural networks. A second major extension is provided by the augmentation of classical (networked) computing devices with sensors and actuators. Many scientist believe that one of the major challenges of computing science is to connect the physical to the digital world, or in other words to use the input provided by the physical surrounding of people to enhance the functionality of computing platforms or to create enhanced experiences of user environments. The cost of sensors may drop to prices below a few Euro cents, and consequently may become abundantly available. They may provide different types of low-level physical inputs including temperature, pressure, movement, acceleration, lighting, humidity, and others. Furthermore, more sophisticated types of sensors may be able to also provide high-level semantic information such as person identification, event detection, case distinction, and others. Both the use of low-level as well as the use of high-level sensor input poses major challenges on the design of intelligent algorithms.

Following the lines of thought imposed by the differences between *more Moore* and *more than Moore*, we can formulate the following challenges for the design of intelligent algorithms.

5.1 More Moore Challenges

- **Small-footprint algorithms**. Small footprint devices allow for the execution of algorithms in mobile systems such as personal digital assistants and personal communication devices. Because the available computational effort is limited, this poses major challenges on general systems functions such as data caching, quality of service, and resource management.
- **Real-time mathematical programming**. Large footprint devices allow real-time execution of algorithms that require large computational efforts for complex decision making processes based on mathematical programming techniques such as dynamic programming, linear programming and quadratic programming. These algorithms call for efficient calculation of incremental input data and effective fallback mechanisms that can handle time-outs.

- **Intentional search**. Large footprint devices allow the use of algorithms the can handle queries based on user intentions. This calls for approximate string search, pattern matching, and classification techniques that can deal with ultra-large databases. These techniques are quite demanding from the point of view of computational effort, because of the growing capacity of storage devices. This type of processing can easily account for another factor of ten in speed of future computing devices following from Moore's law.

- **Feature extraction**. Future databases will extend on the classical functionality of storing and retrieving data items by adding dynamical data manipulation that allows to determine semantic information from the stored data items for the purpose of adding meaning and semantic relations to them. For this we need algorithms that can extract features from audio, video, text, and graphics that enrich the stored data items with meta data.

5.2 More Than Moore Challenges

- **Humanized cost measures**. Algorithms that enhance the user experience should account for human perceived values such as perceptive measures, inconsistency, and feedback. Consequently, rigid physical objective measures resulting from classical measures such as completion time, capacity, speed, and length need to be replaced by more psychological measures that reflect user experiences. These measures are widely unknown and their exploitation in search algorithms requires the quantification of user perceived values, which to a large extent is an unexplored domain of research.

- **Aware algorithms**. The connection of the physical to the digital world calls for algorithms that can deal with large amounts of physical data. The major advantage of the use of physical input is obtained when it is used to provide awareness and contextual information, i.e., environmental input that allows to distinguish between different states of use. Major challenges are robustness, parameter estimation, location detection, and context extraction.

- **Life-long learning algorithms**. True intelligence can only be obtained by developing user models that capture specific knowledge of users in relation to the tasks they may want to perform. As habits of users change over time this involves learning algorithms that can adjust model parameters and autonomously create novel user aspects over time. This requires quite intricate algorithmic features as for instance contained in statistical techniques, and poses major challenges on the design of intelligent algorithms.

- **Collaborative algorithms**. The distributed nature of ambient systems allows for the use of distributed algorithms that consist of autonomous components that run on local devices to perform collaborative tasks. Existing examples are connectionist approaches such as artificial neural networks and artificial life systems. The true challenge is to lift these approaches to a higher semantic level in order to arrive at distributed algorithms that exhibit emergent features. Examples of such features are global control, negotiation, and Internet routing.

Evidently, the formulation of the challenges presented above provides only a coarse guideline for the developments that are needed to arrive at full realization of intelligence in ambient intelligent systems. However, it should be clear from the many new ideas emerging from the detailing and realization of the ambient intelligence vision that this new field poses great challenges on the design of intelligent algorithms.

6 Conclusion

The design of intelligent algorithms can claim an important role in the development of ambient intelligence. Much of the intelligence that is exposed by ambient systems will be generated by techniques that find their roots in computational intelligence. We discussed four examples. Quality-of-experience control maximizes the user-perceived value of adaptive signal processing, by using on-line, stochastic optimization algorithms. Collaborative filtering algorithms for giving personal recommendations can be made private by encrypting profiles and performing calculations on encrypted data. This resulted in an algorithm that is executed distributedly in a network of nodes. Intentional search systems allow users to browse through music databases by applying sophisticated string matching techniques such as dynamic programming and indexing techniques to find approximate matches between input strings and items stored in a database. Sensor fusion using acoustic processing, speech recognition, and computer vision results in natural interaction styles that enable the user to really move freely in his environment while interacting with electronic systems.

A new and interesting field of research is given by systems that connect the physical to the digital world. Sensors and actuators can be used to sense low-level signals in domestic, public, or mobile environments and to control, regulate, and personalize environmental conditions such as temperature, humidity, draft, loudness, and lighting. A next step is to reason about these environmental conditions in order to determine more global environmental states that can be used to personalize settings within ambient systems. A third step then could be to reason about these events and states to deduce pre-responsive actions or suggestions to a user.

It goes without saying that the use of intelligent algorithms within ambient intelligence raises high expectations, resulting in new challenges for the design of intelligent algorithms that extend beyond the use of current algorithms. Especially, the complexity of scale resulting from real-time, on-line and distributed computing gives rise to new problems that need resolution. In addition, the incorporation of human factors and context awareness call for new algorithmic approaches that extend far beyond what current approaches can do in order to connect the physical to the digital world and to the user in the end. Technology will not be the limiting factor in its realization of ambient intelligent environments. The ingredients to let the computer disappear are available, but the true success of the paradigm clearly depends on the ability to come up with concepts that make it possible to interact in a natural way with the digital environments that can be built with the invisible technology of the forthcoming century. The role of intelligent algorithms in this respect is apparent since it is a key enabling factor in realizing natural interaction.

References

1. Verhaegh, W.F.J., E.H.L. Aarts, and J.H.M. Korst (eds.) [2004] *Algorithms in Ambient Intelligence*, Kluwer Academic Publishers, New York.
2. Aarts, E., H. Harwig, and M. Schuurmans [2001], Ambient Intelligence, in: J. Denning (ed.) *The Invisible Future*, McGraw Hill, New York, 235–250.
3. Aarts, E., and S. Marzano (eds.) [2003], *The New Everyday: Visions of Ambient Intelligence*, 010 Publishing, Rotterdam, The Netherlands.
4. National Research Council [2001], *Embedded Everywhere*, National Academy Press, Washington DC.
5. Dertouzos, M. [1999], The Future of Computing, *Scientific American* 281(2), 52–55.
6. Satyanarayanan, M. [2001], Pervasive computing, vision and challenges, *IEEE Personal Communications*, August, 10–17.
7. Weiser, M. [1991], The computer for the twenty-first century, *Scientific American* 265(3), 94–104.
8. Tesler, L.G. [1991], Networked computing in the 1990s, *Scientific American* 265(3), 54–61.
9. Picard, R. [1997], *Affective Computing*, The MIT Press, Cambridge, MA.
10. Beyer, H., and K. Holtzblatt [2002], *Contextual Design: A Customer-centered Approach to Systems Design*, Morgan Kaufmann, New York.
11. Norman, A.M. [1993], *Things That Make Us Smart*, Perseus Books, Cambridge, MA.
12. Reeves, B., and C. Nass [1996], *The Media Equation*, Cambridge University Press, Cambridge, MA.
13. Engelbrecht, A.P. [2002], *Computational Intelligence: An Introduction*, Wiley, Chichester.
14. Sinčák, P., and J. Vaščák (eds.) [2000], *Quo Vadis Computational Intelligence? New Trends and Approaches in Computational Intelligence*, Physica Verlag, Heidelberg.
15. Minsky, M. [1986], *The Society of Mind*, Simon and Schuster, New York.
16. Rich, E., and K. Knight [1991], *Artificial Intelligence (2nd edition)*, McGraw-Hill, New York.
17. Nilsson, N.J. [1998], *Artificial Intelligence: A New Synthesis*, Morgan Kaufmann, San Francisco, CA.

18. Shortliffe, E.H. [1976], *Computer Based Medical Consultations: MYCIN*, Elsevier, New York.
19. Duda, R.O., and E.H. Shortliffe [1983], *Expert Systems Research*, Science 220, 261–268.
20. Meyer, J.A., and S.W. Wilson, (eds.) [1991], *From Animals to Animats*, MIT Press, Cambridge, MA.
21. Brooks, R.A. [2002], *Flesh and Machines*, Pantheon Books, New York.
22. Sejnowski, T., and C. Rosenberg [1987], Parallel networks that learn to pronounce English text, *Complex Systems* 1, 145–168.
23. Liang, T-P., And E. Turban (eds.) [1993], Special Issue on case-based reasoning and its applications, *Expert Systems with Applications*, 6(1).
24. Papadimitriou, C.H., and K. Steiglitz [1982], *Combinatorial Optimization: Algorithms and Complexity*, Prentice-Hall, Englewood Cliffs, NJ.
25. Korte, B., and J. Vygen [2000], *Combinatorial Optimization, Theory and Algorithms*, Springer-Verlag, Berlin.
26. Wolsey, L.A. [1998], *Integer Programming*, Wiley Chichester.
27. Osman, I.H., and J.P. Kelly (eds.) [1996], *Meta-Heuristics: Theory and Applications*, Kluwer Academic Publishers, Boston, MA.
28. Aarts, E.H.L., and J.K. Lenstra (eds.) [1997], *Local Search in Combinatorial Optimization*, Wiley, Chichester.
29. Goldberg, D.E. [1989], *Genetic Algorithms in Search Optimization and Machine Learning*, Addison Wesley, Reading, MA.
30. Tsang, E.P.K. [1993], *Foundations of Constraint Satisfaction*, Academic Press, London.
31. Mitchell, T.M. [1997], *Machine Learning*, McGraw-Hill, New York.
32. Hertz, J., A. Krogh, and R.G. Palmer [1991], *Introduction to the Theory of Neural Computation*, Adison-Wesley, Reading MA.
33. Christianini, N., and J. Shawe-Taylor [2000], *An Introduction to Support Vector Machines*, Cambridge University Press, Cambridge, UK.
34. Pearl, J. [1988], *Probabilistic Reasoning in Intelligent Systems: Networks of Plausible Inference*, Morgan Kaufmann, Dan Mateo, CA.
35. Lenat, D.B. [1990], *Building Large Knowledge Based Systems*, Addison-Wesley, Reading, MA.
36. Berners-Lee, T., J. Hendler, and O. Lassila [2001], The semantic Web, *Scientific American* 284(5), 28–37.
37. Davies, J., D. Fessel, and F. van Harmelen [2003], *Towards the Semantic Web: Onthology-Driven Knowledge Management*, Wiley, Chichester.
38. Maes, P. [1990], *Designing Autonomous Agents*, MIT Press, Cambridge, MA.
39. Wüst, C.C., and W.F.J. Verhaegh [2004a], Dynamic control of scalable media processing applications, In W.F.J. Verhaegh, E. Aarts, and J. Korst (eds.), *Algorithms in Ambient Intelligence,* Kluwer Academic Publishers, New York.
40. Puterman, M.L. [1994], *Markov Decision Processes: Discrete Stochastic Dynamic Programming*, Wiley Series in Probability and Mathematical Statistics. Wiley-Interscience, New York.
41. Sutton, R.S., and A.G. Barto [1998], *Reinforcement Learning: An Introduction.* MIT Press, Cambridge, MA.
42. Wüst, C.C., L. Steffens, R.J. Bril, and W.F.J. Verhaegh [2004b], QoS control strategies for high-quality video processing, Submitted to *Euromicro Conference on Real-Time Systems*.

43. Duijnhoven, A.E.M. van [2003], *Collaborative Filtering with Privacy.* Master's Thesis, Technische Universiteit Eindhoven.
44. Paillier, P. [1999], Public-key cryptosystems based on composite degree residuosity classes, *Advances in Cryptology – EUROCRYPT'99, Lecture Notes in Computer Science*, vol. 1592, pp. 223–238.
45. Peynirçioglu, Z.F., A.I. Tekcan, J.L. Wagner, T.L. Baxter, and S.D. Schaffer [1998], Name or hum that tune feeling of knowing for music, *Memory and Cognition* 26(6), 1131–1137.
46. Pauws, S. [2004], CUBYHUM: Algorithms for query by humming, in: W. Verhaegh, E. Aarts, and J. Korst (eds.), *Algorithms in Ambient Intelligence*, Kluwer Academic Publishers, pp. 71–87.
47. Wagner, R.A., and M.J. Fisher [1974], The string-to-string correction problem, *Journal of the ACM* 21(1), 168–173.
48. Sellers, P.H. [1980], The theory and computation of evolutionary distances: Pattern recognition, *Journal of Algorithms* 1, 359–373.
49. Chang, W., and E. Lawler [1994], Sublinear approximate string matching and biological applications, *Algorithmica* 12(4/5), 327–344.
50. Myers, E. [1994], A sublinear algorithm for approximate keyworkd searching. *Algorithmica* 12(4/5), 345–374.
51. Navarro, G. [2001], A guided tour to approximate string matching, *ACM Computer Surveys* 33(1), 31–88.
52. Vuuren, M. van, S. Egner, J. Korst, and S. Pauws [2003], Approximate string matching: An efficient search method for melody retrieval, *Proceedings TALES of the Disappearing Computer*, June 1–4, Santorini, Greece.
53. Brandstein, M., and D. Ward (Eds.) [2001]. *Microphone Arrays*, Springer.
54. McCowan, I.A. [2001], *Robust Speech Recognition using Microphone Arrays*, Ph.D. Thesis, Queensland University of Technology, Australia, 2001.
55. Yang, J., R. Stiefelhagen, U. Meier, and A. Waibel [1998], Visual trading for multimodal human computer interaction, *Proceedings of the International Conference for Computer-Human Interaction 1998*, pp. 140–147.
56. Darrell, T., J. Fisher, P. Viola, and B. Freeman [2000], Audio-visual segmentation and the cocktail party effect, *Proceedings of the International Conference on Multimodal Interfaces*, Bejing, October 2000.
57. Gatica-Perez, D., G. Lathoud, I. McCowan, J.-M. Odobez, and D. Moore [2003], Audio-visual speaker tracking with importance particle filters, *Proceedings IEEE ICIP*, September 2003.
58. Noyce, R.N. [1977], Microelectronics, *Scientific American* 237(3), 63–69.
59. Berman, F., G.C. Fox, and A.J.G. Hey (eds.) [2003], *Grid Computing: Making the Global Infrastructure Reality*, Wiley, Chichester.

Printing: Krips bv, Meppel
Binding: Litges & Dopf, Heppenheim